MOLECULAR BIOLOGY OF THE ISLETS OF LANGERHANS

T0276142

MOLECULAR BIOLOGY OF THE ISLETS OF LANGERHANS

EDITED BY
HIROSHI OKAMOTO

Department of Biochemistry, Tohoku University School of Medicine, Japan

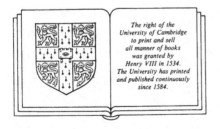

The right of the
University of Cambridge
to print and sell
all manner of books
was granted by
Henry VIII in 1534.
The University has printed
and published continuously
since 1584.

CAMBRIDGE UNIVERSITY PRESS

Cambridge

New York   Port Chester

Melbourne   Sydney

CAMBRIDGE UNIVERSITY PRESS
Cambridge, New York, Melbourne, Madrid, Cape Town, Singapore, São Paulo, Delhi

Cambridge University Press
The Edinburgh Building, Cambridge CB2 8RU, UK

Published in the United States of America by Cambridge University Press, New York

www.cambridge.org
Information on this title: www.cambridge.org/9780521362047

First published 1990
This digitally printed version 2008

A catalogue record for this publication is available from the British Library

Library of Congress Cataloguing in Publication data

Molecular biology of the islets of Langerhans/edited by
Hiroshi Okamoto
    p.   cm.
Includes index.
ISBN 0-521-36204-0
1. Diabetes – Pathophysiology.   2. Islands of Langerhans.
I. Okamoto, Hiroshi, 1939–
RC660.M562 1990
S16.4 62971 – dd20   39-35659

ISBN 978-0-521-36204-7 hardback
ISBN 978-0-521-08800-8 paperback

Contents

List of contributors

GRAEME I. BELL Associate Investigator, *Howard Hughes Medical Institute,* and Associate Professor, *Department of Biochemistry and Molecular Biology, The University of Chicago, Chicago, Illinois 60637, U.S.A.*

STEVEN C. ELBEIN *Division of Endocrinology and Metabolism, University of Utah School of Medicine, Salt Lake City, Utah 84132, U.S.A.*

JOEL F. HABENER Associate Professor, *Laboratory of Molecular Endocrinology, Massachusetts General Hospital,* and *Howard Hughes Medical Institute, Harvard Medical School, Boston, Massachusetts 02114, U.S.A.*

CHIYOKO INOUE *Department of Biochemistry, Tohoku University School of Medicine, Sendai 980, Miyagi, Japan*

NOBUYUKI ITOH Associate Professor, *Division of Biological Engineering, Kyoto Sangyo University, Kyoto 603, Kyoto, Japan*

MOTOO KITAGAWA *Department of Biochemistry, Tohoku University School of Medicine, Sendai 980, Miyagi, Japan*

WILLY J. MALAISSE Professor, *Laboratory of Experimental Medicine, Brussels Free University, B-1000 Brussels, Belgium*

KOJI NATA *Department of Biochemistry, Tohoku University School of Medicine, Sendai 980, Miyagi, Japan*

HIROSHI OKAMOTO Professor, *Department of Biochemistry, Tohoku University School of Medicine, Sendai 980, Miyagi, Japan*

DAVID OWERBACH Assistant Professor, *Endocrinology and Metabolism Section, Department of Pediatrics, Baylor College of Medicine, Houston, Texas 77030, U.S.A.*

M. ALAN PERMUTT Professor, *Division of Endocrinology and Metabolism, Washington University School of Medicine, St Louis, Missouri 63110, U.S.A.*

THUE W. SCHWARTZ Professor, *Laboratory of Molecular Endocrinology, University Hospital, DK-2100 Copenhagen, Denmark*

SUSUMU SEINO *Howard Hughes Medical Institute,* and *Department of Biochemistry and Molecular Biology, The University of Chicago, Chicago, Illinois 60637, U.S.A.*

KIYOTO SHIGA *Department of Biochemistry, Tohoku University School of Medicine, Sendai 980, Miyagi, Japan*

HOWARD S. TAGER Professor, *Department of Biochemistry and Molecular Biology, The University of Chicago, Chicago, Illinois 60637, U.S.A.*

SHIN TAKASAWA *Department of Biochemistry, Tohoku University School of Medicine, Sendai 980, Miyagi, Japan*

KAZUHIKO TATEMOTO Associate Professor, *Department of Psychiatry and Behavioral Sciences, Stanford University School of Medicine, Stanford, California 94305, U.S.A.*

KIMIO TERAZONO *Department of Biochemistry, Tohoku University School of Medicine, Sendai 980, Miyagi, Japan*

MICHAEL D. WALKER *Department of Biochemistry, The Weizmann Institute of Science, Rehovot 76100, Israel*

TAKUO WATANABE *Department of Biochemistry, Tohoku University School of Medicine, Sendai 980, Miyagi, Japan*

TAKASHI YAMAGAMI *Department of Biochemistry, Tohoku University School of Medicine, Sendai 980, Miyagi, Japan*

HIROSHI YAMAMOTO Associate Professor, *Department of Biochemistry, Tohoku University School of Medicine, Sendai 980, Miyagi, Japan*

HIDETO YONEKURA *Department of Biochemistry, Tohoku University School of Medicine, Sendai 980, Miyagi, Japan,* and *Creative Products Research Laboratories, Research Institute, Kissei Pharmaceutical Company Ltd, Matsumoto 399, Nagano, Japan*

YUTAKA YONEMURA Instructor, *Department of Surgery, Kanazawa University School of Medicine, Kanazawa 920, Ishikawa, Japan*

Preface

The islets of Langerhans have been the focus of research on the nature of diabetes mellitus ever since von Mering & Minkowski demonstrated in the 1890s that experimental diabetes can be induced in dogs by pancreatectomy. Following this pioneering work, the next major step was taken in 1922 by Banting & Best, who succeeded in isolating an active preparation of the blood-sugar lowering pancreatic secretion insulin, the body's most abundant peptide hormone and one of the most important in vital processes. In the past decade, advances in molecular techniques have made possible an understanding of the molecular level of the functioning of both the cells distributed at the centre of the islets, which produce insulin, and of the surrounding islet cells, which synthesize glucagon, somatostatin and pancreatic polypeptide. In consequence, new light has been thrown on both the etiology and the course of diabetes itself.

The aim of this book is to provide a contemporary and coherent view of the peptide hormones that are produced by the islets of Langerhans, as well as an explanation, at the molecular level, of defects in the organ that may lead to the pathological condition of insulin-dependent diabetes. The first section of the book is designed to provide a full account of recent information on islet hormone biosynthesis. The second section is devoted to an examination of the effects of deleterious conditions and agents on the functioning of islet cells, especially of the insulin-producing islet B-cells.

The book is intended to provide a detailed picture of the subject for researchers in endocrinology, particularly those interested in islet peptide hormones, and for medical endocrinologists interested in diabetes; it should also be of interest to those many physiologists and biochemists studying peptide hormone biosynthesis and to graduate students in endocrinology.

I express my appreciation to my colleagues, past and present, who have collaborated with me during the past twelve years in investigating the molecular biology of the islets of Langerhans, namely Drs N. Itoh, S. Miyamoto, Y. Noto, H. Mabuchi, R. Takeda, G. Yamazaki, Y. Yone-mura, T. Sei, K. Nose, H. Yamamoto, A. Kawamura, Y. Uchigata, K. Obata, Y. Hayakawa, M. Watanabe, H. Nagai, T. Yamagami, A. Miwa, S. Takasawa, M. Nishizawa, K. Ohsawa, K. Nata, K. Takahashi, C. Inoue, K. Terazono, M. Noguchi, E. Gotoh, K. Shiga, H. Yonekura, M. Kitagawa, T. Sawa, T. Watanabe, J. Aruga, K. Igarashi, I. Kato, T. Abo, M. Ohneda, A. Sugawara, M. Unno and A. Tohgo. This book is one outcome of their endeavors. It is also a pleasure to express my sincere appreciation to Dr Osamu Hayaishi, Dr Yoshimasa Yoneyama, Dr Yoshio Goto and Dr Nakao Ishida, whose encouragement and generous support were the driving force in initiating and continuing this project.

My special thanks are due to Dr G.P.R. McMaster, who provided essential help in the preparation of the contributed manuscripts and in editorial work. It is also a pleasure to express my sincere appreciation to Ms R. Torigoe for typing. The book presents data derived from experiments in my laboratories at Toyama and Sendai, most of which were supported by grants from the Ministry of Education, Science and Culture, Japan, for which I express my sincere gratitude. Thanks are extended to Drs Adam S. Wilkins and Robin C. Smith of Cambridge University Press for their cooperation, and to Drs Hiroshi Yamamoto, Kimio Terazono and Shin Takasawa, who have done very thorough work in the preparation of the subject index. Inevitably, the greatest debt of gratitude is owed to the individual authors whose work has made this book possible.

Sendai, August 1989   Hiroshi Okamoto

Introductory overview

HIROSHI OKAMOTO

Recent advances in molecular biology have made it possible to study the structure and function of the genes that encode the peptide hormones. Among the best characterized of these are insulin, glucagon, somatostatin and pancreatic polypeptide, all of which are produced in a multiendocrine organ, the islet of Langerhans. In addition, an explanation at the molecular level of the principal pathological condition associated with this organ, insulin-dependent diabetes, is now possible.

In the first section of this volume, a full account is offered of state-of-the-art thinking on islet hormone biosynthesis. The second section is devoted to an examination of the effects of deleterious agents and conditions on the functioning of islet cells, particularly of islet B-cells, which produce insulin.

The molecular biology of peptide hormones in islets of Langerhans

Understanding the structure, organization and expression of the genes that encode insulin, glucagon, somatostatin, pancreatic polypeptide, vasoactive intestinal peptide and pancreastatin (Chapters 1, 2, 4, 5, 6, 7 and 8) has been made possible by the cloning and sequencing of both cDNAs and genomic sequences for these genes. From these sequences, the precise amino acid sequences of the peptide precursors have been deduced. In addition, from the genomic clones, it has been possible to map the exons (the encoding sequences) with respect to the introns (the intervening, non-encoding sequences). Finally, cis-acting regulatory sequences that flank the structural genes have been identified, as have, in part, trans-acting factors that interact with the cis-acting sequences.

Ever since Steiner & Oyer's discovery of proinsulin (Steiner & Oyer,

1

1967) it has been recognized that peptide hormones are initially synthesized as larger precursor forms that undergo conversion to one or more biologically active forms as they traverse the secretory pathway. Within the functionally distinct domains of the precursors preproinsulin, preproglucagon, preprosomatostatin and preprovasoactive intestinal peptide, there are significant amino acid sequence homologies among different species, suggesting the existence of strong evolutionary constraints on the evolution of the separate domains. However, in the rat, mouse, guinea pig, canine and human prepropancreatic polypeptides, the carboxy-terminal domains vary greatly in both size and sequence, whereas the amino-terminal domains, which contain the pancreatic polypeptide, exhibit a high degree of homology. It seems likely that, under the pressure of different selective constraints, the exons coding for the amino-terminal domain evolved at different rates from those coding for the carboxy-terminal domain (Chapter 6).

There are several steps in the pathway of peptide hormone biosynthesis at which expression of the genes that encode them can be regulated, namely the transcriptional, translational and posttranslational processing levels. On the whole, studies on the regulation of these genes have been focussed on transcriptional controls, largely because the regulatory sequences that flank the structural genes of insulin, glucagon, somatostatin, pancreatic polypeptide and vasoactive intestinal peptide have been identified. However, in 1980 Itoh & Okamoto demonstrated that it is at the translational level that the primary regulation of insulin biosynthesis takes place (Itoh & Okamoto, 1980). One indication that this must be the case is that in a very short time after the elevation of blood glucose levels, increases as high as 10-fold are observed in the rate of proinsulin biosynthesis. Glucose clearly must act, either directly or indirectly, by affecting the efficiency of the translation of proinsulin mRNA that already exists in islet B-cells. When food is eaten, large amounts of insulin need to be released immediately, and since this rapid release of insulin into the bloodstream depletes the insulin stored in the B-cells of the islets, additional insulin is provided by the B-cell's employment of the immediate mechanism of enhancing the translational efficiency of already existing proinsulin mRNA (Chapter 3).

Processing of the precursor forms is also an essential element in the production of the hormones. Cleavage of islet peptide precursors occurs either at pairs of basic amino acid residues or at single residues, and the structure and processing pattern of the precursors for the individual islet cell hormones are described in Chapter 9. This chapter also deals with the processing of islet hormone precursors in heterologous systems, i.e. in transfected cell lines, yeast and prokaryotes.

The islet of Langerhans is a multiendocrine organ; a given peptide hormone in the islet can be expected to affect the functions of islet cells that synthesize other hormones. Somatostatin, for example, has been reported to exert a direct inhibition on the secretion of insulin and glucagon from the islet; in Chapter 8, Tatemoto reports on a novel pancreatic peptide, pancreastatin, which inhibits insulin secretion and stimulates glucagon secretion.

Molecular aspects of diabetes mellitus

Insulin-dependent (type I) diabetes is caused by a destructive process affecting the insulin-producing B-cells of the islets of Langerhans. Destruction of B-cells may be induced by immunologic abnormalities, infectious agents, inflammatory tissue damage, and B-cytotoxins such as alloxan and streptozotocin. In Chapter 10, it is proposed that although insulin-dependent diabetes can be caused by many different agents, the process by which islet B-cells are destroyed is common (the Okamoto model). This process involves DNA damage, nuclear poly(ADP-ribose) synthetase activation to repair DNA lesions, and NAD depletion in the cells. The B-cells are critically dependent on the maintenance of the NAD pool, and the depletion of NAD by the DNA repair process is sufficient to kill the cell. Both alloxan and streptozotocin cause DNA damage in B-cells through the formation of free radicals such as $OH^{\cdot}$ and $CH_3^{\cdot}$. Since induction of insulin biosynthesis in islet B-cells is achieved at the level of translation (as discussed above and in Chapter 3), an immediate decrease in insulin biosynthetic activity cannot be attributed to DNA damage itself. Rather, the B-cell, it may reasonably be assumed, commits suicide in its attempt to repair the DNA strand breaks.

In humans, the extent of an immune response to individual foreign antigens is controlled by cell-surface molecules encoded by HLA class II genes. Insulin-dependent (type I) diabetes has been demonstrated by restriction fragment length polymorphism (RFLP) and oligonucleotide probe analyses of class II genes to be closely associated with the DQβ locus (Chapter 11). The class II antigens are not expressed on normal pancreatic B-cells but have been observed on islet cells of diabetic patients; they function by presenting antigens to T-helper cells. This ultimately leads to the destruction of the islet B-cells by antigen-specific cytotoxic T-lympho-cytes (Bottazzo et al., 1987). The T-lymphocyte-mediated cytolytic process may also be initiated in the target nucleus, which results in damage to its DNA, as shown in Fig. 10.4. Furthermore, certain islet B-cell antigens, released during islet B-cell injury, are processed and presented by macrophages to T-helper lymphocytes. This can initiate the production of cytokines, of which interleukin-1 (IL-1) is cytotoxic to islet B-cells. To our knowledge, no one has yet determined whether cytokines such as IL-1 and tumor necrosis factor induce DNA damage in target cells; this is of interest since cell-free lymphotoxin-containing supernatants cause the release of DNA from targets, and since murine cytotoxic T-lymphocytes contain a novel cytotoxin, leukalexin, which causes DNA fragmentation in several types of target cell (Chapter 10).

The formation of islet B-cell tumors (insulinomas), following the combined administration of streptozotocin (or alloxan) and a poly(ADP-ribose) synthetase inhibitor, is also related to DNA strand breaks (Chapter 10). In this case, the activation of poly(ADP-ribose) synthetase and the depression of the cellular NAD level are prevented, and islet B-cells therefore are allowed to survive. However, prompt DNA repair will not occur, and B-cells that have experienced some genetic alterations may exhibit abnormal expression of *rig* (rat insulinoma gene) (Chapter 14) and be converted to tumor cells.

As stated above, the DNA repair process can be fatal to islet B-cells; this is the molecular foundation of insulin-dependent (type I) diabetes. From a practical point of view, then, strategies for influencing the replication of islet B-cells and the growth of the B-cell mass, in order to ameliorate not only type I but also type II (non-insulin-dependent) diabetes, may be important. This type is not the result of an ongoing destruction of islet B-cells, but may reflect the inability of the B-cells to meet an increased peripheral demand for insulin. In 1984, Yonemura *et al.* demonstrated that administration of poly(ADP-ribose) synthetase inhibitors such as nicotinamide to 90% depancreatized rats induces regeneration of islet B-cells in the remaining parts of the pancreas, thereby forestalling the development of diabetes (Chapter 10). Recently, Terazono *et al.* have isolated a novel gene, named *reg* (regenerating gene), from the regenerating islets of 90% depancreatized and nicotinamide-treated rats (Chapter 15). The activation of *reg* may be involved in islet B-cell regeneration.

The role of insulin gene mutation in diabetes, especially in type II diabetes, has recently been evaluated through the cloning of the insulin gene and identification of its RFLPs. The available data give no support, however, to the hypothesis that there is a strong linkage between type II diabetes and insulin gene polymorphism (Chapter 12). It is suggested in Chapter 12 that, although insulin gene RFLPs are not routinely used in diabetes diagnosis, they could have a useful role in the analysis of suspected hereditary defects in the insulin gene.

The role of the 5' polymorphic region of the insulin gene has not yet been firmly characterized. However, strong arguments have been offered for the involvement in some forms of diabetes of structurally abnormal and biologically less active insulin molecules, resulting from mutations in the insulin coding sequence. These are discussed in Chapter 13.

A central fact about insulin, of course, is that its synthesis and release are under physiological conditions regulated primarily by glucose. As Malaisse *et al.* (1979) and Matschinsky *et al.* (1989) argue, the role of D-glucose in insulin secretion is intimately related to its capacity to be metabolized and to augment the rate of ATP generation in the pancreatic B-cell. In Chapter 16, anomalous hexose metabolisms, insulin biosynthesis and release in a tumor cell line are described.

References

Bottazzo, G.F., Pujol-Borrell, R. & Gale, E.A.M. (1987). Autoimmunity and type I diabetes: bringing the story up to date. In *The diabetes annual 3* (ed. K.G.M.M. Alberti & L.P. Krall), pp. 15–38. Amsterdam: Elsevier.

Itoh, N. & Okamoto, H. (1980). Translational control of proinsulin synthesis by glucose. *Nature* **283**, 100–2.

Malaisse, W.J., Sener, A., Herchuelz, A. & Hutton, J.C. (1979). Insulin release: the fuel hypothesis. *Metabolism* **28**, 373–86.

Matschinsky, F.M., Corkey, B.E., Prentki, M., Meglasson, M.D., Erecinska, M., Shimizu, T., Ghosh, A. & Parker, J. (1989). Metabolic connectivity and signalling in pancreatic B-cells. In *Diabetes 1988* (ed. R.G. Larkins, P.Z. Zimmet & D.J. Chisholm), pp. 17–26. Amsterdam – New York – Oxford: Excerpta Medica.

Steiner, D.F. & Oyer, P.E. (1967). The biosynthesis of insulin and a probable precursor of insulin by a human islet cell adenoma. *Proceedings of the National Academy of Sciences of the U.S.A.* **57**, 473–80.

I

The molecular biology of peptide
hormones in the islets of Langerhans

1

The organization and structure of insulin genes

GRAEME I. BELL AND SUSUMU SEINO

1.1. Introduction

Insulin is a two-chain disulfide-linked polypeptide hormone which is essential for normal growth and metabolic regulation in higher organisms. It is a member of a superfamily of structurally related peptides which includes relaxin (a two-chain polypeptide hormone that functions in vertebrate reproductive physiology (Bryant-Greenwood, 1982)) and insulin-like growth factors (IGF) I and II. IGF-I and -II are single-chain polypeptides which stimulate growth *in vitro* and *in vivo* and are probably required for normal fetal and postnatal growth and development (Froesch & Zapf, 1985). Insulin is also related to two invertebrate neuropeptides that have recently been isolated and characterized: prothoracicotrophic hormone of the silkworm *Bombyx mori*, which regulates the secretion of the molting hormone ecdysone (Nagasawa *et al.*, 1986; Jhoti *et al.*, 1987); and molluscan insulin-related peptide (MIP) (Smit *et al.*, 1988), a pond snail (*Lymnaea stagnalis*) hormone, which is synthesized by the cerebral light-green cell and is believed to regulate, at least in part, growth of soft body parts and shell and glycogen metabolism. These invertebrate insulin-like proteins are two-chain polypeptide hormones like insulin and relaxin and in contrast to the single-chain IGFs. Although the origins and relationships of the various members of the insulin gene family have not been clarified, it seems likely that they evolved from a common ancestral protein through a process of gene duplication and diversification. Furthermore, the presence of two non-allelic insulin or relaxin genes in some vertebrate species suggests that this process is still ongoing. The primary structures of insulins from about 50 vertebrate species have been determined (reviewed in Hallden *et al.*, 1986; see also Conlon & Thim, 1986; Cutfield *et al.*, 1986; Conlon *et al.*, 1986; Seino *et al.*, 1987; Conlon *et al.*, 1987; Pollock *et al.*, 1987; Plisetskaya *et al.*, 1988). These include representatives of most vertebrate classes, including the hagfish, which diverged from the main line of vertebrate evolution 500 million years ago and is the most primitive vertebrate from which insulin has been isolated, as well as several species of fish, reptile, bird and mammal. Although it is uncertain if insulin or another member of the insulin superfamily represents the ancestral protein, it seems likely that the appearance of insulin antedated that of the vertebrates since it occurs and appears to function similarly in all vertebrate species.

1.2. Structure of the insulin gene

Protein sequencing and molecular cloning studies indicate that insulin is encoded by a gene that has persisted throughout 500–600 million years of vertebrate evolution. The insulin genes of ten species have been isolated and at least partly characterized to date: human (Bell *et al.*, 1980*b*), chimpanzee (unpublished data), African green monkey (an Old World monkey) (unpublished data), owl monkey (a New World monkey) (Seino *et al.*, 1987), dog (Kwok *et al.*, 1983), rat (Cordell *et al.*, 1979; Lomedico *et al.*, 1979), mouse (Wentworth *et al.*, 1986), guinea pig (Chan *et al.*, 1984), chicken (Perler *et al.*, 1980) and hagfish (Steiner *et al.*, 1985; S. Nagamatsu, S.J. Chan & D.F. Steiner, personal communication). In most species there is a single insulin gene per haploid chromosomal complement (or two alleles per diploid cell). However, in several species, including the laboratory rat and mouse and at least two species of fish (tuna and toadfish), two distinct insulin proteins have been isolated, indicating the presence of two non-allelic insulin genes; in fact, two non-allelic insulin genes have been isolated from the rat and mouse. Interestingly, Frank *et al.* (1984) recently identified a variant of bovine proinsulin; the new proinsulin differs from the previously sequenced protein by a single amino acid residue in the C-peptide. They suggest that the presence of a second form of proinsulin in cattle is the result of a gene duplication. Their results raise the intriguing possibility that there may be other species that have duplicated insulin genes and, moreover, suggest that the isolation of a single form of the insulin protein cannot be used as evidence for the presence of a single insulin gene.

Of the insulin genes that have been isolated, the human and rat genes have been studied in the greatest detail. The human insulin gene is a single-copy gene located near the end of the short arm of chromosome 11 in band p15.5 (Harper *et al.*, 1981; Zabel *et al.*, 1985) (Fig. 1.1). The human insulin gene contains three exons (i.e. regions that are also present

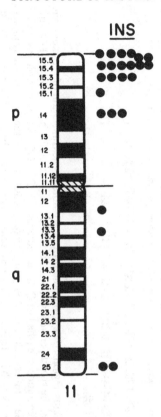

Fig. 1.1. Localization of the insulin gene on human chromosome 11 by *in situ* hybridization. The distribution of silver grains on chromosome 11 is indicated. Letters p and q refer to the short and long arms of the chromosome. The bands observed on giemsa staining are noted. (Adapted from Zabel *et al.* (1985).)

in insulin mRNA) separated by two intervening sequences or introns (Figs. 1.2 and 1.3). This organization is the same as that predicted for the ancestral insulin gene. The intron sequences are present in the gene and the primary RNA transcript and are removed from the mRNA precursor by splicing before the mRNA moves from the nucleus to the cytoplasm. In the insulin genes that have been examined to date, intron A is shorter than intron B (Fig. 1.3). Interestingly, although the sizes and sequences of the introns are quite variable, the positions at which they interrupt the mature mRNA and preproinsulin protein sequences have been conserved; intron A interrupts the region of the gene coding for the 5'-untranslated region of the mRNA at a site 38–43 base pairs (bp) from the 5'-end of the transcript

```
CCCCTTCCTCACTCCCCACTGTCCCACCCCACCACTTTGGCCCCATCATGGGCAATCTTGGCCCTCATGGGGCATC TTGGCCCTCATGGGCAATCTTGGCCCTCATGGGGCATC            -807
Tandem Repeats (VNTR)                                    Hypervariable Region − Variable Number of
GACAGGGGTGTGGGGACAGGGGT                                  (ACAGGGGTGTGGGGACAGGGGTCCGGGGACAGGGGTCTGGG   -686

CTGGGGACAGGGGTGTGGGGACAGGGGTGTGGGGACAGGGGTGTGGGGACAGGGGTGTGGGGACAGGGGTGTGGGGACAGGGT   -565

AGGGGTGTGGGGACAGGGGTGTGGGGATAGGGGTGTGGGACAGGGGTGTGGGGACAGGGGTGTGGGACAGGGT   -444

CTGAGGACAGGGGTGTGGGCCACAGGGGTCCTGGGGACAGGGGTTCTGGGGACAGGGGTCTGGGG) ACAGCAGGGCAAAGAGCCCCGCCCTGCAGCCTCTCC   -323

TGTCTAATGTGGAAAGTGGCCCAGGTGAGGGCTTTGTCTCCTCGTGAGACATTTGCCCCGTGTGACGCAGGGACAGGTCTGGCCACCGGCCCCTTGGTTAAGACTCTAATGACCCGCTG   -202

GTCCTGAGGAAGAGGTGCTGACGACCAAGGAGATCTTCCCACAGACCCAGCACTGGGGCCCAGCCCAGGCCCAGCAGGGCAGGCCCAGCCCAGGCCCCACCCCAGCCCT   -81

AATGGCCAGGCAGGCAGGGGTTGACAGGTAGGGGAGAAGTAATAAAGCCCCAGCCCTCCAGCCCTCCAGGACAGCAGGCTGCATCCAGCA   41
Intron A
G GTCTGTTCCAAGGGCCTTTGCGTCAGGTGGGCTTCAGGTTCCAGGTGGCTGGACCCCAGCCTCGCAGCCAGGGAGGACAGCTGGCCTGGGCTCGTGAGCCATGTGGGGGTGAGCCC   162
                                                                                -20
                                                                  Met Ala Leu Trp Met Arg Leu Leu Pro Leu Leu
AGGGGCCCCAAGGCAGGGCACCTGGCCTTCAGCCTGCCTCAGCCCTGCCTGTCTCCCAG ATCACTGTCCTTCTGCC ATG GCC CTG TGG ATG CGC CTC CTG CCC CTG CTG   271
Exon 2                                                      1                                    10
Ala Leu Ala Leu Trp Gly Pro Asp Pro Ala Ala Ala Phe Val Asn Gln His Leu Cys Gly Ser His Leu Val Glu Ala Leu Tyr Leu
GCG CTG CTG GCC CTC TGG GGA CCT GAC CCA GCC GCA GCC TTT GTG AAC CAA CAC CTG TGC GGC TCA CAC CTG GTG GAA GCT CTC TAC CTA   361
                                                                                                Intron B
Val Cys Gly Glu Arg Gly Phe Phe Tyr Thr Pro Lys Thr Arg Arg Glu Ala Glu Asp Leu Gln V   Gln G   GTGAGCCAACCGCCCATTGCTGCCCCTGGCCGCCCC   461
                30
GTG TGC GGG GAA CGA GGC TTC TTC TAC ACA CCC AAG ACC CGC CGG GAG GCA GAG GAC CTG CAG G
CAGCCACCCCTGCTCCTGGCGCTCCCACCCAGCATGGGCAGAAGGGGCACTTTTTAAAAAGAAGTTCTGTTGGTCAGTCCTAAA   582
GTGACCAGCTCCCTGTGGCCCAGTCAGAATCTCAGCCTGAGGACGGTGTTGGCTTCGGCAGCCCCAGCCATCACAGGGGTGGCACGGTCGCCTCCCTCCACTGCCGTCCTCCCTCAACAAATGC   703
CCCGCAGCCCATTTCTCCACCCTCATTTGATGACCGCAGATTCAGTCTTTGTTTAAGTCTGGGTGACCTCCATACTCAGGAGTCACAGGGTGGGAACAGGTGGGGCCCTGGGAGAG   824
CATCACGCCGGAGGGAAGGGTCTGGGCTGACCCTGCCTGGGGAGACCCCCGGGGGGAAGGAGGTGGGCAGGGGCTGGGAACATGTGGGCGTTGGGGTGTAGGCTCTCCGTGTTCGGAACCT   945
TACTGGGATCACCTGTTCAGGCTCCTCCCTCCTCTGCCTGAGTGGGAGGTGGCTTGGCCTGGGGGCAGGCCCTGGGACGTGTGGGCTGGTTCCTCCCGTGTCCCGGTGTTCGGAACCT   1,066
CTGGGTTCAGCCCCGGCTGGAGATGGGGAGTGGGAGGTGGGACCTTAGGGGTGCAGGGCGCCAGCTGTCCCCTGGCCCTGGCGCTCTCCTCGGTGTCCCGCGTGTTCGGAACCT   1,187
                                                                                                  60
Exon 3                            Gln Val Glu Gly Gly Gly Ala Gly Ser Leu Gln Pro Leu Ala Leu Glu Gly Ser Leu
                                  Gln Val Glu Gly Gly Gly Pro Gly Ala Gly Ser Leu Gln Pro Leu Ala Leu Glu Gly Ser Leu
GCTCTGCGGCACGTCTGCAG TG GGG CAG GTG GAG GTG GAG GGG GGC CCT GGA GGC AGC CTG CAG CCC TTG GCC CTG GAG GGG TCC CTG   1,282
                70                                              80                                    86
Gln Lys Arg Gly Ile Val Glu Gln Cys Cys Thr Ser Ile Cys Ser Leu Tyr Gln Leu Glu Asn Tyr Cys Asn AM
CAG AAG CGT GGC ATT GTG GAA CAA TGC TGT ACC AGC ATC TGC TCC CTC TAC CAG CTG GAG AAC TAC TGC AAC TAG ACGCAGCCTGCAGGCAGCCCC   1,378
Polyadenylation Site
ACACCCGCCGCCTCCTGCACCGAGAGAGATGGAATAAAGCCCTTGAACCAGC CCTGGTCTGGCCGTCTGTGTGTCTGGGGCCCCTGGGAGCCCACCTTCCCGGCACTGTTGTGAGCCC   1,499
```

Fig. 1.2. Sequence of the human insulin gene and flanking regions. The exons and introns are noted. The translation of the exons has been numbered using the proinsulin numbering system: the signal peptide is amino acids −24 to −1; the B-chain, residues 1–30; the C-peptide, 33–63; and the A-chain, 66–86. The polyadenylation site (the last C in the sequence AACCAGC) is the position at which a poly(A) tract is added after synthesis of the mRNA. The number at the end of each line corresponds to that of the last nucleotide in that line; numbering is relative to the first nucleotide of exon 1 which is the site at which transcription of the gene begins. The region between nucleotides −364 and −1 contains sequences which regulate transcription of this gene. The region of length and sequence variation in the 5'-flanking region — the hypervariable region — is noted and enclosed within parentheses. The number and arrangement of the family of sequences related to the 14 bp sequence ACAGGGGTGTGGGG can differ between chromosomes and as a consequence it is often possible to identify the insulin gene derived from the father and mother by using Southern blotting techniques.

(*a*) 'ancestral' vertebrate insulin gene

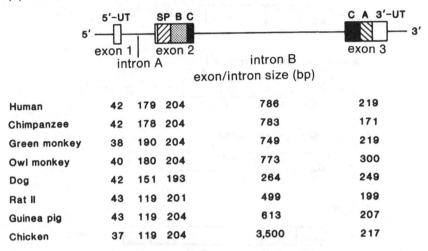

exon/intron size (bp)

	5'-UT	SP	B C	intron B	3'-UT
Human	42	179	204	786	219
Chimpanzee	42	178	204	783	171
Green monkey	38	190	204	749	219
Owl monkey	40	180	204	773	300
Dog	42	151	193	264	249
Rat II	43	119	201	499	199
Guinea pig	43	119	204	613	207
Chicken	37	119	204	3,500	217

(*b*) 'duplicated' murine insulin gene

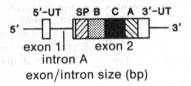

exon/intron size (bp)

Fig. 1.3. Schematic comparison of vertebrate insulin genes. The structures of the two types of insulin gene that have been described are indicated. The sizes of the exons and introns are presented. Intron B in the hagfish is 144 bp (Steiner *et al.*, 1985); the size of intron A has not been determined. The structures of the mouse insulin I and II genes (Wentworth *et al.*, 1986) are similar to the corresponding genes of the rat. The regions of the mRNA and preproinsulin encoded by each exon are noted: 5'- and 3'-UT denote the untranslated region of the mRNA; SP, B, C and A are the signal peptide, B-chain, C-peptide and A-chain, respectively.

and 13–17 bp before the ATG codon at which translation is initiated. The second intron, intron B, interrupts the gene between the first and second nucleotide of the codon for amino acid 7 of the C-peptide, thereby dividing the protein coding portion of the gene into two regions, exons 2 and 3, one of which (exon 2) encodes the signal peptide and A-chain domains of proinsulin together with a portion of the C-peptide; exon 3 encodes the remainder of the C-peptide and the B-chain domain. The close correspondence between exons and functional regions of insulin mRNA

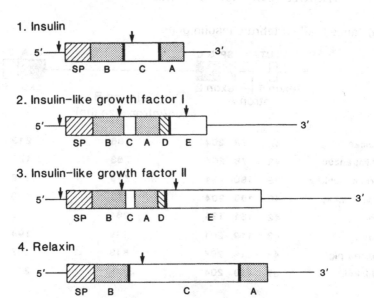

Fig. 1.4. Intron locations in genes encoding members of the insulin superfamily. The sites at which introns interrupt the mRNA or protein sequence are noted. The 5'- and 3'-untranslated regions are indicated as lines and the protein coding region as a box. The domains of the protein are indicated: SP, B, C, A, D and E are the signal peptide, B-chain or domain, C-peptide or domain, A-chain or domain, D- and E-domain, respectively.

or preproinsulin protein suggests that the exons were brought together (exon recruitment) resulting in a gene composed of three exons.

The introns in the genes encoding other members of the insulin gene family are also located in similar positions (Fig. 1.4). The single intron in the human and porcine relaxin genes is located between the first and second nucleotides of the codon for amino acid 14 of the C-peptide (Hudson et al., 1983; Haley et al., 1987). In the IGF-I (Bell et al., 1985; Rotwein et al., 1986) and IGF-II (Dull et al., 1984; Bell et al., 1985; de Pagter-Holthuizen et al., 1987; Hoppener et al., 1988) genes, there are introns in the 5'-untranslated region, between the first and second nucleotides of the codons for amino acids 26 and 29, respectively (these residues are near the COOH-terminus of the B-domain) and in the region encoding the E-domain. The second intron in both the IGF-I and -II genes has shifted slightly (exon sliding) relative to the corresponding introns in the insulin and relaxin genes but is in an analogous position. The similarity in the exon–intron organization of the insulin, relaxin and IGF genes further supports their ancestral relationship. Moreover, the presence of an additional exon encoding much of the E-domain of both the IGF-I and IGF-II precursors supports the notion that exon recruitment has played a role in the evolution of the insulin gene family.

1.3. Structure of a duplicated insulin gene present in laboratory rats and mice

In contrast to humans and most other vertebrates, laboratory rats and mice synthesize and secrete similar amounts of two structurally related insulins (Clark & Steiner, 1969). Insulins I and II are produced by closely related genes that possess over 90% nucleotide sequence homology (Lomedico *et al.*, 1979; Soares *et al.*, 1985; Wentworth *et al.*, 1986). This region of extensive nucleotide sequence homology extends for 500 bp upstream of the transcriptional initiation site and includes the promoter–enhancer region which is responsible for temporal and tissue-specific expression of the insulin gene. As expected from the extensive nucleotide sequence identity in the promoter–enhancer region, the two non-allelic rat insulin genes are coordinately regulated (Giddings & Carnaghi, 1988). Although the sequences of rat insulin I and II mRNAs and preproinsulins are very similar, there is one feature that distinguishes the two genes: the rat–mouse insulin I gene has only one intron rather than the two present in other insulin genes (Fig. 1.3). The intron interrupting the region encoding the C-peptide (intron B) has been precisely deleted in the rat–mouse insulin I gene. As a consequence, the protein coding region of the insulin I gene is contained within a single exon rather than two as in the rat–mouse insulin II gene or the other vertebrate insulin genes. Moreover, as the rat insulin I and II mRNAs are present at similar levels in the insulin-producing B-cell (Giddings & Carnaghi, 1988), it seems unlikely that intron B serves a critical function in *cis* in regulating expression of the insulin gene. The nucleotide sequences at the boundaries where the sequences of the homologous rat–mouse insulin I and II genes diverge have provided a clue to the evolutionary process which generated the insulin I gene. This sequence comparison revealed that the rat–mouse insulin I genes end in a short adenine-rich sequence and are flanked by imperfect 31 bp direct repeats. Such structural features are characteristic of retroposons, a name suggested for dispersed DNA sequences of RNA origin (Rogers, 1983) (in retroposons, the RNA-related sequence terminates in an adenine-rich segment and is flanked by a short 10–30 bp duplication of chromosomal sequences at the insertion site). The similarity of the rat–mouse insulin I gene to a retroposon suggests that it may have arisen by a process involving insertion into the germ line of a DNA copy of a partly spliced insulin gene transcript, i.e. one in which intron B had been removed but which still retained intron A. Since a reverse transcriptase-like enzyme would be required to convert the RNA transcript to a double-stranded DNA molecule, the process generating the insulin I gene may have been facilitated by a retrovirus. Moreover, the extensive homology in the 5'-flanking regions of the rat–mouse insulin I and II genes indicates that the transcript that served as the template for the reverse-transcriptase-like activity was initiated 500 bp upstream of the site at which transcription of the insulin gene normally occurs.

The physiological reasons for the maintenance of two functional insulin genes encoding proteins of slightly different sequence in laboratory rats and mice are still unclear. Perhaps the cloning of the two non-allelic insulin genes of the tuna and toadfish could provide additional clues. Are the duplicated genes in the tuna and toadfish also retroposons, or were they generated by some other molecular mechanism?

1.4. Sequence of the 5'-flanking promoter–enhancer region of the insulin gene

Recent studies from a number of laboratories (Walker et al., 1983; Hanahan, 1985; Edlund et al., 1985; Ohlsson & Edlund, 1986; Bucchini et al., 1986; Selden et al., 1986; Nir et al., 1986; Sample & Steiner, 1987; Karlsson et al., 1987; Alpert et al., 1988) indicate that the insulin gene has a compact promoter and that all the genetic information necessary for both its tissue-specific and temporal regulation are contained within a region of about 300–350 bp upstream from the transcriptional initiation site. Comparison of the sequences of the 5'-flanking regions of the human, dog, rat I, rat II and guinea pig insulin genes indicates that there are short segments of identical sequence in these five genes. It seems likely that these conserved regions are cis-acting transcriptional regulatory sequences and represent binding sites for general transcription factors as well as for specific regulatory proteins (Fig. 1.5). These regulatory proteins include the factors present in the insulin-producing B-cell that facilitate transcription of the insulin gene as well as those that modulate its expression in response to glucose and possibly other molecules. The specific stimulatory effect of glucose on transcription of the insulin gene may be mediated in part by cAMP (Nielsen et al., 1985). In fact, there is a sequence (Fig. 1.5) located 180 bp from the transcriptional initiation site that is similar to the consensus cAMP-responsive element identified within cAMP-responsive genes (Montminy

B-Cell Specific Enhancer and Transcriptional Regulatory Region

```
Human  (HVR) ACA:::GCAGCG:CAAAGAGCCCGCCCTGC:AGCCTC::CAGCTCTCCTGG::TCTAATGTGGAAAGTGGCCCAGGT:::::::::::GAGGG::C
Dog          --G-A--TCAATATTA-GT-CCT:-A-AA-TG--GT:T-C----GGAA-G-T------A:T--T---CCAAAGATG::--G--GG-
Rat I        ---GGGCC--G-A--AAGATACCA-GT-CCCC-A-AA-TG--AC:T-T----GAAA-G-G------A:T--T---CCAAGGAAAAA--G--::--
Rat II       ---GTTCA--G-A--AAGTTACTA-GT-CCCC-A-AA-TG--GC:C-C----GGAA-G-T------AAT--T---CCAAGGACAAA-A--::--
```

```
Human  TTTGCTCTC:CTGGAGACATTGCCCCAGCTGTGAGCAGGACAGGTCTGGCCACCGG::GCCCCTGGTTAAGACTCTAATGACCCGCTGGTCCTGAG
Dog    TCTG-C--C-:G----T-GCCCCCA----CC-G-::::::::::----T-AGG:::-----CGCC--G-C-C----CG----CGAG-:CCGG-
Rat I  CTTA-C-T-TC--:G----A-GATTG::T----TG-A-TGCTTCA::TCA----T-TG:::G-----TGTT--T-A-C---TT----TA:G-TC:TAA-
Rat II CTCA-C-T-TC--:A---A-GTCCCC:T----TG-A-TGGTTCA::TCA----C-CAGGAG-----C:TT--G-C-C---TT----TA:A-GC:TAA-
```

CRE

```
Human      GAAGAGGTGC [TGACGACCA] AGGAGATCTTCCCACAGACCCAGCACCAGGGAAATTGCAGCCTCAGCCCC::AGCCATCTGCGACCCC
Dog        :::::G--C   -G---CT--C -CG----CC----TG--G--A--CCA----T-G--G-C-----T-C::G-------CACCCC
Rat I      T:AGAGT--T  --AC-T--:  -T---CG--T-TG----TT---TA--C--G-:-TTG---T-A--G-T------C-TCTC------CTACCTA
Rat II     T:AGAGG--T  -T---CA--T -TG----CT----CA--C--G--TTG---C--G-G-T------C-TCTG--------TGATCA
Guinea Pig                        -TG------CC----CA--G--A--CCA----T-G--A--C------C-:TG--------TGATGCC
```

Promoter

```
Human      CCCACCCC::AGGCCCTAATGGGCCAGGCGGCAGGGGTTGACAGTAGGGGAGATGGGCTCTGAGAC::
Dog        ::::::::::::::T---C--C-AGGCC-::::::::::::::::T-GG--CG-GAG-::
Rat I      CCCCTCCTAGAGCCT-A--G--C-AAACG-CAAAGTCCAGGGGCAGAGGAG-TG--TT-:GA-::
Rat II     CCC:::::::::T-A--G--A-AAACA-CAAAGTCAGGGTCAGGGGGG-TG--TT-:GA-::
Guinea Pig ACCACCCCCAGGTCCC-A--G--C-TGGTG-CAGAGTTT:::::::GGGAAGAT-GG--CA-:GG-TA
```

```
                                                                    1
           TATAAA  G:CCAGCGGGGGCCCAGCAGCCCTC|A
                   -:CAG:GAGGGTCC--C-G-C-C-|A
                   -:CTAGTGGAGACG--T-A--T-C-|A
                   -:CTAGTGGGATTC--T-A-C-C-|A
                   -TCCAAGGACCTA--:-G-C-C-|A
```

TATA Element

Fig. 1.5. Comparison of the sequences of insulin gene 5'-flanking regions. This region is divided into promoter (RNA polymerase II binding site) and enhancer or cis-acting transcriptional regulatory regions. Sequence identities are indicated by dashes. Gaps are noted by colons. The putative cAMP-responsive element (CRE) is marked, as is the TATA sequence which is responsible for positioning RNA polymerase II in the promoter region, thereby facilitating transcription initiation at nucleotide 1.

et al., 1986); however, the role of this sequence in modulating expression of the insulin gene in response to glucose has not been specifically tested. Interestingly, the 5'-flanking region of the chicken insulin gene (Perler *et al.*, 1980; Soares *et al.*, 1985) cannot be aligned with the corresponding region of any of the mammalian insulin genes. It is not clear if this absence of homology is a consequence of misidentification of the promoter region of the chicken insulin gene (a chicken insulin cDNA clone has not been isolated and thus independent confirmation of the promoter assignment is lacking) or indicates that the chicken gene is regulated in a different fashion from mammalian insulin genes.

1.5. Repeated sequences in the region of the human insulin gene

A region of *c.* 60 kilobase pairs (kbp) in the region of the human insulin gene has been isolated as a series of overlapping DNA fragments (Bell *et al.*, 1985). In addition to the 1430 bp corresponding to the insulin gene and two dispersed middle repetitive DNA Alu sequences of 300 bp (Bell *et al.*, 1980*a*; Bell *et al.*, 1985) (dispersed middle repetitive DNA sequences have also been identified in the region of the rat insulin I gene (Bell *et al.*, 1980*a*; Lakshmikumaran *et al.*, 1985; Winter *et al.*, 1987)), a region consisting of tandem copies arranged in a head-to-tail array of a family of 14 bp oligonucleotides having the consensus sequence ACAGGGGTGTGGGG has been identified (Figs. 1.2 and 1.6) (Bell *et al.*, 1982). This region begins 365 bp before exon 1 of the insulin gene (Fig. 1.2) and extends for a variable distance upstream. Population studies indicate that the size of the hypervariable region (HVR) (an alternative designation for this region is variable numbers of tandem repeats, VNTR) can vary from 500 to 10 000 bp with average sizes of 570, 1320 and 2740 bp, corresponding to 40, 95 and 170 repeats, respectively, of the 14 bp family of GC-rich oligonucleotides (Bell *et al.*, 1981, 1984). The function of the HVR is unknown but it is unlikely to serve a general role in the regulation of insulin gene expression since a similar type of sequence is not present in the corresponding region of either the rat–mouse insulin I and II genes or the insulin genes of the owl monkey or African green monkey (the sequence of the corresponding region of the dog, guinea pig and chicken insulin genes has not been reported). However, as a similar sequence is present flanking the chimpanzee insulin gene, it is possible that it could have a specific role in regulating insulin gene expression in humans, chimpanzees and, possibly, other higher apes. Although the function of the HVR is unknown, it is useful in genetic studies since it can serve as a molecular marker for the adjacent insulin gene allele, thereby

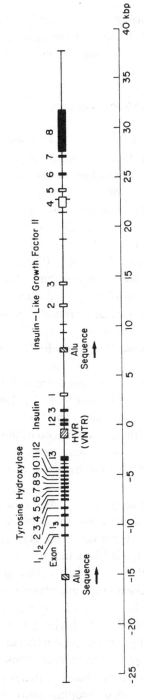

Fig. 1.6. Map of the human insulin, IGF-II and tyrosine hydroxylase gene region of chromosome 11. The exons are indicated, as are the two Alu sequences (the arrows indicate the polarity of these dispersed middle repetitive DNA elements) as well as the hypervariable region (HVR) in the 5'-flanking region of the insulin gene. The zero coordinate on this map corresponds to the transcriptional initiation site of the insulin gene. The unmarked vertical lines represent *EcoRI* sites. The exons denoted by open boxes in the IGF-II gene represent exons which comprise the 5'-untranslated regions of the different IGF-II mRNA transcripts; there are promoters upstream of exons 1, 4 and 5. Adult liver IGF-II mRNA transcripts contain exons 1–3 and 6–8; fetal tissue and adult brain transcripts comprise exons 1–3 and 6–8; and some fetal liver transcripts contain exons 5–8.

facilitating studies of the segregation of the insulin gene in families in which some members are suspected of synthesizing and secreting an abnormal insulin or proinsulin molecule (see, for example, Elbein *et al.*, 1985).

1.6. Insulin, IGF-II and tyrosine hydroxylase genes are contiguous

The isolation and characterization of the human IGF-II gene revealed that the IGF-II gene was adjacent to the 3'-end of the insulin gene and that the insulin and IGF-II genes were transcribed in the same direction (Bell *et al.*, 1985) (Fig. 1.6). Moreover, the promoter which is adjacent to exon 1 of the IGF-II gene and functions in adult liver is only 1.4 kbp from the 3'-end of the insulin gene (de Pagter-Holthuizen *et al.*, 1987). The functional significance of the proximity of the insulin and IGF-II genes is unknown but one intriguing possibility is that IGF-II is also expressed in the insulin-producing B-cell and its expression in this tissue is coordinately regulated with insulin via the insulin gene enhancer. As three different tissue-specific and/or developmentally regulated promoters have been described in the IGF-II gene (Hoppener *et al.*, 1988), this is not an unreasonable hypothesis. As expected, the rat insulin II and IGF-II genes are also adjacent to one another (Soares *et al.*, 1986).

O'Malley & Rotwein (1988) have recently determined that the gene for the catecholamine biosynthetic enzyme, tyrosine hydroxylase, is also contiguous with the insulin and IGF-II genes (Fig. 1.6). The tyrosine hydroxylase gene is located upstream of the insulin gene; the 3'-end of this gene is separated by 2.7 kbp from the 5'-end of the insulin gene. The tyrosine hydroxylase, insulin and IGF-II genes all have the same transcriptional polarity.

The significance of the close proximity of the tyrosine hydroxylase, insulin and IGF-II genes is unclear, but it seems likely that this arrangement allows them to share common regulatory elements. As the IGF-II gene is expressed by most fetal and adult tissues, it is likely that it is also expressed in the fetal B-cell and possibly the adult B-cell as well. Although the tyrosine hydroxylase gene is primarily expressed in specific regions of the brain, sympathetic ganglia and adrenal medulla, it is also expressed together with the insulin gene in the precursor cells which give rise to various cells of the islet as well as in the developing nervous system (Alpert *et al.*, 1988). Thus, the proximity of the insulin and tyrosine hydroxylase genes would allow them to share common regulatory elements, thereby facilitating coordinate transcriptional activation in developing neurons and islet cells. Similarly, IGF-II gene transcription could be integrated with expression of the insulin and/or tyrosine hydroxylase genes.

1.7. Conclusion

The recent demonstration that the synthesis and secretion of an abnormal insulin protein can result in glucose intolerance and diabetes mellitus has stimulated interest in the role of the insulin gene in the development of this disorder (Haneda *et al.*, 1984; Nanjo *et al.*, 1987). In addition, the isolation of cDNA clones encoding a molluscan insulin-related peptide (Smit *et al.*, 1988) has also renewed interest in the genes encoding insulin and other members of the insulin superfamily and raised several questions regarding the origin and evolution of insulin and related peptides. Will the molluscan insulin-related peptide gene have the three-exon organization characteristic of vertebrate insulin genes? It is likely that the characterization of the molluscan insulin-related peptide gene as well as the isolation of cDNAs and genes encoding other invertebrate insulin or insulin-related peptides will provide valuable insight into the evolution of the insulin superfamily. In addition, the isolation of insulin-related peptides from invertebrate neurosecretory cells addresses an important issue regarding the presence of insulin or related peptides in the vertebrate central nervous system (CNS). Present data indicate that the insulin gene is not expressed in the adult CNS (Baskin *et al.*, 1988), although recent data suggest that it may be transiently expressed during embryogenesis in the cells of the neural tube and the neural crest (Alpert *et al.*, 1988). The physiological relevance of transient expression of insulin in the developing CNS remains to be determined, but none the less such expression suggests that insulin or proinsulin could play a role in this process. By exploiting the experimental foundation provided by studies on the insulin, relaxin and IGF genes as well as the invertebrate insulin-related neuropeptides, it may be possible to develop novel strategies which can be used to determine the function of insulin–proinsulin in the CNS as well as to identify unrecognized insulin-related molecules expressed in this tissue. If the current interest in the genes encoding insulin and related peptides can be a guide, it seems likely that our understanding of the evolution of insulin and related peptides will increase dramatically over the next few years. It also seems likely that nature will continue to surprise us regarding the structural and functional diversity of the insulin superfamily.

We thank Ms Julie Dicig for her invaluable assistance in the preparation of this manuscript. In addition, we thank Dr T. Shapiro and Dr D. F. Steiner for their comments on this manuscript, and Dr T. B. Shows for providing Fig. 1.1.

References

Alpert, S., Hanahan, D. & Teitelman, G. (1988). Hybrid insulin genes reveal a developmental lineage for pancreatic endocrine cells and imply a relationship with neurons. *Cell* **53**, 295–308.

Baskin, D.G., Wilcox, B.J., Figlewicz, D.P. & Dorsa, D.M. (1988). Insulin and insulin-like growth factors in the CNS. *Trends in Neurosciences* **11**, 107–11.

Bell, G.I., Gerhard, D.S., Fong, N.M., Sanchez-Pescador, R. & Rall, L.B. (1985). Isolation of the human insulin-like growth factor genes: Insulin-like growth factor II and insulin genes are contiguous. *Proceedings of the National Academy of Sciences of the U.S.A.* **82**, 6450–4.

Bell, G.I., Horita, S. & Karam, J.H. (1984). A polymorphic locus near the human insulin gene is associated with insulin-dependent diabetes mellitus. *Diabetes* **33**, 176–83.

Bell, G.I., Karam, J.H. & Rutter, W.J. (1981). A polymorphic region adjacent to the 5' end of the human insulin gene. *Proceedings of the National Academy of Sciences of the U.S.A.* **78**, 5759–63.

Bell, G.I., Pictet, R.L. & Rutter, W.J. (1980a). Analysis of the regions flanking the human insulin gene and sequence of an Alu family member. *Nucleic Acids Research* **8**, 4091–109.

Bell, G.I., Pictet, R.L., Rutter, W.J., Cordell, B., Tischer, E. & Goodman, H.M. (1980b). Sequence of the human insulin gene. *Nature* **284**, 26–32.

Bell, G.I., Selby, M.J. & Rutter, W.J. (1982). The highly polymorphic region near the human insulin gene is composed of simple tandemly repeating sequences. *Nature* **295**, 31–5.

Bryant-Greenwood, G.D. (1982). Relaxin as a new hormone. *Endocrine Reviews* **3**, 62–90.

Bucchini, D., Ripoche, M.-A., Stinnakre, M.G., Besbois, P., Lores,P., Monthioux, E., Absil, J., Lepesaut, J.-A., Pictet, R. & Jami, J. (1986). Pancreatic expression of human insulin gene in transgenic mice. *Proceedings of the National Academy of Sciences of the U.S.A.* **83**, 2511–15.

Chan, S.J., Episkopou, V., Zeitlin, S., Karathanasis, S.K., MacKrell, A., Steiner, D.F. & Efstratiadis, A. (1984). Guinea pig preproinsulin gene: an evolutionary compromise? *Proceedings of the National Academy of Sciences of the U.S.A.* **81**, 5046–50.

Clark, J.L. & Steiner, D.F. (1969). Insulin biosynthesis in the rat: demonstration of two proinsulins. *Proceedings of the National Academy of Sciences of the U.S.A.* **62**, 278–85.

Conlon, J.M., Dafgard, E., Falkmer, S. & Thim, L. (1986). The primary structure of ratfish insulin reveals an unusual mode of proinsulin processing. *FEBS Letters* **208**, 445–50.

Conlon, J.M., Davis, M.S. & Thim, L. (1987). Primary structure of insulin and glucagon from the flounder (*Platichthys flesus*). *General and Comparative Endocrinology* **66**, 203–9.

Conlon, J.M. & Thim, L. (1986). Primary structure of insulin and a truncated C-peptide from an elasmobranchian fish, *Torpedo marmorata*. *General and Comparative Endocrinology* **64**, 199–205.

Cordell, B., Bell, G., Tischer, E., DeNoto, F.M., Ullrich, A., Pictet, R., Rutter, W.J. & Goodman, H.M. (1979). Isolation and characterization of a cloned rat insulin gene. *Cell* **18**, 533–43.

Cutfield, J.F., Cutfield, S.M., Carne, A., Emdin, S.O. & Falkmer, S. (1986). The isolation, purification and amino acid sequence of insulin from the teleost fish *Cottus Scorpius* (daddy sculpin). *European Journal of Biochemistry* **158**, 117–23.

de Pagter-Holthuizen, P., Jansen, M., van Schaik, F.M.A., van der Kammen, R., Oosterwijk, C., Van den Brande, J.L. & Sussenbach, J.S. (1987). The human insulin-like growth factor II gene contains two development-specific promoters. *FEBS Letters* **214**, 259–64.

Dull, T.J., Gray, A., Hayflick, J.S. & Ullrich, A. (1984). Insulin-like growth factor II precursor gene organization in relation to insulin gene family. *Nature* **310**, 777–81.

Edlund, T., Walker, M.D., Barr, P.J. & Rutter, W.J. (1985). Cell-specific expression of the rat insulin gene: evidence for role of two distinct 5' flanking elements. *Science* **230**, 912–16.

Elbein, S.C., Gruppuso, P., Schwartz, R., Skolnick, M. & Permutt, M.A. (1985). Hyperproinsulinemia in a family with a proposed defect in conversion is linked to the insulin gene. *Diabetes* **34**, 821–4.

Frank, B.H., Pekar, A.H., Pettee, J.M., Schirmer, E.M., Johnson, M.G. & Chance, R.E. (1984). Isolation and characterization of a genetic variant of bovine proinsulin. *International Journal of Peptide Protein Research* **23**, 506–15.

Froesch, E.R. & Zapf, J. (1985). Insulin-like growth factors and insulin: comparative aspects. *Diabetologia* **28**, 485–93.

Giddings, S.J. & Carnaghi, L.R. (1988). The two nonallelic rat insulin mRNAs and pre-mRNAs are regulated coordinately *in vivo*. *Journal of Biological Chemistry* **263**, 3845–9.

Haley, J., Crawford, R., Hudson, P., Scanlon, D., Tregear, G., Shine, J. & Niall, H. (1987). Porcine relaxin: gene structure and expression. *Journal of Biological Chemistry* **262**, 11940–6.

Hallden, G., Gafvelin, G., Mutt, V. & Jornvall, H. (1986). Characterization of cat insulin. *Archives of Biochemistry and Biophysics* **247**, 20–7.

Hanahan, D. (1985). Hertiable formation of pancreatic β-cell tumours in transgenic mice expressing recombinant insulin-simian virus 40 oncogenes. *Nature* **315**, 115–22.

Haneda, M., Polonsky, K.S., Bergenstal, R.M., Jaspan, J.B., Shoelson, S.E., Blix, P.M., Chan, S.J., Kwok, S.C.M., Wishner, W.B., Zeidler, A., Olefsky, J.M., Freidenberg, G., Tager, H.S., Steiner, D.F. & Rubenstein, A.H. (1984). Familial hyperinsulinemia due to a structurally abnormal insulin. *New England Journal of Medicine* **310**, 1288–94.

Harper, M.E., Ullrich, A. & Saunders, G.F. (1981). Localization of the human insulin gene to the distal end of the short arm of chromosome 11. *Proceedings of the National Academy of Sciences of the U.S.A.* **78**, 4458–60.

Hoppener, J.W.M., Mosselman, S., Roholl, P.J.M., Lambrechts, C., Slebos, R.J.C., de Pagter-Holthuizen, P., Lips, C.J.M., Jansz, H.S. & Sussenbach, J.S. (1988). Expression of insulin-like growth factor-I and -II genes in human smooth muscle tumors. *The EMBO Journal* **7**, 1379–85.

Hudson, P., Haley, J., John, M., Cronk, M., Crawford, R., Haralambidis, J., Tregear, G., Shine, J. & Niall, H. (1983). Structure of a genomic clone encoding biologically active human relaxin. *Nature* **301**, 628–31.

Jhoti, H., McLeod, A.N., Blundell, T.L., Ishizaki, H., Nagasawa, H. & Suzuki, A. (1987). Prothoracicotropic hormone has an insulin-like tertiary structure. *FEBS Letters* **219**, 419–25.

Karlsson, O., Edlund, T., Moss, J.B., Rutter, W.J. & Walker, M.D. (1987). A mutational analysis of the insulin gene transcriptional control region: Expression in beta cells is dependent on two related sequences within the enhancer. *Proceedings of the National Academy of Sciences of the U.S.A.* **84**, 8819–23.

Kwok, S.C.M., Chan, S.J. & Steiner, D.F. (1983). Cloning and nucleotide sequence analysis of the dog insulin gene. *Journal of Biological Chemistry* **258**, 2357–63.

Lakshmikumaran, M.S., D'Ambrosio, E., Laimins, L.A., Lin, D.T. & Furano, A.V. (1985). Long interspersed repeated DNA (LINE) causes polymorphism at the rat insulin 1 locus. *Molecular and Cellular Biology* **5**, 2197–203.

Lomedico, P., Rosenthal, N., Efstratiadis, A., Gilbert, W., Kolodner, R. & Tizard, R. (1979). The structure and evolution of two nonallelic rat preproinsulin genes. *Cell* **18**, 545–58.

Montminy, M.R., Sevarino, K.A., Wagner, J.A., Mandel, G. & Goodman, R.H. (1986). Identification of a cyclic-AMP-responsive element within the rat somatostatin gene. *Proceedings of the National Academy of Sciences of the U.S.A.* **83**, 6682–6.

Nagasawa, H., Kataoka, H., Isogai, A., Tamura, S., Suzuki, A., Mizoguchi, A., Fujiwara, Y., Suzuki, A., Takahashi, S.Y. & Ishizaki, H. (1986). Amino acid sequence of a prothoracicotropic hormone of the silkworm *Bombyx mori*. *Proceedings of the National Academy of Sciences of the U.S.A.* **83**, 5840–3.

24   G. I. BELL AND S. SEINO

Nanjo, K., Miyano, M., Kondo, M., Sanke, T., Nishimura, S., Miyamura, K., Inouye, K., Given, B.D., Chan, S.J., Polonsky, D.S., Tager, H.S., Steiner, D.F. & Rubenstein, A.H. (1987). Insulin Wakayama: familial mutant insulin syndrome in Japan. *Diabetologia* **30**, 87–92.
Nielsen, D.A., Welsh, M., Casadaban, M.J. & Steiner, D.F. (1985). Control of insulin gene expression in pancreatic β-cells and in an insulin-producing cell line, RIN-5F cells. *Journal of Biological Chemistry* **260**, 13585–9.
Nir, U., Walker, M.D. & Rutter, W.J. (1986). Regulation of rat insulin 1 gene expression: evidence for negative regulation in nonpancreatic cells. *Proceedings of the National Academy of Sciences of the U.S.A.* **83**, 3180–4.
Ohlsson, H. & Edlund, T. (1986). Sequence-specific interactions of nuclear factors with the insulin gene enhancer. *Cell* **45**, 35–44.
O'Malley, K.L. & Rotwein, P. (1988). Human tyrosine hydroxylase and insulin genes are contiguous on chromosome 11. *Nucleic Acids Research* **16**, 4437–46.
Perler, F., Efstratiadis, A., Lomedico, P., Gilbert, W., Kolodner, R. & Dodgson, J. (1980). The evolution of genes: the chicken preproinsulin gene. *Cell* **20**, 555–66.
Plisetskaya, E.M., Pollock, H.G., Elliott, W.M., Youson, J.H. & Andrews, P.C. (1988). Isolation and structure of lamprey (*Petromyzon marinus*) insulin. *General and Comparative Endocrinology* **69**, 46–55.
Pollock, H.G., Kimmel, J.R., Hamilton, J.W., Rouse, J.B., Ebner, K.E., Lance, V. & Rawitch, A.B. (1987). Isolation and structures of alligator gar (*Lepisosteus spatula*) insulin and pancreatic polypeptide. *General and Comparative Endocrinology* **67**, 375–82.
Rogers, J. (1983). Retroposons defined. *Nature* **301**, 460.
Rotwein, P., Pollock, K.M., Didier, D.K. & Krivi, G.G. (1986). Organization and sequence of the human insulin-like growth factor I gene. *Journal of Biological Chemistry* **261**, 4828–32.
Sample, C.E. & Steiner, D.F. (1987). Tissue-specific binding of a nuclear factor to the insulin gene promoter. *FEBS Letters* **222**, 332–6.
Seino, S., Steiner, D.F. & Bell, G.I. (1987). Sequence of a New World primate insulin having low biological potency and immunoreactivity. *Proceedings of the National Academy of Sciences of the U.S.A.* **84**, 7423–7.
Selden, R.F., Skoskiewicz, M.J., Howie, K.B., Russell, P.S. & Goodman, H.M. (1986). Regulation of human insulin gene expression in transgenic mice. *Nature* **321**, 525–8.
Smit, A.B., Vreugdenhil, E., Ebberink, R.H.M., Geraerts, W.P.M., Klootwijk, J. & Joose, J. (1988). Growth-controlling molluscan neurons produce the precursor of an insulin-related peptide. *Nature* **331**, 535–8.
Soares, M.B., Schon, E., Henderson, A., Kathanasis, S.K., Cate, R., Zeitlin, S., Chirgwin, J. & Efstratiadis, A. (1985). RNA-mediated gene duplication: the rat preproinsulin I gene is a functional retroposon. *Molecular and Cellular Biology* **5**, 2090–103.
Soares, M.B., Turken, A., Ishii, D., Mills, L., Episkopou, V., Cotter, S., Zeitlin, S. & Efstratiadis, A. (1986). Rat insulin-like growth factor II gene. *Journal of Molecular Biology* **192**, 737–52.
Steiner, D.F., Chan, S.J., Welsh, J.M. & Kwok, S.C.M. (1985). Structure and evolution of the insulin gene. *Annual Review of Genetics* **19**, 463–84.
Walker, M.D., Edlund, T., Boulet, A.M. & Rutter, W.J. (1983). Cell-specific expression controlled by the 5'-flanking region of insulin and chymotrypsin genes. *Nature* **306**, 557–61.
Wentworth, B.M., Schaefer, I.M., Villa-Komaroff, L. & Chirgwin, J.M. (1986). Characterization of two nonallelic genes encoding mouse preproinsulin. *Journal of Molecular Evolution* **23**, 305–12.
Winter, W.E., Maclaren, N.K., Bell, G.I., Beppu, H., Cooper, D.L. & Wakeland, E.K. (1987). Restriction-fragment-length polymorphisms of the 5'-flanking region of insulin I gene in BB and other rat strains: absence of association with IDDM. *Diabetes* **36**, 193–8.

Zabel, B.U., Kronenberg, H.M., Bell, G.I. & Shows, T.B. (1985). Chromosome mapping of genes on the short arm of human chromosome 11: parathyroid hormone gene is at 11p15 together with the genes for insulin, c-Harvey-*ras* 1, and β-hemoglobin. *Cytogenetics and Cell Genetics* **39**, 200–5.

2

The regulation of insulin gene expression

MICHAEL D. WALKER

2.1. Summary

Insulin biosynthesis in adult vertebrates is strictly confined to the B-cells of the endocrine pancreas. The molecular mechanisms underlying this selective expression of the insulin gene, and of other genes whose expression is also regulated in cell-specific fashion, are beginning to be revealed, largely through the application of recombinant DNA and gene transfer methodologies. A critical point of control in insulin gene expression is at the stage of transcription initiation. The effects are mediated through specific DNA sequences located in the 5'-flanking region of the gene, and appear to involve recognition of these sequences by protein factors present in B-cells. The characterization of such sequence-specific DNA binding proteins and the genes which encode them will contribute decisively to our understanding of differentiation and development of the endocrine pancreas.

2.2. Introduction

As a multicellular organism develops, dramatic morphological differences appear among the various cell types, culminating in the generation of the final differentiated phenotype. These changes are associated with, and to a large extent caused by, alterations in gene expression. Thus a molecular understanding of development requires the elucidation of the process of selective gene expression, both at the level of the final differentiated cell and at earlier stages of development where a commitment (or determination) to differentiate has occurred but no morphological changes are yet apparent. In general, the process of differentiation is not accompanied by extensive loss or rearrangement of the genetic material (Gurdon, 1974). Indeed, under appropriate experi-

27

mental circumstances, the differentiated cell exhibits considerable 'plasticity' or ability to express genes characteristic of an alternative cell type, for example following cell fusion (Blau *et al.*, 1985). Thus the inability of a particular cell type to express a given gene cannot be attributed to irreversible changes at the DNA level (e.g. loss of the gene). If so, what mechanisms direct the characteristic pattern of gene expression of a differentiated cell?

Although the biosynthesis of proteins is a complex multistage process, it has become increasingly clear that a critical control step in determining the final level of a protein is the first stage in the process, namely initiation of transcription (Darnell, 1982). In prokaryotic systems, where powerful genetic tools have permitted a detailed understanding of regulatory mechanisms, transcription initiation is controlled by the interaction of specific proteins with DNA sequences located immediately upstream of the transcribed DNA (Jacob & Monod, 1961). This interaction can lead to activation or repression of transcription. For complex multicellular organisms comparable genetic tools are not available. However, an alternative form of genetic analysis, 'reverse genetics', has in recent years permitted detailed analysis of transcriptional control mechanisms. This approach, rather than relying on the isolation of rare mutant phenotypes and subsequent identification of the genetic basis for the phenotype, uses the cloned gene as starting point, and examines the consequence on phenotype of introduction to live cells of wild type DNA or DNA mutated *in vitro*. Such analyses of transcription control regions of protein-encoding genes of mammalian cells have led to the identification of two major classes of control element (reviewed in McKnight & Tjian, 1986; Maniatis *et al.*, 1987). The promoter is a DNA element located proximal to the transcription start site that is capable of initiating transcription: its activity is dependent on position and orientation relative to the transcribed sequence. In many genes, the promoter contains a consensus TATA sequence located about 20–25 bp upstream from the transcription initiation site (Breathnach & Chambon, 1981). This sequence is necessary for selection of the appropriate start site and maximal rates of initiation. In addition, promoters often contain 'upstream promoter elements' which are required for efficient initiation. Enhancers were initially identified in genes of animal viruses and subsequently in cellular genes. In contrast to promoters, enhancers are incapable of initiating transcription, but rather increase the activity of homologous or heterologous promoters relatively independently of orientation or precise positioning with respect to the transcription unit.

Both promoters and enhancers are believed to function by interacting directly with sequence-specific DNA binding proteins. For some eukaryotic transcription control elements, the cognate binding factors have been identified and molecularly cloned (Kadonaga *et al.*, 1987; Evans, 1988). This has permitted analysis of structure–function relationships. The important observation has been made that these proteins can be divided into distinct domains, one capable of recognizing the appropriate DNA sequence and the other capable of transcription activation (Keegan *et al.*, 1986). This supports the idea that transcription activation occurs as a result of direct protein–protein interactions between DNA binding proteins and components of the transcription machinery, possibly RNA polymerase (Sigler, 1988).

2.3. Experimental approaches

The unambiguous definition of transcription control elements requires a functional assay in which putative regulatory DNA sequences are introduced to live cells and their ability to direct expression measured. In principle, the issue of selective expression of the insulin gene could be addressed by introducing (by transfection) an intact insulin gene to cultured cells of either pancreatic or non-pancreatic origin, followed by measurement of expression of the transfected gene. However, this approach has a number of drawbacks.

(*a*) The expression of the exogenous gene at either the RNA or protein level may be below detectable levels, especially when assayed following transient transfection assay (where only a small percentage of cells contribute to production of a signal (see below)).

(*b*) Discrimination between expression of the endogenous gene and the exogenous one is likely to be complicated by the close similarity at both the nucleic acid level and protein level of vertebrate insulin genes (reviewed in Steiner *et al.*, 1985; see also Chapter 1).

(*c*) Such experiments do not immediately distinguish between control elements located in transcribed DNA sequences (exons and introns) as opposed to elements contained within non-transcribed regions (5'- and 3'-flanking DNA).

An alternative approach based on the construction of hybrid genes (reviewed in Kelly & Darlington, 1985) provides an elegant and convenient solution to these problems. In analyzing the activity of transcription control regions, hybrid genes have been examined utilizing three major types of expression protocol (Fig. 2.1):

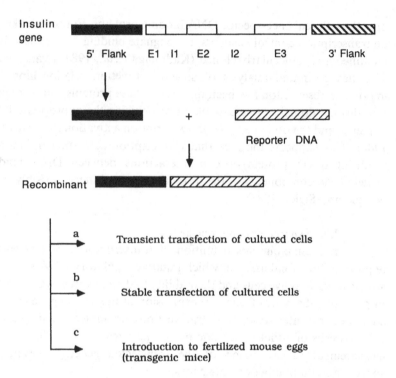

Fig. 2.1. Use of reporter functions using three different experimental protocols. A hybrid gene is constructed by fusing the 5'-flanking DNA sequence from the gene of interest (here insulin) to the DNA encoding an appropriate reporter function. E1–3 represent the exons and I1–2 the introns of the insulin gene. For transient assays the use of an easily assayable enzyme as reporter is common. For stable assays, it is necessary to use an enzyme whose expression can be selected for. However, the gene of interest can be one for which no selective criteria are available if transfection is performed in the presence of an additional marker for which selection conditions are available (Gorman, 1986). To produce transgenic mice, DNA is introduced to fertilized mouse eggs using glass micropipettes, and the eggs implanted to foster mothers and allowed to develop to maturity. The reporter function in this case can be an oncoprotein, enzyme or any gene product whose expression can be readily assayed.

(a) Transient assay (reviewed in Gorman, 1986): Hybrid genes are constructed consisting of the putative regulatory region of the gene in question, linked to a so-called 'reporter' gene, which generally encodes an easily assayable protein, often an enzyme. The genes are introduced to cultured cells by exposure to DNA in the presence of calcium phosphate, DEAE dextran, or by direct electroporation. A small percentage of cells take up and express the DNA. However, expression is transient since the majority of the DNA does not integrate and therefore cannot replicate. After a few days expression levels become undetectable. Therefore expression is usually measured within about 48 h of exposure of the cells to DNA. An obvious advantage of this method is the rapidity with which information can be obtained.

(b) Stable assay (reviewed in Gorman, 1986): Occasionally the exogenous DNA in a transfected cell becomes integrated into the chromosome of the host. If this DNA contains a gene encoding a selectable marker, for example an enzyme required for survival in the presence of antibiotic or antimetabolites (Mulligan & Berg, 1981; Southern & Berg, 1982), then application of this selection will permit isolation of those cells containing and expressing the exogenous gene. Although the selection procedure takes considerably longer to carry out than a typical transient transfection (2–4 weeks, depending on the growth rate of the cells) an advantage is that a stable line is generated which can subsequently be characterized in detail.

(c) Transgenic animals (reviewed in Palmiter & Brinster, 1986): Established cell lines are valuable models for the corresponding cell type in intact animals. However, they differ significantly from the normal counterpart, perhaps most strikingly in their ability to grow indefinitely in culture, and usually also in their more limited expression of the differentiated phenotype. Are transcription control elements defined by transfection of established cells equally important in the natural *in vivo* setting? The production of transgenic animals (most commonly mice) permits this issue to be addressed directly. DNA is injected by glass micropipette into fertilized one-cell embryos. Injected embryos are implanted into the oviduct of a foster mother, and are permitted to develop to term. The expression of the injected DNA (or transgene) is then measured in the cell types of interest. Since every cell in a given transgenic mouse contains the transgene in an identical chromosomal position, every possible cell type is theoretically available for comparison of expression, a situation which does not apply for cell culture experiments. Furthermore, the transgene is present in the cell from the one cell stage onward and is therefore exposed to similar influences during development as is the normal endogenous gene.

2.4. Identification of control sequences

2.4.1. Role of 5'-flanking DNA

In our initial experiment using the hybrid gene approach, we chose to use a transient transfection procedure based on the use of chloramphenicol acetyltransferase (CAT), a reporter function developed by Gorman et al. (1982b). This enzyme is particularly well suited as a reporter molecule since mammalian cells show negligible endogenous activity and enzymatic assay is rapid and sensitive. We constructed hybrid genes containing 5'-flanking sequences of the rat insulin I gene, the human insulin gene, the chymotrypsin gene (a gene expressed in the exocrine pancreas only) and the Rous sarcoma virus LTR (a promoter–enhancer complex which is active in a wide range of cell types (Gorman et al., 1982a)). These sequences were fused to the DNA encoding CAT. The hybrid genes were introduced to cultured cell lines of either pancreatic endocrine, pancreatic exocrine, or non-pancreatic origin, using the calcium phosphate method. Constructions containing 5'-flanking sequences of the rat or human insulin gene were from 50 to 200-fold more active in a line of cells derived from the endocrine pancreas than in cells derived from fibroblasts or pancreatic exocrine cells (Walker et al., 1983) (Fig. 2.2). This result demonstrated several important points.

(a) At least a portion of the information responsible for determining cell-specific expression is present in a relatively short length of DNA (in the case of the rat insulin I gene, 410 bp of 5'-flanking DNA sequence).

(b) A rapid and convenient experimental protocol can be used to identify these sequences.

(c) Since under the conditions of these experiments the transfected DNA is not located in its normal location in the genome, and in fact does not integrate into the chromosome to any significant extent, the positional requirements for reconstituting at least a portion of the cell-specific effect are relatively non-stringent.

The important role of 5'-flanking sequence in cell-specific expression has been confirmed using two other hybrid gene approaches. Episkopou et al. (1984) generated stable cell lines using a retrovirus-based approach in which the 5'-flanking region of the rat insulin II gene was linked to the DNA encoding the xanthine–guanine phosphoribosyl transferase (gpt) gene. Infection of cells with this marker, followed by application of appropriate selection conditions, led to the conclusion that

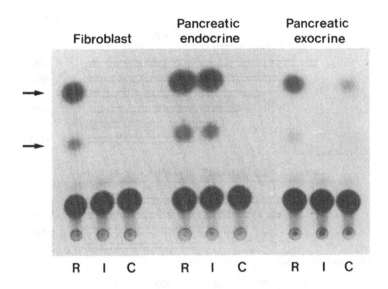

Fig. 2.2. Chloramphenicol acetyltransferase (CAT) activity directed by eukaryotic 5′-flanking sequences in different cells. Recombinant plasmids containing 5′-flanking sequences were applied to tissue culture cells. After 48 h, CAT activity from extract containing 25 μg of protein was determined. The photograph shows an autoradiogram of the thin layer chromatography separation of CAT reaction products. The cell types used were CHO fibroblasts, HIT pancreatic endocrine cells and AR4-2J pancreatic exocrine cells (Jessop & Hay, 1980). The upper two spots, indicated by arrows, correspond to the two isomers of [14C]chloramphenicol monoacetate and the lower intense spot corresponds to unreacted [14C]chloramphenicol. The 5′ sequences are: R, Rous sarcoma virus LTR; I, rat insulin; C, chymotrypsin.

efficient expression occurs in insulin-producing cell lines but not in fibroblasts. Hanahan (1985) prepared transgenic mice containing the promoter of the rat insulin II gene linked to the DNA encoding an oncoprotein, the SV40 T-antigen. Such mice developed tumors of the endocrine B-cells, and in no other cell, indicating that expression of the transgene was limited to this cell type.

The convenience of the transient assay for identification of DNA control elements permitted extensive deletion analysis to be performed on the flanking DNA regions (Walker *et al.*, 1983). The analysis (Fig. 2.3) revealed that, for the rat and human insulin genes, sequences upstream of about − 300, including the polymorphic region present on the insulin gene (Bell *et al.*, 1982; see also Chapter 12) had no effect on

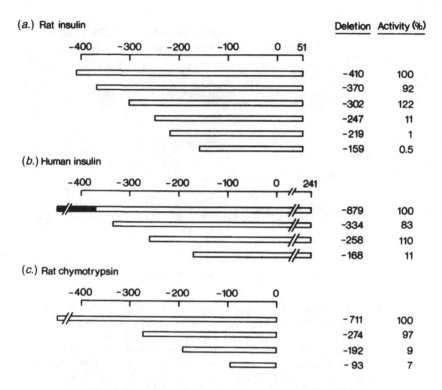

Fig. 2.3. Relative activity of insulin (*a*, *b*) and chymotrypsin (*c*) gene fragments containing 5' deletions. Deletions from the 5'-ends of the above fragments were produced and the precise end points of the deletions determined by DNA sequencing. The end points of all shortened fragments are indicated in base pairs relative to the transcription start site. Activities were determined following transfection of appropriate cells (endocrine cells for insulin gene and exocrine cells for chymotrypsin gene). The relative activities are expressed as a percentage of the activity of the largest fragment tested for each gene. The location of the polymorphic region of the insulin gene (Bell *et al.*, 1982) is indicated by the shaded bar.

expression of the linked gene. However, removal or sequences located downstream of − 302 led to dramatic losses in activity.

2.4.2. *Cell-specific enhancer and promoter*

The type of regulatory signals involved were further examined in a second series of experiments (Edlund *et al.*, 1985) designed to identify putative enhancer sequences in the 5'-flanking DNA. The analysis revealed that the 5'-flanking DNA of the rat insulin I gene does contain a cell-specific enhancer located between about − 103 and − 333. This enhancer is capable of strongly activating (40-fold) the thymidine kinase (TK) promoter in B-cells. No activation was seen with other cell types, indicating that the enhancer represents an important component of the machinery determining cell-specific expression of the insulin gene.

We observed, in addition, that DNA sequences downstream of this region, comprising the insulin promoter, displayed negligible enhancer activity despite quite significant conservation of sequence among the characterized mammalian genes (Bell *et al.*, 1980; Chan *et al.*, 1984). In order to test whether these sequences are involved in cell specificity, we substituted the enhancer region of the insulin gene with the enhancer of the Moloney sarcoma virus (MSV) which is known to display activity in a wide range of cell types (Laimins *et al.*, 1982). In this 'enhancer swap' situation, the novel construction retains 10-fold preferential expression in the insulin-producing cell type, indicating that the promoter region *per se* retains significant ability to direct B-cell specific expression despite the absence of enhancer activity.

2.4.3. Positive and negative control

Based on the behavior of prokaryotic control systems, we considered the possibility of control mechanisms operating through either positive or negative control. Experiments based on fusion of cell types of different transformed phenotype have given evidence for both types of control (Lewin, 1980). In general when such cell fusion experiments are performed the outcome is loss of expression or 'extinction' of the differentiated phenotype (Killary & Fournier, 1984). For example, when liver cells are fused to fibroblasts the expression of characteristic liver genes is lost. When such hybrid cells are cultured over time they lose chromosomes and the liver phenotype can be regained. Interestingly, this reacquisition of phenotype correlates with the loss of a specific fibroblast chromosome. These results are consistent with the notion that fibroblasts contain transcriptional repressor molecules which reduce expression of genes normally expressed only in other differentiated cell types. On the other hand, cell fusion can occasionally also result in activation of a previously silent locus. This situation is best characterized in cases where muscle cells are a partner in the fusion. Production of heterokaryons can lead to activation of previously silent muscle genes in the fusion partner (Blau *et al.*, 1985). In our deletion analysis discussed above, we observed loss of expression following deletion of sequences, a result consistent with the loss of a binding site for a positively acting transcription control factor. We do not observe any increase in expression following such deletions in either expressing or non-expressing cells. We therefore designed experiments to test directly the possible involvement of negatively acting factors in reducing expression in heterologous cells.

Our procedure was based on the principle of competition *in vivo* between plasmids and involved introduction to COS fibroblast cells (Gluzman, 1981) of two plasmids, one containing insulin enhancer sequences linked to the CAT reporter gene (test gene) and the other (competitor plasmid) containing the same enhancer sequences without CAT sequences but containing the SV40 origin of replication (to permit replication in COS cells). Upon introduction of these plasmids to COS cells the competitor plasmid (but not the test plasmid) will replicate to high copy number and titrate out putative repressor molecules, leading to derepression of the test plasmid. Indeed, when such experiments are performed, the presence of insulin enhancer sequences on both plasmids leads to a 5-fold elevation of expression as compared with a situation where the enhancer is absent from either or both plasmids (Nir *et al.*, 1986). Effectively, the presence of the competitor plasmids reveals cryptic activity of the insulin enhancer in an inappropriate cell type. The role of repressor molecules may be to reduce such 'leaky' expression. The putative repressor may have a less stringent sequence requirement for binding than the positive factors since the rat amylase enhancer (Boulet *et al.*, 1986), when present at high copy number, was able to activate cryptic enhancer activity in COS cells. If the function of the repressor is to repress a subset of cell-specific enhancers it may be present also in insulin-producing cells, where its effects are presumably overridden by the dominant activating effects of positive *trans*-activators.

Evidence for the existence of additional negatively acting DNA elements in the rat insulin I gene was obtained by Laimins *et al.* (1986), who identified such 'silencer' regions between 2 and 4 kb upstream of the rat insulin I gene.

2.4.4. Regulation by glucose

Insulin plays a central role in integrating carbohydrate metabolism. In order to fulfill this role the output of insulin from the pancreas must respond appropriately to fluctuating levels in the bloodstream of several metabolites, most importantly glucose (Hedeskov, 1980). This is achieved by regulation at multiple levels, including regulation of stability (Welsh *et al.*, 1985) and translation (Welsh *et al.*, 1986) of insulin mRNA. By measuring the rate of transcription of the insulin gene, an effect of glucose on the rate of transcription initiation was detected (Nielsen *et al.*, 1985). To confirm this effect at the transcription level, and to begin to address the issue of the underlying mechanism, we measured the ability of glucose to increase transcription from a hybrid gene including 410 bp of the rat insulin I gene 5'-flanking DNA linked to CAT, upon introduction to the glucose-responsive HIT cell line (Santerre *et al.*, 1981). The presence of 20 mM glucose reproducibly led to a 5-fold increase in CAT expression levels using such constructions (L.G. Moss, W.J. Rutter & M.D. Walker, unpublished). Control plasmids containing unrelated control sequences did not respond significantly to glucose. Thus glucose effects on transcription are mediated at least in part through 5'-flanking DNA sequences.

2.4.5. Systematic mutagenesis

Given the apparent complexity of the insulin regulatory region, we set out to define the precise location of the various types of control element. The procedure used is based conceptually on the elegant 'linker scanning' approach developed by McKnight and colleagues (McKnight *et al.*, 1981). The availability of automated oligonucleotide synthesizers prompted us to use a synthetic oligonucleotide approach to generate a systematic series of replacement mutants. For convenience, the DNA was constructed as a series of short (*c.* 20 base) overlapping oligonucleotides. The synthetic fragments were ligated together into an expression plasmid and the structure of the regulatory region verified by sequencing. For mutant sequences oligonucleotides were synthesized containing blocks of non-complementary transversions (i.e. A↔C, G↔T) spanning an average of 10 bp. All plasmids were introduced to cultured cells and expression levels determined by measurement of CAT enzymatic activity and steady state levels of CAT mRNA (Karlsson *et al.*, 1987). Although mutation of several regions leads to marked reduction in activity, relatively few mutations cause dramatic loss of activity (Fig. 2.4). The most severely affected regions are (-23 to -32), which spans the TATA box, mutant (-104 to -112) and mutant (-233 to -241). The TATA homology, which is found in most pol II genes, is unlikely to be contributing to cell-specific expression and is probably involved in interacting with constitutive components of the transcription machinery. However, the two other mutants implicate regions which are conserved well among the various insulin genes (Chan *et al.*, 1984) but are not found in identical form in other genes (Moss *et al.*, 1988). These regions are therefore likely to be involved in cell-specific expression. Interestingly, the -104 to

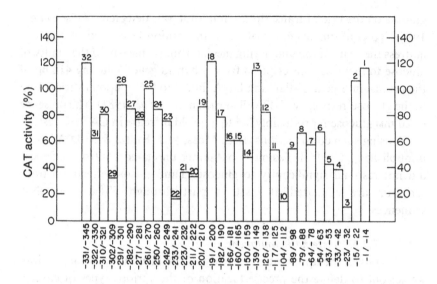

Fig. 2.4. CAT activities directed by mutant plasmids. Plasmids bearing mutations in the insulin 5'-flanking DNA were transfected into HIT cells and CAT activity was measured. Activities are expressed relative to that directed by the wild type plasmid.

-112 and -233 to -241 regions are identical to each other in 8 out of 9 positions, raising the possibility that they are involved in interactions with similar or identical transcription factors. We have tested the ability of these sequences on their own (in the absence of any additional sequence from the rat insulin gene) to elevate expression from a heterologous promoter. As a single copy the sequence stimulates expression c. 2–3-fold in insulin-producing cells. When two copies are combined the expression increases to 5–10-fold (O. Karlsson, M.D. Walker, W.J. Rutter & T. Edlund, unpublished). Thus, although the regulatory region is complex, a measure of cell specificity can be obtained using a very small unit possibly corresponding to the binding site of a single protein. This observation will significantly simplify analysis of control mechanisms.

2.4.6. Role of other sequences in cell specificity

The experiments described above utilizing cultured cells define 5'-flanking sequences which play an important role in cell-specific expression. However, these experiments do not directly address the issue of whether other sequences, located elsewhere in the gene, also play a decisive role in vivo. The transgenic mouse experiment involving insulin–T antigen fusion genes (Hanahan, 1985) suggested that expression of the transgene is appropriate in a qualitative sense. However, given the substantial effects of expression of T antigen on cell metabolism, this protein cannot be regarded as a neutral reporter, and hence levels of T antigen mRNA and protein in B-cells cannot be used as a quantitative measure of transcription initiation from the insulin control region. In order to address the issue of whether 5'-flanking sequences are sufficient to quantitatively direct cell specificity, we designed a construction containing the rat insulin I flanking DNA sequences linked to the human placental lactogen (hPL) gene, a much more neutral reporter function. In preliminary results we observed that levels of pancreatic hPL mRNA were comparable to those of the endogenous mouse insulin mRNA (M.D. Walker, W.J. Rutter & D. Hanahan, unpublished) indicating that the 5'-flanking sequences constitute a very significant component of the cell-specific transcription machinery and may represent the only sequences involved in determining cell specificity. Direct measurement of transcription rate will be required to establish this.

2.4.7. Biochemical approaches

Given the presumed role of sequence-specific transcription factors in activation of promoters and enhancers, much effort has been directed towards purification of these factors biochemically and characterization of their activities. Considerable progress has been made especially in identification of factors involved in transcription of viral genes (summarized in Jones et al., 1988).

In vitro analysis of the interaction of transcription factors with DNA is commonly performed using the techniques of DNase I footprinting (Galas & Schmitz, 1978) and gel mobility shift (Fried & Crothers, 1981). In both of these methods, radioactively labeled DNA containing the sequence of interest is allowed to interact with cell extracts, and the binding of proteins is monitored by determining the ability of the protein to protect the DNA from nuclease digestion (footprinting assay) or the ability of the protein to bind specifically to the DNA, thereby altering its mobility on polyacrylamide gel electrophoresis (mobility shift assay). In the case of the rat insulin I gene, analysis by DNase I footprinting revealed that three distinct regions, E1 (-290 to -335), E2 (-205 to -222) and E3 (-154 to -169) within the enhancer are protected from DNase I cleavage by nuclear extracts prepared from insulin-producing cells (Ohlsson & Edlund, 1986). The protection of E1 was observed with extracts from insulin-producing cells only, whereas protection of E2 and

E3 was also observed with extracts from other cell types. By using the more sensitive gel mobility shift assay, with either whole cell extracts (Moss *et al.*, 1988) or nuclear extracts (Ohlsson *et al.*, 1988), it was possible to show that the two most important elements of the insulin enhancer, namely (− 104 to − 112) and (− 233 to − 241), also interact *in vitro* with nuclear proteins present in insulin-producing cells. Furthermore, nuclear extracts from other cell types show greatly reduced binding activity (Ohlsson *et al.*, 1988). Since these two sequences compete with one another for binding activity, it seems that they bind to a similar protein(s) (designated IEF1) which plays an important role in determining cell-specific expression. In an independent study (Sample & Steiner, 1987), gel mobility shift analysis showed interaction of a nuclear factor found preferentially in insulin-producing cells with a fragment encompassing the rat insulin II gene promoter (from − 86 to + 47).

2.5. Expression during development

Analysis of the expression of insulin hybrid genes during embryonic development in transgenic mice has begun to reveal some exciting new insights into the development of pancreatic endocrine cells, and also into the regulatory mechanisms underlying selective gene expression (Alpert *et al.*, 1988). When transgenes containing the rat insulin 5′-flanking region are used, expression of the linked gene is observed in all hormone-producing cell types of the endocrine pancreas, and indeed this unexpected pattern of expression is true also for the endogenous insulin gene, raising the possibility that the various endocrine cells of the pancreas have a common precursor. This conclusion is consistent with the observation made by several groups that cell lines derived from pancreatic endocrine origin are capable of producing more than one hormone (Gazdar *et al.*, 1980; Madsen *et al.*, 1986). Whatever mechanisms permit expression of multiple hormones in a single cell type, they are clearly operating via a short portion of insulin 5′-flanking DNA, since the behavior of the transgene mimics that of the endogenous gene.

In neuronal tissue, on the other hand, the behavior of the transgene appears to differ from that of the endogenous insulin gene: the transgene is expressed during embryogenesis in cells of the neural tube and neural crest, yet there is no evidence for expression of the insulin gene in vertebrate brain (Baskin *et al.*, 1987). The authors suggest that these results could be due to a combinatorial interaction between elements of the insulin regulatory region and cryptic elements within the reported molecule, as has been suggested in other systems (Evans *et al.*, 1986).

Alternatively, sequence elements outside of the flanking region used may be responsible for negatively controlling expression of the insulin gene in non-pancreatic cell types. Based on the fact that developing endocrine cells express the characteristic neuronal enzyme tyrosine hydroxylase and that in developing neurons a transferred insulin promoter is active, it is suggested that these cell types may share a common precursor. The human tyrosine hydroxylase gene has recently been shown to be located adjacent to the 5'-end of the human insulin gene, at a distance of less than 3 kb (O'Malley & Rotwein, 1988). The unexpected expression of the tyrosine hydroxylase in the pancreas may therefore be related to its proximity to the insulin gene enhancer. Whether the insulin gene is influenced by control elements of the tyrosine hydroxylase gene is unknown.

2.6. Prospects

Given the powerful molecular genetic and biochemical tools currently available, one can anticipate relatively rapid progress toward definition of the mechanism underlying the differential expression of the insulin gene. Some of the more promising lines of approach include the following.

2.6.1. Biochemical

As described above, biochemical purification of cell extracts can be used to isolate transcription factors. An important technical advance has been the recent development of an effective procedure for affinity chromatographic purification of sequence-specific DNA-binding proteins (Kadonaga & Tjian, 1986). The purified material can be assayed for binding activity to the cognate DNA sequence and also for ability to increase transcription activity in *in vitro* transcription systems. The approach has been applied successfully to several viral transcription factors. However, progress has been slower for cell-specific mammalian genes. Gorski *et al.* (1986) have shown that extracts from liver cells are capable of effective *in vitro* transcription from the mouse albumin promoter, whereas extracts from other cell types are much less effective. Bodner & Karin (1987) were able to show that partly purified protein fractions isolated from pituitary cells were capable of stimulating the ability of fibroblast extracts to transcribe *in vitro* the growth hormone gene. One limitation of the approach based on purification by affinity chromatography is the requirement for specific DNA recognition by the purified protein. Several heteromeric transcription factors have recently been isolated (Hahn & Guarente, 1988; Chodosh *et al.*, 1988). Since DNA binding activity can be lost upon dissociation of the complex, modified isolation procedures will be required in such cases.

2.6.2. Genetic

In this approach, DNA from a differentiated cell is used to transfect cultured fibroblast cells and the expression of genes characteristic of the differentiated cell is either screened for or selected for. Such a scheme has been successfully applied in the past to isolate oncogenes (Shih et al., 1981) and has led recently to the identification of two genes which appear to play an important role in the development of differentiated muscle cells (Davis et al., 1987; Pinney et al., 1988). The method used for this involved introduction to fibroblast cells of cloned DNA sequences followed by screening for transformants exhibiting the desired differentiated phenotype. This is based on the observation that 10T1/2 fibroblasts can be converted to myoblasts by transfection with genomic DNA derived from myoblasts, implying that a single genetic locus is sufficient to mediate the conversion (Lassar et al., 1986). In practice, cloned cDNA or cosmid sequences were shown to be able to effect the transfection at high frequency. Given their ability to alter the differentiated phenotype, including the production of dramatic differences in gene expression, the isolated genes may not simply encode transcription factors which activate expression of genes characteristic of the fully differentiated phenotype but may instead be controlling genes located at a 'higher' position in the regulatory hierarchy. This highlights a striking advantage of the genetic approach as compared to the biochemical approach, whose utility is limited to isolating transcription factors relevant to the expression of a specific gene. On the other hand, the genetic approach is limited in that an appropriate choice of recipient cell may be critical: 10T1/2 cells show an inherent predisposition towards muscle cell differentiation (Taylor & Jones, 1979). Furthermore, the transfection approach is unlikely to succeed if simultaneous expression of multiple unlinked genes is essential for bringing about the differentiative transition. Finally, transfection efficiency must be high in order to be able to monitor the rare differentiated colony following transfer of DNA, thereby limiting the range of cell types which can be used as recipients. This latter limitation may be overcome if retroviral expression cDNA libraries are used (Murphy & Efstratiadis, 1987); such an approach can be used with any cell type which can be efficiently infected, including derivatives of B-cells which have lost the ability to express the insulin gene, possibly because of lack of expression of a key transcription factor (Episkopou et al., 1984).

2.6.3. Expression library screening

Cloned DNA sequences can be routinely isolated from plasmid or bacteriophage libraries by *in situ* screening with nucleic acid probes complementary in sequence to that of the desired gene or with antibody probes directed toward the protein encoded by the desired gene. An additional possibility for genes encoding sequence-specific DNA binding proteins, is to probe the library using a DNA probe corresponding in sequence to the binding site of the particular protein. The feasibility of this approach was demonstrated by Singh *et al.* (1988), who identified a recombinant clone from a λGT11 expression library using a DNA sequence corresponding to the binding site of the mammalian transcription factors H2TF1 and NF-κB which are involved in expression of histocompatibility genes and immunoglobulin genes respectively.

The identification of genes encoding transcription factors will allow isolation in quantity of both the wild type protein and site-directed mutants to permit detailed analysis of the structure and function of the protein. Furthermore, the availability of DNA probes for these genes will permit analysis of the intriguing question of how the genes which encode cell specific transcription factors are themselves controlled. In the case of the NF-κB factor, considerable experimental evidence has accumulated to suggest that the factor is present in many cell types but active only in lymphoid cells (Sen & Baltimore, 1986*a*, *b*; Bauerle & Baltimore, 1988). Indeed, the gene identified as encoding the H2TF1 or NF-κB factor is transcribed in both lymphoid and non-lymphoid cells (Singh *et al.*, 1988).

2.7. Concluding remarks

Substantial progress has been made in recent years toward defining the molecular basis underlying the expression of genes characteristic of terminally differentiated cell types. In the case of the insulin gene, multiple control elements lie within the 5′-flanking DNA region responsible for cell-specific expression. A dominant component is attributable to an 8 bp sequence which is found twice within the region. The identification of these sequences, combined with evolving methodologies for isolating transcription control factors and their genes, will lead to further rapid progress in understanding of regulation of the insulin gene in the normal state and the possible role of inappropriate expression in disease states such as diabetes. In type I diabetes, endogenous insulin production is deficient because of partial or complete destruction of the B-cells caused by autoimmune or viral damage. Although the primary defect is not insulin gene expression *per se*, the prospect of treating affected persons with gene therapy has been considered (Selden *et al.*, 1987*b*). Some promising steps have been taken: the production of transgenic mice bearing an intact human insulin gene results in healthy animals in which the transgene is expressed and regulated in normal fashion, implying that the DNA fragment used contained all the appropriate signals and that these are capable of responding following gene transfer (Selden *et al.*, 1986). However, formidable technical problems remain. In particular, an appropriate recipient cell line must be identified: pancreatic B-cells would be subject to immune attack in the patient. Thus, an alternative cell must be identified in which the insulin gene can be expressed and conversion of proinsulin to insulin effected, possibly a non-pancreatic endocrine cell which is capable of proteolytic processing of prohormones (Moore *et al.*, 1983). Selden *et al.* (1987*a*, *b*) have described a protocol, 'transkaryotic implantation', which involves introduction of the desired gene into appropriate cells in culture, followed by characterization of the expression of the introduced gene and implantation into animals. Clearly, if such an approach is to be used in humans, it is essential that insulin output by such a modified cell type be regulated appropriately. A thorough understanding of control of expression of the normal gene is a prerequisite for this.

It is a pleasure to acknowledge the major contributions of several colleagues with whom I have been fortunate to collaborate at various stages during my work on insulin gene regulation. In particular Dr Thomas Edlund and Dr William J. Rutter were closely involved throughout. In addition I thank Dr Uri Nir, Dr Anne Boulet, Dr Olof Karlsson, Dr Larry Moss, Dr Jennifer Barnett Moss and Dr Doug Hanahan, with whom I enjoyed valuable collaborations during various phases of the work. I gratefully acknowledge support from the Juvenile Diabetes Foundation International (Career Development Award). Work carried out in my laboratory was supported by grants from the NIH (GM 38817) and the Israel Academy of Sciences and Humanities.

References

Alpert, S., Hanahan, D. & Teitelman, G. (1988). Hybrid insulin genes reveal a developmental lineage for pancreatic endocrine cells and imply a relationship with neurons. *Cell* **53**, 295–308.

Baskin, D.G., Figlewicz, D.P., Woods, S.C., Porte, D. Jr. & Dorsa, D.M. (1987). Insulin and the brain. *Annual Review of Physiology* **49**, 335–47.

Bauerle, P.A. & Baltimore, D. (1988). Activation of DNA-binding activity in an apparently cytoplasmic precursor of the NF-κB transcription factor. *Cell* **53**, 211–17.

Bell, G.I., Pictet, R.L., Rutter, W.J., Cordell, B., Tischer, E. & Goodman, H.M. (1980). Sequence of the human insulin gene. *Nature* **284**, 26–32.

Bell, G.I., Selby, M.J. & Rutter, W.J. (1982). The highly polymorphic region near the human insulin gene is composed of simple tandemly repeating sequences. *Nature* **295**, 31–5.

Blau, H.M., Pavlath, G.K., Hardeman, E.C., Chiu, C.-P., Dilberstein, L., Webster, S.G., Miller, S.C. & Webster, C. (1985). Plasticity of the differentiated state. *Science* **230**, 758–66.

Bodner, M. & Karin, M. (1987). A pituitary-specific trans-acting factor can stimulate transcription from the growth hormone promoter in extracts of non-expressing cells. *Cell* **50**, 267–75.

Boulet, A.M., Erwin, C.R. & Rutter, W.J. (1986). Cell specific enhancers in the rat exocrine pancreas. *Proceedings of the National Academy of Sciences of the U.S.A.* **83**, 3599–603.

Breathnach, R. & Chambon, P. (1981). Organization and expression of eukaryotic split genes coding for proteins. *Annual Review of Biochemistry* **50**, 349–83.

Chan, S.J., Episkopou, V., Zeitlin, S., Karathanasis, S.K., Mackrell, A., Steiner, D.F. & Efstratiadis, A. (1984). Guinea pig preproinsulin gene: an evolutionary compromise. *Proceedings of the National Academy of Sciences of the U.S.A.* **81**, 6046–51.

Chodosh, L.A., Baldwin, A.S., Carthew, R.W. & Sharp, P.A. (1988). Human CCAAT-binding proteins have heterologous subunits. *Cell* **53**, 11–24.

Darnell, J.E. (1982). Variety in the level of gene control in eukaryotic cells. *Nature* **297**, 365–71.

Davis, R.L., Weintraub, H. & Lassar, A.B. (1987). Expression of a single transfected cDNA converts fibroblasts to myoblasts. *Cell* **51**, 987–1000.

Edlund, T., Walker, M.D., Barr, P.J. & Rutter, W.J. (1985). Cell-specific expression of the rat insulin gene: evidence for role of two distinct 5'-flanking elements. *Science* **230**, 912–16.

Episkopou, V., Murphy, A.J.M. & Efstratiadis, A. (1984). Cell-specified expression of a selectable hybrid gene. *Proceedings of the National Academy of Sciences of the U.S.A.* **81**, 4657–61.

Evans, R.M. (1988). The steroid and thyroid hormone receptor superfamily. *Science* **240**, 889–95.

Evans, R.M., Weinberger, C., Hollenberg, S., Swanson, L., Nelson, C. & Rosenfeld, M.G. (1986). Inducible and developmental control of neuroendocrine genes. *Cold Spring Harbor Symposium on Quantitative Biology* **50**, 389–97.

Fried, M. & Crothers, M.D. (1981). Equilibria and kinetics of lac repressor operator interactions by polyacrylamide gel electrophoresis. *Nucleic Acids Research* **9**, 6505–25.

Galas, D. & Schmitz, A. (1978). DNase footprinting: A simple method for detection of protein-DNA binding specificity. *Nucleic Acids Research* **5**, 3157–70.

Gazdar, A.F., Chick, W.L., Herbert, K.O., King, D.L., Gordon, C.W. & Lauris, V. (1980). Continuous, clonal, insulin- and somatostatin-secreting cell lines established from a transplantable rat islet cell tumor. *Proceedings of the National Academy of Sciences of the U.S.A.* **77**, 3519–23.

Gluzman, Y. (1981). SV-40 transformed simian cells support the replication of early SV40 mutants. *Cell* **23**, 175–82.

Gorman, C. (1986). High efficiency gene transfer into mammalian cells. In *DNA cloning: a practical approach* vol. 2 (ed. D.M. Glover), pp. 143–90. Oxford: IRL Press.

Gorman, C.M., Merlino, G.T., Willingham, M.C., Pastan, T.R. & Howard, B.H. (1982a). The Rous sarcoma virus long terminal repeat is a strong promoter when introduced into a variety of eukaryotic cells by DNA-mediated transfection. *Proceedings of the National Academy of Sciences of the U.S.A.* **79**, 6777–81.

Gorman, C.M., Moffat, L.F. & Howard, B.H. (1982b). Recombinant genomes which express chloramphenicol acetyltransferase in mammalian cells. *Molecular and Cellular Biology* **2**, 1044–51.

Gorski, K., Carnciro, M. & Schibler, U. (1986). Tissue-specific in vitro transcription from the mouse albumin promoter. *Cell* **47**, 767–76.

Gurdon, J.B. (1974). The control of gene expression in animal development. Cambridge, Massachusetts: Harvard University Press.

Hahn, S. & Guarente, L. (1988). Yeast HAP2 and HAP3: transcriptional activators in a heteromeric complex. *Science* **240**, 317–20.

Hanahan, D. (1985). Heritable formation of pancreatic β-cell tumours in transgenic mice expressing recombinant insulin/simian virus 40 oncogenes. *Nature* **315**, 115–22.

Hedeskov, C.J. (1980). Mechanism of glucose-induced insulin secretion. *Physiological Reviews* **60**, 442–509.

Jacob, F. & Monod, J. (1961). Genetic regulatory mechanisms in the synthesis of proteins. *Journal of Molecular Biology* **3**, 318–56.

Jessop, N.W. & Hay, R.J. (1980). Characteristics of two rat pancreatic exocrine cell lines derived from transplantable tumors. *In Vitro* **16**, 212.

Jones, N.C., Rigby, P.W.J. & Ziff, E.B. (1988). Trans-acting protein factors and the regulation of eukaryotic transcription: lessons from studies on DNA tumor viruses. *Genes and Development* **2**, 267–81.

Kadonaga, J.T., Carner, K.R., Masiarz, F.R. & Tjian, R. (1987). Isolation of a cDNA encoding transcription factor Sp1 and functional analysis of the DNA binding domain. *Cell* **51**, 1079–90.

Kadonaga, J.T. & Tjian, R. (1986). Affinity purification of sequence-specific DNA binding proteins. *Proceedings of the National Academy of Sciences of the U.S.A.* **83**, 5889–93.

Karlsson, O., Edlund, T., Moss, J.B., Rutter, W.J. & Walker, M.D. (1987). A mutational analysis of the insulin gene transcription control region: Expression in beta cells is dependent on two related sequences within the enhancer. *Proceedings of the National Academy of Sciences of the U.S.A.* **84**, 8819–23.

Kelly, J.H. & Darlington, G.J. (1985). Hybrid genes: molecular approaches to tissue specific gene regulation. *Annual Review of Genetics* **19**, 273–96.

Killary, A.M. & Fournier, R.E.K. (1984). A genetic analysis of extinction: trans-dominant loci regulate expression of liver specific traits in hepatoma hybrid cells. *Cell* **38**, 523–34.

Laimins, L., Holmhren-Konig, M. & Khoury, G. (1986). Transcriptional 'silencer' element in rat repetitive sequences associated with the rat insulin I gene locus. *Proceedings of the National Academy of Sciences of the U.S.A.* **83**, 3151–5.

Laimans, L.A., Khoury, G., Gorman, C., Howard, B. & Gruss, P. (1982). Host-specific activation of transcription by tandem repeats from simian virus 40 and Moloney murine sarcoma virus. *Proceedings of the National Academy of Sciences of the U.S.A.* **79**, 6453–7.

Lassar, A.B., Paterson, B.M. & Weintraub, H. (1986). Transfection of a DNA locus that mediates the conversion of 10T1/2 fibroblasts to myoblasts. *Cell* **47**, 649–56.

Lewin, B. (1980). *Gene expression*, vol. 2: *Eucaryotic chromosomes*. New York: Wiley and Sons.

Madsen, O.D., Lersson, L.I., Rehfeld, J.F., Schwartz, T.W., Lernmark, A., Labrecque, A.D. & Steiner, D.F. (1986). Cloned cell lines from a transplantable islet cell tumor are heterogeneous and express cholecystokinin in addition to islet hormones. *Journal of Cell Biology* **103**, 2025–34.

Maniatis, T., Goodbourn, S. & Fischer, J.A. (1987). Regulation of inducible and tissue-specific gene expression. *Science* **236**, 1237–45.

McKnight, S.L., Gavis, E.R., Kingsbury, R. & Axel, R. (1981). Analysis of transcriptional regulatory signals of the HSV thymidine kinase gene: identification of an upstream control region. *Cell* **25**, 385–98.

McKnight, S. & Tjian, R. (1986). Transcriptional selectivity of viral genes in mammalian cells. *Cell* **46**, 795–805.

Moore, H.-P., Walker, M.D., Lee, F. & Kelly, R.B. (1983). Expressing a human proinsulin cDNA in a mouse ACTH-secreting cell. Intracellular storage, proteolytic processing, and secretion on stimulation. *Cell* **35**, 531–8.

Moss, L., Moss, J.B. & Rutter, W.J. (1988). Systematic binding analysis of the insulin gene transcription control region: insulin and immunoglobulin enhancers utilize similar transactivators. *Molecular and Cellular Biology* **8**, 2620–7.

Mulligan, R.D. & Berg, P. (1981). Selection for animal cells that express the *E. coli* gene coding for xanthine-guanine phosphoribosyltransferase. *Proceedings of the National Academy of Sciences of the U.S.A.* **78**, 2072–6.

Murphy, A.J.M. & Efstratiadis, A. (1987). Cloning vectors for expression of cDNA libraries in mammalian cells. *Proceedings of the National Academy of Sciences of the U.S.A.* **84**, 8277–81.

Nielsen, D.A., Welsh, M., Casadaban, M.J. & Steiner, D.F. (1985). Control of insulin gene expression in pancreatic β-cells and in an insulin-producing cell line, RIN-5F cells. I. Effects of glucose and cyclic AMP on the transcription of insulin mRNA. *Journal of Biological Chemistry* **260**, 13585–9.

Nir, U., Walker, M.D. & Rutter, W.J. (1986). Regulation of rat insulin 1 gene expression: evidence for negative regulation in non-pancreatic cells. *Proceedings of the National Academy of Sciences of the U.S.A.* **83**, 3180–4.

Ohlsson, H. & Edlund, T. (1986). Sequence-specific interactions of nuclear factors with the insulin gene enhancer. *Cell* **45**, 35–44.

Ohlsson, H., Karlsson, O. & Edlund, T. (1988). A beta-cell-specific protein binds to the two major regulatory sequences of the insulin gene enhancer. *Proceedings of the National Academy of Sciences of the U.S.A.* **85**, 4228–31.

O'Malley, K.L. & Rotwein, P. (1988). Human tyrosine hydroxylase and insulin genes are contiguous on chromosome 11. *Nucleic Acids Research* **16**, 4437–45.

Palmiter, R.D. & Brinster, R.L. (1986). Germ-line transformation of mice. *Annual Review of Genetics* **20**, 465–99.

Pinney, D.F., Pearson-White, S.H., Konieczny, S.F., Lattham, K.E. & Emerson, C.P. Jr. (1988). Myogenic lineage determination and differentiation: Evidence for a regulatory gene pathway. *Cell* **53**, 781–93.

Sample, C.E. & Steiner, D.F. (1987). Tissue-specific binding of a nuclear factor to the insulin gene promoter. *FEBS Letters* **222**, 332–6.

Santerre, R.F., Cook, R.A., Crisel, R.M.D., Sharp, J.D., Schmidt, R.J., Williams, D.C. & Wilson, C.P. (1981). Insulin synthesis in a clonal cell line of simian virus 40-transformed hamster pancreatic beta cells. *Proceedings of the National Academy of Sciences of the U.S.A.* **78**, 4339–43.

Selden, R.F., Skoskiewictz, M.J., Howie, K.B., Russell, P.S. & Goodman, H.M. (1986). Regulation of human insulin gene expression in transgenic mice. *Nature* **321**, 525–8.

Selden, R.F., Skoskiewictz, M.J., Howie, K.B., Russell, P.S. & Goodman, H.M. (1987*a*). Implantation of genetically engineered fibroblasts into mice: implications for gene therapy. *Science* **236**, 714–18.

Selden, R.F., Skoskiewictz, M.J., Russell, P.S. & Goodman, H.M. (1987*b*). Regulation of insulin gene expression: implications for gene therapy. *New England Journal of Medicine* **317**, 1067—76.

Sen, R. & Baltimore, D. (1986*a*). Multiple nuclear factors interact with the immunoglobulin enhancer sequences. *Cell* **46**, 705–16.

Sen, R. & Baltimore, D. (1986*b*). Inducibility of κ immunoglobulin enhancer-binding protein NF-κB by a posttranslational mechanism. *Cell* **47**, 921–8.

Shih, C., Padhy, L.C., Murray, M. & Weinberg, R.A. (1981). Transforming genes of carcinomas and neuroblastomas introduced into mouse fibroblasts. *Nature* **290**, 261–4.

Sigler, P.B. (1988). Acid blobs and negative noodles. *Nature* **333**, 210–12.

Singh, H., LeBowitz, J.H., Baldwin, A.S., Jr. & Sharp, P.A. (1988). Molecular cloning of an enhancer binding protein: isolation by screening of an expression library with a recognition site DNA. *Cell* **52**, 415 23.

Southern, P.J. & Berg, P. (1982). Transformation of mammalian cells to antibiotic resistance with a bacterial gene under control of the SV40 early region promoter. *Journal of Molecular and Applied Genetics* **1**, 327–41.

Steiner, D.F., Chan, S.J., Welsh, J.M. & Kwok, S.C.M. (1985). Structure and evolution of the insulin gene. *Annual Review of Genetics* **19**, 463–84.

Taylor, S.M. & Jones, P.A. (1979). Multiple new phenotypes induced in 10T1/2 and 3T3 cells treated with 5-azacytidine. *Cell* **17**, 771–9.

Walker, M.D., Edlund, T., Boulet, A.M. & Rutter, W.J. (1983). Cell-specific expression controlled by the 5' flanking region of insulin and chymotrypsin genes. *Nature* **306**, 557–61.

Welsh, M., Nielsen, D.A., MecKrell, A.J. & Steiner, D.F. (1985). Control of insulin gene expression in pancreatic β-cells and in an insulin-producing cell line, RIN-5F cells. II. Regulation of insulin mRNA stability. *Journal of Biological Chemistry* **260**, 13590–4.

Welsh, M., Scherberg, N., Gilmore, R. & Steiner, D.F. (1986). Translational control of insulin biosynthesis. Evidence for regulation of elongation, initiation and signal-recognition-particle-mediated translation arrest by glucose. *Biochemical Journal* **235**, 459–67.

3

The translational control of proinsulin synthesis by glucose

NOBUYUKI ITOH

3.1. Introduction

Insulin has long been known to be synthesized in the pancreatic B-cells of islets of Langerhans. However, an additional protein, similar to but larger than insulin, was discovered by Steiner & Oyer (1967). These workers used human islet-cell tumors, but it was not until a method for isolating actual islets from the pancreas was established that further progress could be made in the study of insulin biosynthesis. Using islets thus isolated from rat pancreases, Steiner and others were able to show the larger protein to be a precursor of insulin and proposed for it the name proinsulin (Steiner *et al.*, 1972; Steiner & Tager, 1979; Permutt, 1981).

Both proinsulin synthesis and insulin secretion in the islets of Langerhans are shown to be highly stimulated by glucose, in marked contrast to non-proinsulin proteins, the synthesis of which is only slightly stimulated. Thus, there must be in the islets a highly selective mechanism for proinsulin synthesis (Steiner *et al.*, 1972; Steiner & Tager, 1979; Permutt, 1981), either at the transcription level of the proinsulin gene or at the translation level of the proinsulin mRNA or both. Permutt & Kipnis (1972) demonstrated that actinomycin D had little effect on the stimulation of proinsulin synthesis by glucose in the early phase (within 45 min). This result suggested that the stimulation of proinsulin synthesis was regulated at the post-transcription level. In order to clarify the mechanism by which glucose stimulates proinsulin synthesis, a direct determination of the proinsulin mRNA level and the transcriptional activity of the proinsulin gene in the islets during the induction of proinsulin synthesis by glucose seemed desirable. Recent advances in molecular biology techniques made it possible to resolve this problem. In this chapter, the present knowledge of the mechanism by which glucose stimulates proinsulin synthesis will be described.

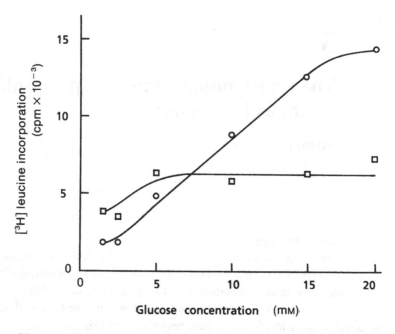

Fig. 3.1. Effect of glucose on proinsulin and non-proinsulin protein synthesis (M. Watanabe, N. Itoh & H. Okamoto, unpublished data). Rat pancreatic islets were incubated for 60 min at various glucose concentrations in the presence of [³H]leucine. The [³H]leucine-labeled product in the islets was analyzed by SDS–polyacrylamide gel electrophoresis and the amounts of synthesized proinsulin (circles) and non-proinsulin protein (squares) were determined by summing the [³H]leucine radioactivity of gel slices corresponding to the respective fractions.

3.2. Effect of glucose on proinsulin and non-proinsulin protein synthesis

Glucose was shown to stimulate proinsulin synthesis greatly in the pancreatic islets of Langerhans. Isolated rat pancreatic islets were incubated for 60 min at various concentrations of glucose in the presence of [³H]leucine. The labeled islet proteins were analyzed by SDS–polyacrylamide gel electrophoresis. The amounts of proinsulin and non-proinsulin protein synthesized were determined. As shown in Fig. 3.1, glucose stimulated proinsulin synthesis 9-fold, although non-proinsulin protein synthesis in the islets was stimulated only slightly (1.5-fold) (M. Watanabe, N. Itoh & H. Okamoto, unpublished data). The effect of glucose on the non-proinsulin protein synthesis was considered to be non-specific, since glucose has been shown to stimulate protein synthesis 1- to 2-fold in a large number of tissues (Davis & Morris, 1963). At 20 mM glucose, proinsulin synthesis composed 70% of total islet protein synthesis but only 28% at 1.5 mM when [³H]leucine labeling was used (Fig. 3.1). The content of leucine in rat proinsulin is 15.9% (Lomedico et al., 1979) and the average content of leucine in the total pancreatic protein was shown to be 9.3% (Müting & Wortmann, 1954). Thus, we can estimate that proinsulin synthesis accounted for 41% and 17% of the total islet protein at 20 mM and 1.5 mM glucose, respectively.

Using isolated rat islets, Permutt & Kipnis (1972) examined the mechanism by which glucose stimulates proinsulin synthesis. Islets incubated at 15.3 mM glucose incorporated 7- to 8-fold more [³H]leucine into proinsulin than those at 2.8 mM glucose. There was a linear incorporation of [³H]leucine into proinsulin at 2.8 mM glucose over a 2 h incubation. At 15.3 mM glucose, the rate of incorporation of [³H]leucine into proinsulin in the islets increased during the first 45 min and then continued linearly for the rest of the incubation; this indicated a lag in the attainment of the maximal rate of proinsulin synthesis at high concentrations of glucose which was not present at low concentrations of glucose. Actinomycin D, an inhibitor of RNA synthesis, had for the first 45 min little effect on the stimulation of proinsulin synthesis by glucose, which it subsequently did inhibit for the rest of the 2 h incubation. These results suggested that glucose could increase the proinsulin synthesis using previously existing mRNA (a post-transcriptional effect) and that glucose stimulated the proinsulin mRNA synthesis to increase proinsulin synthesis further (a transcriptional effect). In order to clarify the mechanism by which glucose stimulates proinsulin synthesis, direct quantification of proinsulin mRNA in the islets during the stimulation of proinsulin synthesis by glucose seemed desirable, as many side effects of actinomycin D had been documented (Lodish, 1976).

3.3. Effect of glucose on proinsulin mRNA level

It is possible to quantify proinsulin mRNA by assaying preproinsulin synthesized in a cell-free protein synthesizing system of wheatgerm or reticulocytes (Itoh et al., 1978; Okamoto et al., 1979). When nucleic acid extracted from rat pancreatic islets was incubated at 25 °C in the cell-free system, [³H]leucine was incorporated into protein at a linear rate with an incubation time of up to 90 min. SDS–polyacrylamide gel electrophoresis of the [³H]leucine-labeled product revealed that the cell-free translation product consisted of preproinsulin. The amount of radioactivity incorporated into the preproinsulin fraction was increased proportionally to the amount of nucleic acid added to the cell-free system. Using the cell-free translational assay system, Itoh *et al.* (1978) compared the proinsulin synthesis in isolated rat pancreatic islets with the corresponding mRNA level. When glucose was increased from 3.3 to 20 mM, proinsulin synthesis in isolated rat islets was stimulated. In contrast, the level of proinsulin mRNA in the islets remained essentially unchanged, irrespective of glucose concentrations (Itoh *et al.*, 1978; Okamoto *et al.*, 1979). This study, the first quantification of proinsulin mRNA in pancreatic islets, suggested that the induction of proinsulin by glucose was achieved at the translational level (translational control).

Investigations of the mechanisms by which transcription and translation are controlled in eukaryotic cells were facilitated by isolation of a specific mRNA and construction of the corresponding cDNA. Several investigators tried to purify proinsulin mRNA from mammalian and fish pancreatic islets or from X-ray-induced islet tumors (Duguid *et al.*, 1976; Rapoport *et al.*, 1976; Permutt *et al.*, 1978). However, because of the limited availability of these materials or the low content of proinsulin mRNA, proinsulin mRNA could not be obtained in a purified form. Itoh & Okamoto (1977) used rat pancreatic B-cell tumors induced by streptozotocin and nicotinamide as a material from which to purify proinsulin mRNA, since the tumors were found to contain a large amount of proinsulin mRNA and could easily be produced in rats (Itoh & Okamoto, 1977; Itoh *et al.*, 1979). Purification was achieved through the use of oligo(dT)-cellulose chromatography and sucrose gradient centrifugation (Itoh *et al.*, 1979). As judged from the translational activity in a cell-free protein synthesizing system of wheatgerm, proinsulin mRNA was purified 362-fold from the total nucleic acid of the tumors (Table 3.1). Purified proinsulin mRNA migrated as a single symmetrical peak on sucrose gradient centrifugation and as a single band on polyacrylamide gel electrophoresis in the presence of formamide. The translation product of purified proinsulin mRNA in the cell-free system was found to be only preproinsulin, as analyzed by SDS–polyacrylamide gel electrophoresis and immunoprecipitation. Proinsulin cDNA was then synthesized from the purified proinsulin mRNA. The fidelity of the hybridization reaction with proinsulin mRNA and its cDNA was ascertained by the following evidence: (i) the hybridization occurred within two orders of magnitude, as expected of a pseudo-first-order reaction; and (ii) the experimental $Cot_{1/2}$ value (2.1×10^{-4} mol l^{-1} s; the value required for half-reassociation) was in good agreement with the expected $Cot_{1/2}$ value (2.7×10^{-4})

Table 3.1. *Purification of proinsulin mRNA from rat B-cell tumors (Itoh et al., 1979).*

Proinsulin mRNA activity is given as the [^{3}H]leucine incorporated into preproinsulin synthesized in the wheatgerm system.

fraction	nucleic acid (μg)	proinsulin mRNA activity		
		(counts min^{-1} μg RNA^{-1})	-fold	yield (%)
total nucleic acid	8560	1450	1.0	100
LiCl-precipitable RNA	4880	1940	1.3	76
poly(A)-containing RNA	55.5	120 000	82.8	54
first sucrose gradient	10.2	507 000	350	42
second sucrose gradient	5.6	525 000	362	24

calculated from the Cot$_{\frac{1}{2}}$ value of the hybridization with mouse globin mRNA and its cDNA (Itoh *et al.*, 1979). A trace amount of proinsulin mRNA (0.05–0.5 ng) was able to be quantified with the cDNA hybridization method. As it was difficult to prepare many islets from the rat pancreas, hybridization was a very useful method for the quantification of the proinsulin mRNA in the islets (Itoh & Okamoto, 1980).

Itoh & Okamoto (1980) confirmed the previous result that the amount of proinsulin mRNA in the glucose-stimulated (25 mM glucose) or unstimulated (2.8 mM glucose) islets remained unchanged during the period of proinsulin induction for up to 60 min by the hybridization method, clearly indicating that proinsulin synthesis was regulated at the translation level (translational control) (Fig. 3.2). Recently, Fukumoto *et al.* (1986) examined whether the acute increase in insulin release induced by oral glucose ingestion in rats was associated with alterations in pancreatic proinsulin mRNA levels. Rats were given glucose orally in a conscious and unrestrained state. Pancreatic RNA was extracted and the relative level of proinsulin mRNA was determined by dot–blot analysis with a cloned rat proinsulin cDNA probe. Proinsulin mRNA levels showed no significant change after acute glucose administration, whereas blood glucose and plasma insulin increased rapidly to maximum values 15 min after glucose administration. This result indicated that proinsulin synthesis was also regulated at the translational level by glucose *in vivo*. The amount of

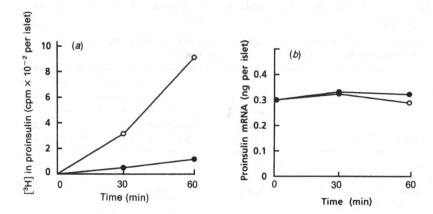

Fig. 3.2. Proinsulin mRNA levels and the time course of induction of proinsulin synthesis by glucose (Itoh & Okamoto, 1980). Rat pancreatic islets were incubated at 37° C in the presence of 2.8 mM (filled circles) or 25 mM (open circles) glucose and [³H]leucine. (a) [³H]Leucine-labeled product in the islets was analyzed by SDS–polyacrylamide gel electrophoresis and the amount of synthesized proinsulin was determined by summing [³H]leucine radioactivity of the gel slices corresponding to the proinsulin peak. (b) Aliquots of the nucleic acid from the incubated islets were used for the quantification of proinsulin mRNA by hybridization with proinsulin cDNA.

proinsulin mRNA was shown to be approximately 0.3 ng per islet. The number of molecules of proinsulin mRNA per rat pancreatic B-cell, the insulin-producing cell in the islet, was estimated to be approximately 2.7×10^5 from the following evidence (Okamoto, 1981): (i) the relative molecular mass was estimated to be approximately 2×10^5; (ii) the amount of DNA in the rat pancreas was shown to be 7.4 pg per cell and that in rat pancreatic islets was found to be 20 ng per islet; and (iii) B-cells made up 82% of the total cell population of rat pancreatic islets. Several investigators estimated the number of some specific mRNA molecules that are most abundant in certain types of eukaryotic cells. For example, the numbers of globin mRNA molecules in a reticulocyte and of ovalbumin mRNA molecules in an oviduct cell were reported to be 1.4 to 1.5×10^5 per cell (Humphries et al., 1976; Harris et al., 1975), figures that are thought to represent the highest possible mRNA content of eukaryotic cells. The number of proinsulin mRNA molecules per rat pancreatic B-cell was comparable to those of globin mRNA in a reticulocyte and of ovalbumin mRNA in an oviduct cell. Giddings et al. (1982) examined the long-term effect of glucose on the proinsulin mRNA level in pancreatic

Table 3.2. *Effect of glucose on synthesis of proinsulin, total RNA and proinsulin mRNA in rat pancreatic islets (Itoh et al., 1982).*

Rat pancreatic islets were incubated at 37 °C for 60 min in the presence of 2.8 mM glucose, 0 or 4 μg α-amanitin, and incubated further at 37 °C for 60 min by adding [³H]uridine for proinsulin mRNA synthesis, or [³H]leucine for proinsulin synthesis, and glucose (the final concentration of glucose was 2.8 mM or 25 mM). Synthesis values are in ³H disintegrations per minute per islet.

glucose	total RNA synthesis	proinsulin mRNA synthesis	proinsulin synthesis
2.8 mM	3.57×10^4	24.4	5.05×10^2
2.8 mM (+ α-amanitin)	3.08×10^4	4.9	4.44×10^2
25 mM	5.14×10^4	46.5	42.5×10^2
25 mM (+ α-amanitin)	4.99×10^4	11.4	36.1×10^2

islets of rats which were starved for 3 days. Rats starved for 3 days were injected with glucose or saline for 24 h. In starved rats, the proinsulin mRNA level was 15–29% that of fed controls. Glucose injection produced a specific 3- to 4-fold increase in the proinsulin mRNA level relative to the total pancreatic RNA within 24 h. Giddings *et al.* (1982) estimated that the proinsulin synthesis was regulated at least in part by the proinsulin mRNA level. But the significance of the long-term effect of glucose on the proinsulin mRNA level under non-physiological conditions remained to be elucidated.

3.4. Effect of glucose on proinsulin mRNA synthesis

As described above, the induction of proinsulin synthesis by glucose is achieved by enhancing the availability of proinsulin mRNA for translation rather than by simply increasing the total amount of proinsulin mRNA. On the other hand, as actinomycin D partly inhibited the stimulation of proinsulin synthesis by glucose, it was speculated that transcriptional control might also participate in the stimulation of proinsulin synthesis (Permutt & Kipnis, 1972). However, in order to clarify the participation of transcriptional control in the glucose stimulation of proinsulin synthesis, it seemed to be necessary to examine directly the effect of glucose on the transcriptional activity of the proinsulin gene in islets. Itoh *et al.* (1982) examined the effect of glucose on the transcriptional activity of the proinsulin gene in rat pancreatic islets using a cloned rat proinsulin cDNA probe. [³H]Uridine was incorporated into

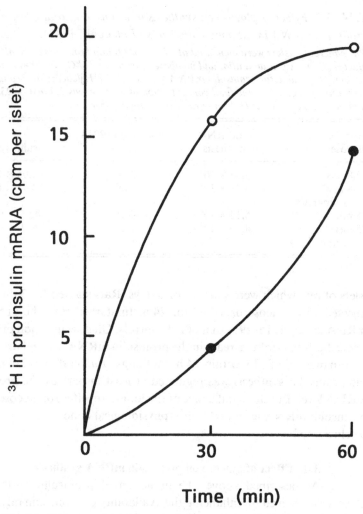

Fig. 3.3. Kinetics of newly synthesized proinsulin mRNA
accumulation in the nucleus and cytoplasm of rat pancreatic islets
(Itoh *et al.*, 1982). Rat pancreatic islets were incubated at 37° C for the
indicated time in the presence of [³H]uridine and 25 mM glucose.
Newly synthesized proinsulin mRNA was determined by the filter
hybridization method. Open circles, nucleus; filled circles, cytoplasm.

total RNA in isolated rat pancreatic islets at a linear rate during a period
of induction of up to 60 min. Total RNA synthesis in the islets incubated
for 60 min was only slightly stimulated by glucose (1.4-fold). Newly
synthesized proinsulin mRNA was determined by a filter hybridization
method with a cloned rat proinsulin cDNA probe. Proinsulin mRNA

synthesis was stimulated only 1.9-fold by glucose (Table 3.2). α-Amanitin, an RNA polymerase II inhibitor, strongly inhibited proinsulin mRNA synthesis but only slightly inhibited total RNA synthesis and proinsulin synthesis. These results also indicated that the stimulation of proinsulin synthesis by glucose was mainly regulated at the translational level. The accumulation kinetics of newly synthesized proinsulin mRNA in the nucleus and the cytoplasm of rat pancreatic islets was also studied (Fig. 3.3). A significant amount of newly synthesized proinsulin mRNA was detected in the cytoplasm of islets incubated for 30 min. From the cytoplasm of islets incubated for 60 min, nearly half of the newly synthesized proinsulin mRNA was recovered. Glucose stimulation at the transcription level may thus play a part in the maintenance of the proinsulin mRNA level in the islets (Itoh et al., 1982). Nielsen et al. (1985) examined the long-term effect of glucose on the transcriptional activity of the proinsulin gene in isolated rat pancreatic islets by incubating islets at low (3.3 mM) or high (17 mM) glucose for 24 h. Glucose stimulated the transcriptional activity of the proinsulin gene 5.6-fold in contrast with a 1.9-fold increase of total RNA synthesis in the islets. Nielsen et al. (1985) concluded that glucose exerted a specific stimulatory effect on the transcription of the proinsulin gene in the islets.

3.5. Mechanism of translational control

Permutt (1974) reported that glucose slightly stimulated the translation initiation efficiency of total protein synthesis in rat pancreatic islets. However, he did not determine the effect of glucose on the translation initiation and the polypeptide elongation efficiency of proinsulin synthesis in the islets. On the other hand, Lomedico & Saunders (1977) speculated, from the results of an experiment in which fetal bovine pancreatic poly(A)$^+$ RNA containing proinsulin mRNA was translated in a wheatgerm cell-free protein synthesizing system, that the initiation efficiency might play an important role in proinsulin synthesis. However, the role of the initiation efficiency in the glucose-induced proinsulin synthesis in pancreatic islets was not examined. Watanabe et al. examined the initiation efficiency of proinsulin mRNA translation and of non-proinsulin mRNA translation in glucose-stimulated or -unstimulated islets by two different methods, one being partial inhibition of polypeptide chain elongation with low doses of cycloheximide (Watanabe et al., 1982) and the other a hypertonic initiation block (M. Watanabe, N. Itoh & H. Okamoto, unpublished data).

Cycloheximide, which inhibits polypeptide chain elongation, provides

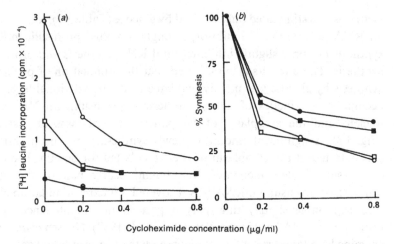

Cyclohexamide concentration (μg/ml)

Fig. 3.4. Effect of cycloheximide on proinsulin and non-proinsulin protein synthesis (Watanabe *et al.*, 1982). Rat pancreatic islets were incubated at 1.5 mM or 20 mM glucose for 60 min in the presence of [³H]leucine and various concentrations of cycloheximide. (*a*) The amounts of proinsulin (open circles, at 20 mM glucose; filled circles, at 1.5 mM glucose) and non-proinsulin protein (open squares, at 20 mM glucose; filled squares, at 1.5 mM glucose) synthesized at each concentration of cycloheximide were determined by summing [³H]leucine radioactivity incorporated into the respective fractions. (*b*) Proinsulin and non-proinsulin protein synthesis at each concentration of cycloheximide were plotted as a percentage of synthesis without cycloheximide.

one means of distinguishing between mRNAs with different initiation efficiencies through the preferential reduction of highly efficient mRNA translation (Lodish, 1971). Isolated rat pancreatic islets were incubated at 1.5 mM or 20 mM glucose in the presence of [³H]leucine and low doses of cycloheximide. Newly synthesized polypeptides were analyzed by SDS–polyacrylamide gel electrophoresis. The amounts of proinsulin and non-proinsulin protein synthesized at each concentration of cycloheximide were determined (Fig. 3.4*a*) (Watanabe *et al.*, 1982). As shown in Fig. 3.4*b*, there was no difference in the relative sensitivities of proinsulin and non-proinsulin protein synthesis to cycloheximide at 1.5 mM or 20 mM glucose. This result indicated that there was no difference in the initiation efficiencies of proinsulin and non-proinsulin mRNA translation at either glucose concentration. Under treatment with low doses of cycloheximide, the amount of total non-proinsulin protein (Fig. 3.4*a*) synthesized at 20 mM glucose was equal to that at 1.5 mM glucose. In contrast, the amount of proinsulin synthesized at 20 mM glucose was 5–6 times more than that at 1.5 mM glucose in spite of the application of the same treatment (Fig. 3.4a). According to Lodish (1976), if the initiation step is rate-limiting for protein synthesis, the relative amount of polypeptide synthesized should be proportional only to the relative amount of mRNA when polypeptide chain elongation is slowed by cycloheximide. The amounts of translatable proinsulin mRNA (Itoh & Okamoto, 1980) and non-proinsulin mRNA (Watanabe et al., 1982) were shown to be unaffected by glucose. Therefore, Watanabe et al. estimated that the glucose effect on non-proinsulin protein synthesis was achieved only because of an enhancement of the initiation efficiency and that the specific induction of proinsulin synthesis by glucose was not regulated at the initiation step (Watanabe et al., 1982). Hypertonicity, most commonly 100–150 mM excess of NaCl, which selectively inhibits the translational initiation, also provides another means of distinguishing between mRNAs with different initiation efficiencies (Jen et al., 1978; Ignotz et al., 1981). Isolated rat pancreatic islets were incubated at 1.5 mM or 20 mM glucose in the presence of [^{3}H]leucine and 100–150 mM excess of NaCl. There was no difference in the relative sensitivities of proinsulin and non-proinsulin protein synthesis to hypertonicity at 1.5 mM or 20 mM glucose (Fig. 3.5) (M. Watanabe, N. Itoh & H. Okamoto, unpublished data). This result was consistent with the result of the cycloheximide experiment that the effect of glucose on the initiation efficiency of proinsulin mRNA translation was equal to that on non-proinsulin mRNA translation. From these results, the induction of proinsulin synthesis by glucose was suggested to be regulated by the stimulation of polypeptide elongation efficiency or recruitment of proinsulin mRNA from an untranslatable pool. Glucose-stimulated proinsulin synthesis was not accompanied by an increase in the amount of proinsulin mRNA on the membrane-bound polysomes where proinsulin was shown to be synthesized (Table 3.3) (Itoh & Okamoto, 1980; Watanabe et al., 1982). Therefore, the induction of proinsulin synthesis by glucose was considered to be regulated mainly by the enhancement of polypeptide elongation efficiency (Watanabe et al., 1982).

There are two possible mechanisms by which proinsulin synthesis is stimulated at the polypeptide elongation step: (i) in glucose-unstimulated islets, the translation of proinsulin mRNA is selectively blocked at the elongation step, and in the stimulated islets, the blockage is released; or (ii) in the stimulated islets, the translation of proinsulin mRNA is selectively enhanced at the elongation step. As described above, the proinsulin synthesis accounted for 41% and 17% of total islet protein synthesis in glucose-stimulated and -unstimulated islets, respectively. RNA from rat

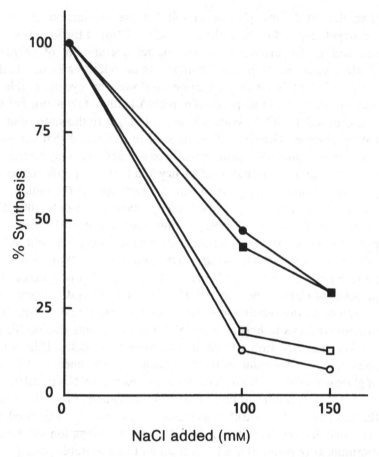

Fig. 3.5. Relative sensitivity of proinsulin and non-proinsulin synthesis
to hypertonicity (M. Watanabe, N. Itoh & H. Okamoto, unpublished
data). Rat pancreatic islets were incubated at 1.5 mM or 20 mM glucose
in the presence of [³H]leucine and 100 mM or 150 mM excess of NaCl.
The amounts of proinsulin and non-proinsulin protein synthesized
were determined as described in the legend for Fig. 3.4 and plotted as a
percent of synthesis without excess NaCl. (Symbols as for Fig. 3.4.)

pancreatic islets was translated in a reticulocyte cell-free system using
[³⁵S]methionine, and the translation products were analyzed by SDS–
polyacrylamide gel electrophoresis (M. Watanabe, N. Itoh & H.
Okamoto, unpublished observations). The amount of [³⁵S]methionine
incorporated into preproinsulin was found to be 23% of the amount
totally incorporated into islet protein. The average content of methionine
in rat preproinsulin I and II was 2.6% (Lomedico *et al.*, 1979), and the

Table 3.3. *Subcellular distribution of proinsulin mRNA in rat pancreatic islets (Itoh & Okamoto, 1980).*

Rat pancreatic islets were incubated at 37 °C for 60 min in the presence of 2.8 mM or 25 mM glucose. Subcellular fractions were prepared from the incubated islets. Aliquots of the nucleic acids from the subcellular fractions were used for the quantification of proinsulin mRNA by hybridization with proinsulin cDNA. Values are expressed as ng proinsulin mRNA per islet. The numbers in parentheses give the percentage of the postnuclear supernatant.

glucose	membrane-bound polysome	free polysome	post-polysomal supernatant
expt 1			
2.8 mM	0.099 (39.4)	0.033 (13.1)	0.199 (47.5)
25 mM	0.136 (57.9)	0.028 (11.9)	0.071 (30.2)
expt 2			
2.8 mM	0.109 (41.4)	0.031 (11.8)	0.123 (46.8)
25 mM	0.145 (58.4)	0.036 (14.5)	0.067 (27.1)

average content of methionine in total pancreatic protein had been shown to be 2.1% (Müting & Wortmann, 1954). Thus, proinsulin mRNA was estimated to make up 19% of the total islet mRNA. At 19%, proinsulin mRNA bore almost the same relation to total islet mRNA as proinsulin protein did to total islet protein (17%), suggesting that in the unstimulated islets both proinsulin mRNA and non-proinsulin mRNA were translated at the same translational efficiency. Therefore, the induction of proinsulin synthesis was assumed to be achieved by the selective enhancement of the translation of proinsulin mRNA at the elongation level and not by the release of selective blockage of proinsulin mRNA translation (N. Itoh, M. Watanabe & H. Okamoto, unpublished observations). Direct evidence for this assumption can be obtained by measurement of the ribosomal transit time for proinsulin synthesis.

Welsh *et al.* (1986) also examined the mechanism by which proinsulin synthesis is regulated by glucose and reported that glucose regulated the translation of proinsulin mRNA at three distinct levels: glucose (i) may have stimulated elongation rates of nascent proinsulin at concentrations up to 5.6 mM; (ii) stimulated the initiation of protein synthesis involving both proinsulin and non-proinsulin mRNA at concentrations above 3.3 mM; and (iii) increased the transfer of initiated proinsulin mRNA molecules from the cytoplasm to endoplasmic reticulum membranes by a signal-recognition particle (SRP)-mediated mechanism that involved the modification of interactions between SRP and its receptor.

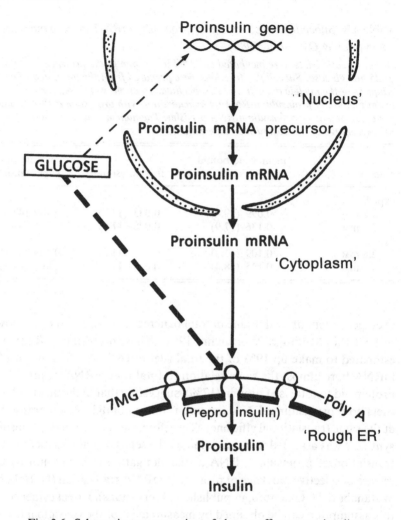

Fig. 3.6. Schematic representation of glucose effect on proinsulin synthesis.

3.6. Concluding remarks

As described in this chapter, the induction of proinsulin synthesis by glucose in pancreatic islets is mainly achieved by enhancement of the translation efficiency of proinsulin mRNA on the membrane-bound polysomes (translational control) (Fig. 3.6). Since the effect of glucose stimulation on protein synthesis in pancreatic islets is highly specific to proinsulin synthesis, it may be assumed that there is a factor specific to proinsulin mRNA translation. Whether metabolites of or receptors for glucose exist to account for the stimulus of proinsulin synthesis remains to be determined.

Numerous observations have indicated that the rate of protein synthesis, such as of ovalbumin (Chan *et al.*, 1973), tyrosine aminotransferase (Nickol *et al.*, 1976), tryptophan oxygenase (Roewekamp *et al.*, 1976), growth hormone (Martial *et al.*, 1977), phosphoenolpyruvate carboxykinase (Iynedjian & Hanson, 1977), immunoglobulin (Honjo *et al.*, 1977) and phosphogluconate dehydrogenase (Hutchison & Holten, 1978), was closely correlated with the amount of mRNA coding for these proteins, presumably through a changed rate of transcription of specific genes (transcriptional control). On the other hand, there are some well-documented examples of the translational control of specific gene products. In such examples as ornithine aminotransferase (Mueckler *et al.*, 1983), ornithine decarboxylase (White *et al.*, 1987) and ferritin (Rogers & Munro, 1987), translational control is regulated at the translation initiation level. The translational control of proinsulin synthesis, which seems to be regulated mainly at the elongation level, is a unique regulation of gene expression in eukaryotic cells. As insulin regulates the glucose level in the blood, the rates of both insulin secretion and insulin synthesis are required to change rapidly according to the change in the blood glucose level. This regulation of proinsulin synthesis is the means by which the rapid regulation of insulin synthesis by glucose is ensured.

References

Chan, L., Means, A.R. & O'Malley, B.W. (1973). Rates of induction of specific translatable messenger RNAs for ovalbumin and avidin by steroid hormones. *Proceedings of the National Academy of Sciences of the U.S.A.* **70**, 1870–4.

Davis, J.R. & Morris, R.N. (1963). Effect of glucose on incorporation of L-lysine-U-C[14] into testicular proteins. *American Journal of Physiology* **205**, 833–6.

Duguid, J.R., Steiner, D.F. & Chick, W.L. (1976). Partial purification and characterization of the mRNA for rat preproinsulin. *Proceedings of the National Academy of Sciences of the U.S.A.* **73**, 3539–43.

Fukumoto, H., Seino, Y., Koh, G., Takeda, J., Tsuji, K., Kurose, T., Kitano, N., Tsuda, K., Taminato, T. & Imura, H. (1986). Effect of oral glucose administration on preproinsulin mRNA in rats *in vivo*. *Biochemical and Biophysical Research Communications* **141**, 1201–6.

Giddings, S.J., Chirgwin, J. & Permutt, A.M. (1982). Effect of glucose on proinsulin messenger RNA in rats *in vivo*. *Diabetes* **31**, 624–9.

Harris, S.E., Rosen, J.M., Means, A.R. & O'Malley, B.W. (1975). Use of a specific probe for ovalbumin messenger RNA to quantitate estrogen-induced gene transcript. *Biochemistry* **14**, 2072–81.

Honjo, T., Shimizu, A., Tsuda, M., Natori, S., Kataoka, T., Dohmoto, C. & Mano, Y. (1977). Accumulation of immunoglobulin messenger ribonucleic acid in immunized mouse spleen. *Biochemistry* **16**, 5764–9.

Humphries, S., Windass, J. & Williamson, R. (1976). Mouse globin gene expression in erythroid and non-erythroid tissues. *Cell* **7**, 267–77.

Hutchison, J.S. & Holten, D. (1978). Quantitation of messenger RNA levels for rat liver 5-phosphogluconate dehydrogenase. *Journal of Biological Chemistry* **253**, 52–7.

Ignotz, G.G., Hokari, S., DePhilip, R.M., Tsukada, K. & Lieberman, I. (1981). Lodish model and regulation of ribosomal protein synthesis by insulin-deficient chick embryo fibroblasts. *Biochemistry* **20**, 2550–8.

Itoh, N., Nose, K. & Okamoto, H. (1979). Purification and characterization of proinsulin mRNA from rat B-cell tumor. *European Journal of Biochemistry* **97**, 1–9.

Itoh, N., Ohshima, Y., Nose, K. & Okamoto, H. (1982). Glucose stimulates proinsulin synthesis in pancreatic islets without a concomitant increase in proinsulin mRNA synthesis. *Biochemistry International* **4**, 315–21.

Itoh, N. & Okamoto, H. (1977). Synthesis of preproinsulin with RNA preparation of streptozotocin–nicotinamide induced B-cell tumor. *FEBS Letters* **80**, 111–14.

Itoh, N. & Okamoto, H. (1980). Translational control of proinsulin synthesis by glucose. *Nature* **283**, 100–2.

Itoh, N., Sei, T., Nose, K. & Okamoto, H. (1978). Glucose stimulation of the proinsulin synthesis in isolated pancreatic islets without increasing amount of proinsulin mRNA. *FEBS Letters* **93**, 343–7.

Iynedjian, P.B. & Hanson, R.W. (1977). Increase in level of functional messenger RNA coding for phosphoenolpyruvate carboxykinase (GTP) during induction by cyclic 3′: 5′-monophosphate. *Journal of Biological Chemistry* **252**, 655–62.

Jen, G., Birge, C.H. & Thach, R.E. (1978). Comparison of initiation rates of encephalomyocarditis virus and host protein synthesis in infected cells. *Journal of Virology* **27**, 640–7.

Lodish, H.F. (1971). Alpha and beta globin messenger ribonucleic acid. Different amounts and rates of initiation of translation. *Journal of Biological Chemistry* **246**, 7131–8.

Lodish, H.F. (1976). Translational control. *Annual Review of Biochemistry* **45**, 39–72.

Lomedico, P., Rosenthal, N., Efstratiadis, A., Gilbert, W., Kolodner, R. & Tizard, R. (1979). The structure and evolution of the two nonallelic rat preproinsulin genes. *Cell* **18**, 545–58.

Lomedico, P.T. & Saunders, G.F. (1977). Cell-free modulation of proinsulin synthesis. *Science* **198**, 620–2.

Martial, J.A., Baxter, J.D., Goodman, H.M. & Seeburg, P.H. (1977). Regulation of growth hormone messenger RNA by thyroid and glucocorticoid hormones. *Proceedings of the National Academy of Sciences of the U.S.A.* **74**, 1816–20.

Mueckler, M.M., Merrill, M.J. & Pitot, H.C. (1983). Translational and pretranslational control of ornithine aminotransferase synthesis in rat liver. *Journal of Biological Chemistry* **258**, 6109–14.

Müting, D. & Wortmann, V. (1954). Zum Aminosäurenaufbau der Serum- und Gewebeeiweisskörper gesunder Menschen und Tiere. *Biochemische Zeitschrift* **325**, 448–58.

Nickol, J.M., Lee, K.L., Hollinger, T.G. & Kenney, F.T. (1976). Translation of messenger RNA specific for tyrosine aminotransferase in oocytes of *Xenopus laevis*. *Biochemical and Biophysical Research Communications* **72**, 687–93.

Nielsen, D.A., Welsh, M., Casadaban, M.J. & Steiner, D.F. (1985). Control of insulin gene expression in pancreatic B-cells and in an insulin-producing cell line, RIN-5F cells. I. Effect of glucose and cyclic AMP on the transcription of insulin mRNA. *Journal of Biological Chemistry* **260**, 13585–9.

Okamoto, H. (1981). Regulation of proinsulin synthesis in pancreatic islets and a new aspect to insulin-dependent diabetes. *Molecular and Cellular Biochemistry* **37**, 43–61.

Okamoto, H., Nose, K., Itoh, N., Sei, T. & Yamamoto, H. (1979). The translational control of proinsulin synthesis in pancreatic islets. In *Proinsulin, insulin, C-peptide* (ed. S. Baba, T. Kaneko & N. Yanaihara), pp. 27–35. Amsterdam: Excerpta Medica.

Permutt, M.A. (1974). Effect of glucose on initiation and elongation rates in isolated rat pancreatic islets. *Journal of Biological Chemistry* **248**, 2738–42.

Permutt, M.A. (1981). Biosynthesis of insulin. In *The Islets of Langerhans* (ed. C.J. Cooperstein & D.T. Watkins), pp. 75–95. New York: Academic Press.

Permutt, M.A., Boime, I., Chyn, R. & Goldford, M. (1978). Isolation and partial purification of catfish pancreatic islet messenger RNA. *Biochemistry* **17**, 537–43.

Permutt, M.A. & Kipnis, D.M. (1972). Insulin biosynthesis I. On the mechanism of glucose stimulation. *Journal of Biological Chemistry* **247**, 1194–9.

Rapoport, T.A., Hohne, W.E., Klatt, D., Prehn, S. & Hahn, V. (1976). Evidence for the synthesis of a precursor of carp proinsulin in a cell free translation system. *FEBS Letters* **69**, 32–6.

Roewekamp, W.G., Hofer, E. & Sekeris, C.E. (1976). Translation of mRNA from rat-liver polysomes into tyrosine aminotransferase and tryptophan oxygenase in a protein-synthesizing system from wheat germ. *European Journal of Biochemistry* **70**, 259–68.

Rogers, J. & Munro, H. (1987). Translation of ferritin light and heavy subunit mRNAs is regulated by intracellular chelatable iron levels in rat hepatoma cells. *Proceedings of the National Academy of Sciences of the U.S.A.* **84**, 2277–81.

Steiner, D.F., Kemmler, W., Clark, J.L., Oyer, P.E. & Rubenstein, A.H. (1972). The biosynthesis of insulin. In *Handbook of Physiology*, section 7, vol. 1 (ed. D.F. Steiner & N. Freinkel), pp. 175–98. Washington, D.C.: American Physiological Society.

Steiner, D.F. & Oyer, P.E. (1967). The biosynthesis of insulin and a probable precursor of insulin by a human islet cell adenoma. *Proceedings of the National Academy of Sciences of the U.S.A.* **57**, 473–80.

Steiner, D.F. & Tager, H.S. (1979). Biosynthesis of insulin and glucagon. In *Endocrinology*, vol. 2 (ed. L.J. DeGroot, G.F. Cahill, Jr., W.D. Odell, L. Martini, J.T. Potts, Jr, D.H. Nelson, E. Steinberger & A.I. Winegrad), pp. 921–34. New York: Grune & Stratton.

Watanabe, M., Itoh, N. & Okamoto, H. (1982). Translational control of proinsulin synthesis by glucose. In *Endocrinology*, International Congress Series no. 598 (ed. K. Shizume, H. Imura & N. Shimizu), pp. 266–70. Amsterdam: Elsevier.

Welsh, M., Scherberg, N., Gilmore, R. & Steiner, D.F. (1986). Translational control of insulin synthesis. Evidence for regulation of elongation, initiation and signal-recognition-particle-mediated translational arrest by glucose. *Biochemical Journal* **235**, 459–67.

White, M.W., Kameji, T., Pegg, A.E. & Morris, D.R. (1987). Increased efficiency of translation of ornithine decarboxylase mRNA in mitogen-activated lymphocytes. *European Journal of Biochemistry* **170**, 87–92.

4

The structure and regulation of the glucagon gene

JOEL F. HABENER

4.1. Introduction

Glucagon is a peptide hormone of 29 amino acids produced and secreted by the A-cells of the pancreatic islets (Unger & Orci, 1981). The hormone is one of a family of several peptides with similar primary structures that includes secretin (Mutt *et al.*, 1970), vasoactive intestinal peptide (Mutt & Said, 1974), gastric inhibitory peptide (Brown, 1971), and growth hormone releasing hormone (Spiess *et al.*, 1982). The secretion of glucagon is regulated by blood levels of glucose (Gerich *et al.*, 1974) and amino acids (Assan *et al.*, 1977), as well as by a variety of hormonal stimuli (Samols *et al.*, 1983). The action of glucagon on its target tissues, particularly the liver, is an important factor in protein and carbohydrate metabolism (Aoki *et al.*, 1974; Cherrington *et al.*, 1976). Abnormal regulation of glucagon gene expression has been implicated in the pathogenesis of diabetes mellitus (Dobbs *et al.*, 1975).

Peptides related immunologically to glucagon are produced in several extrapancreatic tissues such as brain (Tager *et al.*, 1980), salivary glands (Lawrence *et al.*, 1977) and intestine (Conlon, 1980; Parker *et al.*, 1984; Hoshino *et al.*, 1984). The principal hormonal function of glucagon is to regulate carbohydrate, fat and protein metabolism, but it is also possible that glucagon and glucagon-related peptides function as paracrine agents by way of communicating with adjacent cells, for example within pancreatic islet cells and as neurotransmitters within the nervous system. Glucagon and glucagon-related peptides have been isolated and characterized from catfish (Andrews & Ronner, 1985) and anglerfish (Andrews *et al.*, 1986) endocrine pancreases.

PRE-PROGLUCAGON cDNAs

5'—AAGCAGAGGAACTAACAGCACTATTTGAGGGAGAAAAGAATAAATACGGTTGTAAAC
GAAGCTCAAACA

SIGNAL PEPTIDE

```
        Met Lys Arg Ile His Ser Leu Ala Gly Ile Leu Leu Val Leu Gly Leu Ile Gln Ser Ser
AFG I   ATG AAA CGC ATC CAC TCC CTG GCT GGT ATC CTT CTG GTG CTT GGT TTA ATC CAG AGC AGC
AFG II  ATG ACA AGT CTT CAC TCT CTC GCT GGA CTC CTG CTC CTC ATG --- ATC ATC CAA AGC AGC
        Met Thr Ser Leu His Ser Leu Ala Gly Leu Leu Leu Leu Met     Ile Ile Gln Ser Ser
```

NH₂ PEPTIDE

```
        Cys Arg Val Leu Met Gln Glu Ala Asp Pro Ser Ser Ser Leu Glu Ala Asp Ser Thr Leu
AFG I   TGC CGG GTT CTT ATG CAG GAG GCT GAT CCC AGC TCA AGT TTG GAG GCA GAC AGC ACA CTG
AFG II  TGG CAG ATG CCT GAC CAG GAC CCA GAC CGG AAC TCT ATG CTT CTG AAT GAA AAC TCC ATG
        Trp Gln Met Pro Asp Gln Asp Pro Asp Arg Asn Ser Met Leu Leu Asn Glu Asn Ser Met
```

Val 5

```
        Lys Asp Glu Pro Arg Glu Leu Ser Asn Met Lys Arg His Ser Glu Gly Thr Phe Ser Asn
AFG I   AAG GAC GAG CCG AGA GAG CTT TCA AAC ATG AAG AGA CAC TCG GAG GGA ACT TTC TCC AAC
AFG II  TTG ACC GAA CCC ATC GAG CCC CTT AAC ATG AAG AGA CAC TCA GAG GGA ACT TTC TCC AAC
        Leu Thr Glu Pro Ile Glu Pro Leu Asn Met Lys Arg His Ser Glu Gly Thr Phe Ser Asn
```

10   15   ANGLERFISH GLUCAGON   25

```
        Asp Tyr Ser Lys Tyr Leu Glu Asp Arg Lys Ala Gln Glu Phe Val Arg Trp Leu Met Asn
AFG I   GAC TAC AGC AAA TAC CTG GAG GAC AGG AAG GCA CAG GAG TTT GTT CGG TGG CTG ATG AAC
AFG II  GAC TAT AGT AAA TAC CTG GAG ACA AGA AGA GCA CAA GAT TTT GTC CAG TGG CTG AAG AAC
        Asp Tyr Ser Lys Tyr Leu Glu Thr Arg Arg Ala Gln Asp Phe Val Gln Trp Leu Lys Asn
```

29   INTERVENING PEPTIDE   5   10

```
        Asn Lys Arg Ser Gly Val Ala Glu Lys Arg His Ala Asp Gly Thr Phe Thr Ser Asp Val
AFG I   AAC AAG AGG AGC GGT GTG GCA GAA AAG CGT CAC GCT GAT GGG ACC TTC ACC AGC GAT GTC
AFG II  TCA AAA AGA AAT GGT TTA TTT --- AGA CGC CAT GCA GAC GGC ACC TAC ACC AGT GAC GTG
        Ser Lys Arg Asn Gly Leu Phe     Arg Arg His Ala Asp Gly Thr Tyr Thr Ser Asp Val
```

15   GLUCAGON-RELATED COOH PEPTIDE   25   30

```
        Ser Ser Tyr Leu Lys Asp Gln Ala Ile Lys Asp Phe Val Asp Arg Leu Lys Ala Gly Gln
AFG I   AGC TCC TAC CTC AAA GAC CAG GCA ATC AAA GAC TTT GTG GAC AGG CTC AAG GCT GGA CAA
AFG II  AGC TCC TAC CTG CAG GAC CAG GCA GCA AAA GAC TTT GTG TCC TGG CTC AAC GCC GGC CGA
        Ser Ser Tyr Leu Gln Asp Gln Ala Ala Lys Asp Phe Val Ser Trp Leu Lys Ala Gly Arg
```

34

```
        Val Arg Arg Glu STOP
AFG I   GTC AGA AGA GAG TAG  GCTGCCCTCCATACTTCTCCTACAGTACCAAAAAGCTTAATGGCCGCCTTCATGTGTCTGATTCCATCATGAC
AFG II  GGC AGA AGA GAG TAA  ACAACAGCTTTCTGAATCATACAACGGTTTACACTGATTCTGTTCTGTTATATGTCTCACAGTCTGTACTACG
        Gly Arg Arg Glu STOP
```

AFG I   GAGAGAGAGCCGAGAGCCGAGAGAGATCTCGAATCAAAGAGCTAATCAACTGCTTCTTGTGACAACTTTATCTCCTTTCTTTGCGTTTCTTGATGTTCCC

AFG II CTCTTCTTCTTGGTTATGCAATTAAACCCCCCCTCATAAG—Poly(A)

AFG I   TGTGGTTGTGCTAAAAATAAATCACCTGAACTTTT—Poly(A)

Fig. 4.1. Nucleotide and corresponding deduced amino acid sequences of cDNAs encoding two anglerfish preproglucagons (AFG I and AFG II). The cDNA sequences shown were obtained by complete sequencing of both strands of each of five separate cDNAs. Dots between the nucleotide sequences indicate identical nucleotides (two single codon gaps, shown as dashes, were introduced into the AFG II sequence to maximize homology). The amino acid sequences of the two precursors, derived from the nucleotide sequences of the cDNAs, are shown above the nucleotide sequences. Boxed regions indicate regions of identity of the amino acid sequences in the two precursors. Arrows represent predicted prohormone cleavage sites (heavy arrows predict sites of cleavage by trypsin-like activity; light arrows predict sites of cleavage by carboxypeptidase B-like activity). Horizontal lines delineate peptides formed by potential cleavages at the arrowed sites. The line indicating the NH₂-terminal peptide is dashed to indicate uncertainty as to the site of cleavage of the signal sequence. The underlined DNA sequences AATAAA and AATTAAA at the 3'-ends of AFG I and AFG II cDNAs, respectively, are characteristic of sites involved in addition of the poly(A) tract to eukaryotic mRNAs.

4.2. Recombinant cDNAs encoding preproglucagons

Analysis of mRNAs isolated from anglerfish islets has shown that two independent glucagon genes encode two separate and distinct mRNAs (Lund *et al.*, 1983). Nucleotide sequence analyses of cloned cDNAs corresponding to these mRNAs reveal that the biosynthetic precursors of the two anglerfish preproglucagons each contain a glucagon and a second peptide related in structure to glucagon located in the carboxy-terminal sequences of the prohormones of M_r 12 500 and 14 500, referred to as glucagon-related carboxyl terminal peptides (Fig. 4.1).

The nucleic acid sequences have been determined for cloned cDNAs corresponding to the mRNAs encoding rat (Heinrich *et al.*, 1984*b*), bovine (Lopez *et al.*, 1983), Syrian hamster (Bell *et al.*, 1983), and guinea pig (Seino *et al.*, 1986) preproglucagon (Fig. 4.2). The mammalian glucagon precursors of M_r 18 000 encoded by these mRNAs are longer than the anglerfish preproglucagons, and contain the sequences of glucagon and two, rather than one, glucagon-like peptides arranged in tandem in the prohormone (Fig. 4.3).

4.3. Structure of the glucagon gene

Knowledge of the nucleotide sequence of rat preproglucagon cDNA (Heinrich *et al.*, 1984*b*) was used to isolate subfragments of the genomic subclones for nucleotide sequence analysis. This approach ensured that most sequencing would be started within exons of the glucagon gene (Heinrich *et al.*, 1984*a*). Exon and intron junctions, and the deduced amino acid sequence of the encoded glucagon precursor, were assigned according to the known rat glucagon cDNA nucleotide sequence. The gene consists of six exons and five introns. Exon 1 contains most of the 5′-untranslated region of glucagon mRNA. Exon 6 contains the entire 3′-untranslated region and the last four nucleotides of the coding region. Exon 2 encodes the signal peptide and a portion of the NH_2-terminal extension of glucagon (Figs. 4.4, 4.5). Exons 3, 4 and 5 contain glucagon, glucagon-like peptide I (GLP-I) and glucagon-like peptide II (GLP-II), respectively. Thus, each of the peptide-encoding, and presumably functional, domains of preproglucagon is encoded by a separate exon. Introns B, C and D in the coding sequence are located in intervening or connecting peptide regions (see Figs. 4.2 and 4.5). The sizes of the introns were estimated by restriction endonuclease mapping and blot hybridization. Intron A is the longest and consists of approximately 3.0 kb. Introns B, C and D are all of similar length of approximately 1.6 kb. Intron E is the shortest and consists of about 600–700 bp (base pairs). Introns B, C and D

5'
```
                                                                    -20       Asn Ile
                                                                    Met Lys Thr Val Tyr
aaaggagctccacctgtctacacctcctctcagctcagtcccacaaggcagaataaaaaaATG AAG ACC GTT TAC  75
                                                                            A   A
```
————————— SIGNAL PEPTIDE —————————
```
              Phe     Cys Gly Ala Gly                          -1 +1 Ser Leu
Ile Val Ala Gly Leu Phe Val Met Leu Val Gln Gly Ser Trp Gln His Ala Pro Gln Asp  135
 T                    T       TGT GGT GCT GT              T       T   A
```
————————— NH₂-PEPTIDE —————————
```
        Lys Ser                        15              Asp                    25
Thr Glu Glu Asn Ala Arg Ser Phe Pro Ala Ser Gln Thr Glu Pro Leu Glu Asp Pro Asp  195
            A   T                              C       C   G
```
—————   35   ————— GLUCAGON —————
```
                                                                             45
Gln Ile Asn Glu Asp Lys Arg His Ser Gln Gly Thr Phe Thr Ser Asp Tyr Ser Lys Tyr  255
 A           T   CGC
                 G
```
```
                           55
Leu Asp Ser Arg Arg Ala Gln Asp Phe Val Gln Trp Leu Met Asn Thr Lys Arg Asn Arg  315
 G               A   C       G                   C                        A
```
— IP-I —┐   ┌—   75   ————— GLP I —————
```
                                                                           85
Asn Asn Ile Ala Lys Arg His Asp Glu Phe Glu Arg His Ala Glu Gly Thr Phe Thr Ser  375
                 C   C       G               C                               C
```
```
                           95              Ala Lys Glu Phe Ile Ala Trp Leu Val Lys
Asp Val Ser Ser Tyr Leu Glu Gly Gln Gly  105
GAT GTG AGT TCT TAC TTG GAG GGC CAG GCA GCA AAG GAA TTC ATT GCT TGG CTG GTG AAA  435
             C                       T
```
—————   ————— IP-II —————
```
                            115   Thr          Val                       125
Gly Arg Gly Arg Arg Asp Phe Pro Glu Glu Val Ala Ile Ala Glu Glu Leu Gly Arg Arg  495
 A              G                        A        A   T   T   A        C   C
```
—————   ————— GLP II —————
```
                            135                        Ser         145
His Ala Asp Gly Ser Phe Ser Asp Glu Met Asn Thr Ile Leu Asp Asn Leu Ala Thr Arg  555
         G   C   C       C                             GT              G
```
```
                            155
Asp Phe Ile Asn Trp Leu Ile Gln Thr Lys Ile Thr Asp Lys Lys End
GAC TTC ATC AAC TGG CTG ATT CAA ACC AAG ATC ACT GAC AAG AAA TAG gaatatttcaccatt  618
                                                      A
```
```
cacaaccatcttcacaacatctcctgccagtcacttgggatgtacatttgagagcatatccgaagctatactgctttgc  697

atgcggacgaatacatttccctttagcgttgtgtaacccaaaggttgtaaatggaataaagtttttccagggtgttgat  776

aaagtaacaactttacagtatgaaaatgctggattctcaaattgtctcctcgtttttgaagttaccgccctgagattact  855

tttctgtggtataaattgtaaattatcgcagtcacgacacctggattacaacaacagaagacatggtaacctggtaacc  933

gtagtggtgaacctggaaagagaacttcttccttgaacccctttgtcataaatgcgctcagctttcaatgtatcaagaat 1012

agatttaaataaatatctcat                                                           1024
            3'
```

are located in positions corresponding to those in the human glucagon gene (Bell *et al.*, 1983).

4.4. Tissue-specific post-translational processing of proglucagon

The decoding of cloned cDNAs for the preproglucagons indicates that a single prohormone encoded by a unique gene and mRNA contains multiple different peptides that are liberated by posttranslational processing in a tissue-specific manner. Analysis of the processing of proglucagon in intact pancreas and intestine (Mojsov *et al.*, 1986; Ørskov *et al.*, 1986; George *et al.*, 1985), as well as in established pancreatic endocrine cell lines (Drucker *et al.*, 1986; Philippe *et al.*, 1986) confirms earlier studies (Tager & Markese, 1979) that multiple peptides are generated by cell-specific processing of proglucagon. It is now certain that the mammalian proglucagons are encoded by a single gene and the identical prohormones are expressed in pancreatic A-cells and in the intestinal neuroendocrine L-cells (Mojsov *et al.*, 1986; Novak *et al.*, 1987). Thus, the remarkable diversity in the expression of the proglucagon gene occurs at the level of differential posttranslational processing resulting in the formation of specific peptides which have distinctly different bioactivities.

The A-cells of the pancreatic islets process proglucagon predominantly to glucagon, glicentin-related pancreatic peptide, and to a lesser extent, to the glucagon-like peptides I, GLP-I(1–37) and GLP-I(7–37) (Fig. 4.6). The glucagon-like peptide II and intervening peptide II (IP-II), as well as a large major proglucagon fragment consisting of a single carboxy-terminal fragment of GLP-I, IP-II and GLP-II remain as biologically inactive 'discard' fragments resulting from the processing of proglucagon. In contrast, in the intestinal L-cells, cleavage of proglucagon results in the

Fig. 4.2. Composite nucleotide sequence of the rat preproglucagon cDNA, with the derived amino acid sequence. Two recombinant cDNAs hybridizing to the Syrian hamster probe were selected for sequence analysis. In combination, the two cDNAs were sequenced on both strands. Restriction endonuclease fragments were prepared, labeled at the 3'- and 5'-ends with ^{32}P, and sequenced. The signal sequence includes amino acids -20 to -1, the amino terminal peptide (glicentin) of the proglucagon amino acids $+1$ to $+33$, glucagon amino acids $+33$ to $+61$, and the COOH-terminal peptide, which contains glucagon-like peptides I and II, as well as corresponding intervening peptides I and II, amino acids $+62$ to $+160$. Within the coding region, the appended nucleotides and amino acids shown below and above the continuous rat sequences represent the comparative changes in the corresponding sequences of the preproglucagon cDNA obtained from the Syrian hamster.

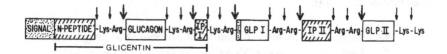

Fig. 4.3. Schematic diagram of the preproglucagon encoded in the rat
pancreatic mRNA. The two amino acids linking the boxed peptide
domains are indicated. Stippled box, signal peptide; hatched box,
NH₂-terminal peptide and two intervening peptides (IP-I and II); open
boxes, glucagon and two glucagon-like peptides (GLP-I and II).

highly efficient formation of the glucagon-like peptides and a COOH-
amidated form of the intervening peptide II. Glucagon is not formed in the
intestine. Rather, the glucagon remains in the form of the incompletely
processed, bioinactive precursors, glicentin (Thim & Moody, 1981) and
oxyntomodulin (Fig. 4.6). Therefore, as a result of the posttranslational
processing from the identical prohormone, the pancreatic A-cells produce
large amounts of glucagon and small amounts of the GLP-Is and the
intestinal L-cells produce large amounts of the glucagon-like peptides and
no glucagon. This situation of the remarkable cell-specific utilization of
alternative patterns of posttranslational processing of proglucagon in
these two tissues raises interest about the potential bioactivities of the
glucagon-like peptides.

4.5. Discovery of a unique insulinotropic activity of glucagon-like peptide I

The close similarities of the structures of the anglerfish COOH-
terminal peptides and the glucagon-like peptide I(7–37) of mammals,
representing a conservation of amino acid sequences over an evolutionary
time of approximately 400 million years, strongly suggested (Lund *et al.*,
1982, 1983) that glucagon-like peptide I may be a bioactive peptide (Fig.
4.7).

Much experimental evidence now indicates that GLP-I(7–37), and the
C-terminally modified peptide GLP-I(7–36) des Gly, Arg amide, provide
the most potent insulinotropic activities of any insulinotropic factors yet
recognized to date. At concentrations as low as 10^{-12} M the GLP-I
peptides stimulate pancreatic B-cells to release insulin, increase levels of
proinsulin mRNA and elevate the content of cAMP in B-cells (Drucker *et
al.*, 1987c). Infusion of GLP-I into subjects at amounts to achieve just

RAT   GLUCAGON   GENE

```
E-1   5' gatgacgagagtgggcgagtgaaatcatttgaacaaaaccccattatttacagatgagaaatttatattgtcagcgtaatatctgcaaggctaaacagcttggagacta tataa a   -19

      gccacagcacctttggtgc AGAAGGGCAGAGCTTGGGCGCAGAACACACTCAAAGTTCCCAAAGGAGCTCCACCTGTCTACACCTCCTCTCAGCTCAGTCCCCACAAG gtaagaggca   98

      cactgtggggtggggatgtctggaggacattggaca...


                              Intron A (3.0kb)


                                      -20                          -10
                                      Met Lys Thr Val Tyr Ile Val Ala Gly Leu Phe Val Met Leu Val Gln Gly Ser
E-2   ...cccctacccccactctgtgtccaacag GCAGAATAAAAA ATG AAG ACC GTT TAC ATC GTG GCT GGA TTG TTT GTA ATG CTG GTA CAA GGC AGC   3200

      -1 +1
      Trp Gln His Ala Pro Gln Asp Thr Glu Glu Asn Ala Ar
      TGG CAG CAT GCC CCT CAA GAC ACG GAG GAG AAC GCC AG gtactaaacagtagagcagcctgtcctgtggttagagtgag...   3278


                              Intron B (1.6kb)


            ...ttataaagcgtatactcctggagtatttaccctttgtgtcctgctcctacagttgtacttatgctacttaattactgcctctcaca   4800

                          20                          30                              40
            g Ser Phe Pro Ala Ser Gln Thr Glu Pro Leu Glu Asp Pro Asp Gln Ile Asn Glu Asp Lys Arg His Ser Gln Gly Thr Phe Pro Ser
E-3         agA TCA TTC CCA GCT TCC CAG ACA GAA CCA CTT GAA GAC CCT GAT CAG ATA AAC GAA GAC AAA CGC CAT TCA CAG GGC ACA TTC ACC AGT   4890

            50 ——— GLUCAGON ——— 60
            Asp Tyr Ser Lys Tyr Leu Asp Ser Arg Arg Ala Gln Asp Phe Val Gln Trp Leu Met Asn Thr Lys Arg Asn Ar
            GAC TAC AGC AAA TAC CTA GAC TCC CGC CGT GCT CAA GAT TTT GTG CAG TGG TTG ATG AAC ACC AAG AGG AAC CG gtaggagtctgaagttgtt   4983

            gtacacaaatgcacattaagaaattctcctgactctcctaggttatcatgctcaccaaacaggcttattcattttgggtgttcgtgaggttgt...   5076


                              Intron C (1.6kb)


                                        ...gagatgaact   6700

                          70              80
            g Asn Asn Ile Ala Lys Arg His Asp Glu Phe Glu Arg His Ala Glu Gly Thr Phe Thr Ser Asp
            tcattcaacacaactcccacatttctttcag G AAC AAC ATT GCC AAA CGT CAT GAT GAA TTT GAG AGG CAT GCT GAA GGG ACC TTT ACC AGT GAT   6794

E-4         90 —————— GLP-I ——————— 100          110
            Val Ser Ser Tyr Leu Glu Gly Gln Ala Ala Lys Glu Phe Ile Ala Trp Leu Val Lys Gly Arg Gly Arg Arg As
            GTG AGT TCT TAC TTG GAG GGC CAG GCA GCA AAG GAA TTC ATT GCT TGG CTG GTG AAA GGC CGA GGA AGG CGA GA gtaagtctctggtttaa   6885

            tatctgatgattttcctagatgagaaattacaaagaaaaaaagactatcctatggaaagctatactaaaaacttaatattt...   6967


                              Intron D (1.6kb)


                                                  120
            p Phe Pro Glu Gly Val Ala Ile Ala Glu Glu Leu Gly Arg Arg
            ...cagcagttcacacgaatttatttaagataaacactgactccaagccttctttgcttag C TTC CCG GAA GAA GTC GCC ATA GCT GAA GAA CTT GGG CGC AGA   8700

            130 ————— GLP II ——————— 150
E-5         His Als Asp Gly Ser Phe Ser Asp Glu Met Asn Thr Ile Leu Asp Asn Leu Ala Thr Arg Asp Phe Ile Asn Trp Leu Ile Gln Thr Lys
            CAT GCT GAT GGA TCC TTC TCT GAT GAG ATG AAC ACG ATT CTC GAT AAC CTT GCC ACC AGA GAC TTC ATC AAC TGG CTG ATT CAA ACC AAG   8790

            Ile Thr Asp Ly
            ATC ACT GAC AA gtaagggtttttagttgattttgaaatctagtaaaacttt...   8842


                              Intron E (0.6kb)


                                        ...aattattctgaacaattcctgcttttgacc   9500

            —160—
            s Lys END
            tcatag G AAA TAG GAATATTTCACCATTCACAACCATCTTCACAACATCTCCTGCCAGTCACTTGGGATGTACATTTGAGAGCATATACCGAAGCTATACTGCTTGGCATGCG   9613

E-6         GACGAATACATTTCCCTTTAGCGTTGTGTAACCCAAAGGTTGTAAATGGAATAAAGTTTTTCCAGGGTGTTGATAAAGTAACAACTTTACAGTATGAAAATGCTGGATTCTCAAATTGT   9732

            CTCCTCGTTTTGAAGTTACCGCCCTGAGATTACTTTTCTGTGGTATAAATTGTAAATTATCGCAGTCACGACACCTGGATTACAACAACAGAAGACATGGTAACCTGGTAA.CGTAGTG   9851

            GTGAACCTGGAAAGAGAACTTCTTCCTTGAACCCTTTGTCATAAATGCGCTCAGCTTTCAAATGTATCAAGAATAGATTTAAATAAATATCTCAtctttgttattatgctcctttctc   9970

            ttttattan... 3'   9979
```

Fig. 4.4. Nucleotide sequence of specific regions of the rat glucagon gene. The nucleotide sequence of 5'-flanking region, exons, and exon–intron junctions of the gene were determined by chemical sequencing. Exon sequences are in capital letters; flanking and intronic sequences are in lower case letters and are given from 5'- to 3'-ends of the genomic insert, as indicated. Nucleotides are numbered starting with + 1 at the most 5' nucleotide of exon 1, and running through each intron whose length, i.e. number of nucleotides, was estimated by restriction enzyme and blot hybridization analysis. The nucleotides of the 5'-flanking region are numbered from − 1 beginning with the nucleotide immediately 5' to exon 1, and continuing in the 5' direction.

(*Caption continues over page*)

twice normal physiological concentrations of the measurable circulating concentrations of GLP-I has a profound effect on insulin secretion and blood glucose levels (Kreymann et al., 1987). Because GLP-I is liberated from proglucagon at a high efficiency in the intestinal L-cells it is tempting to speculate that GLP-I may be one of the 'incretins', that is, factors that are postulated to originate from the intestine in response to the ingestion of oral nutrients and to augment insulin release to a greater extent than does the administration of an equivalent amount of nutrients by the systemic (intravenous) route (Creutzfeldt, 1979).

The discovery of this highly potent insulinotropic peptide co-encoded with glucagon in the proglucagon gene raises intriguing possibilities regarding the possible role of the peptide in the pathogenesis of non-insulin-dependent diabetes, a disease manifested by the development of both insulin resistance and defective glucose-mediated regulation of insulin secretion. Perhaps, either under- or overproduction of GLP-I by the intestine (and pancreas) may alter the sensitivity of B-cells to glucose and thereby impair the coupling of blood glucose levels to the dynamic cellular processes of insulin secretion.

4.6. Regulation of glucagon gene expression

The endocrine secretions of the pancreatic islets in mammals are governed by a complex number of controlling factors that include blood-borne metabolites (for example, glucose, amino acids and catecholamines) and local neural and paracrine influences (Samols et al., 1983). The major islet hormones (insulin, glucagon, somatostatin) interact among their specific cell types (B-, A- and D-cells, respectively) to modulate secretory responses mediated by the metabolites (Samols et al., 1983). Pancreatic glucagon secretion, and presumably expression of the glucagon gene at the level of transcription and hormone biosynthesis, appears to be regulated by a combination of nutrient and hormonal factors (Pipeleers et al., 1985). Secretion from the A-cell of the pancreatic islet is stimulated by amino acids, which in turn is augmented by analogs of cyclic AMP and epinephrine, and is suppressed by glucose, somatostatin, and insulin.

Caption to Fig. 4.4 (cont.)

> The promoter of the glucagon gene is boxed; the two polyadenylation sites are underlined. Amino acids encoded by exons are indicated above each codon, and are numbered − 20 to − 1 for the signal peptide, and + 1 to + 160 for the remainder of proproglucagon. Sites of known and presumptive proteolytic processing of preproglucagon are indicated by arrows.

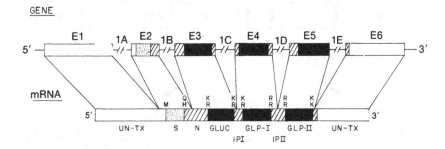

Fig. 4.5. Structure of the encoded rat preproglucagon mRNA. The mRNA encoded by the glucagon gene, and the regions of preproglucagon corresponding to a given exon of the gene, are indicated by lines between gene and mRNA diagrams to illustrate the correspondence between exon and polyprotein functional domain structures. Exons (boxed) are designated E1–E6, and intervening sequences (interrupted lines) IA–IE. Open boxes represent untranslated regions of preproglucagon mRNA (UN-TX). The stippled boxes represent signal peptides (S); hatched boxes represent N-terminal extension (N) and C-terminal extension (IPI) of glucagon and the intervening peptides (IPI and IPII). Closed boxes represent glucagon and two glucagon-like peptides, I and II (GLP-I and GLP-II).

Studies of peptide secretion from the intestinal L-cell, although less extensively pursued than studies of the islet A-cell, indicate that the release of glicentin is stimulated by the presence of nutrients (triglycerides, glucose) in the gastrointestinal lumen (Böttger *et al.*, 1973), by bowel resection (Sagor *et al.*, 1982), and by bombesin (Matsuyama *et al.*, 1980), an amphibian gastrin-releasing peptide that acts through the phosphoinositol pathway (Sutton & Martin, 1982). At the present time, no definitive information is available bearing directly upon the mechanisms of glucagon gene expression, either in the A- or in the L-cell. However, some preliminary observations have been made regarding potential cellular pathways of signal transduction that may be involved in the metabolic induction of glucagon gene expression and in the identification of DNA control elements that may determine the cell-specific expression of the glucagon gene with the A-cells of the pancreatic islets.

4.7. Cellular signaling pathways in glucagon gene expression
Endocrine regulation of gene expression occurs through at least two different mechanisms. Specific intracellular receptor proteins, identified for both steroid and thyroid hormones, interact directly with chroma-

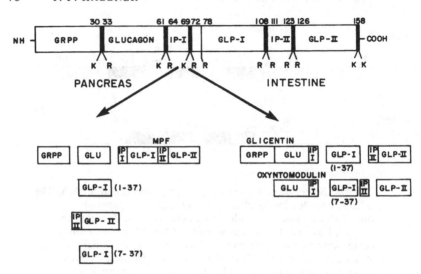

Fig. 4.6. Tissue-specific processing of proglucagon in pancreas and large intestine. Peptides present in the pancreas are glucagon, GLP-I, and a variety of carboxy-terminal extension peptides. Peptides present in the intestine are glicentin, oxyntomodulin, GLP-I, GLP-II, and IPII. GRPP, glicentin-related pancreatic peptides; MPF, major proglucagon fragment; K, lysine, R, arginine. Partly processed peptides are indicated under the predominant peptides.

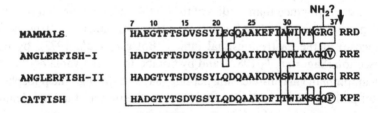

Fig. 4.7. Similarities in glucagon-like peptide I sequences. Amino acids that are conserved or functionally interchangeable, e.g. F and Y, E and D, are enclosed in the boxed area. Numbering of the mammalian peptides begins at position 7, which is preceded by an arginine at position 6 in the mammalian proglucagons; this provides optimal alignment of the sequence similarities to the fish sequences corresponding to the histidine immediately following paired basic amino acids. Some question remains regarding the structure of the COOH-termini of the peptides. Sequences of peptides of human (Bell et al., 1983), rat (Heinrich et al., 1984), bovine (Lopez et al., 1983), guinea pig (Seino et al., 1986) and hamster (Bell et al., 1983) are identical. Sequence of fish peptides: anglerfish and catfish (Lund et al., 1983; Andrews & Ronner, 1985) differ considerably in the COOH-region.

tin. This interaction has been further defined and involves specific DNA sequences in a number of genes known to be regulated by steroid hormones (Doehmer *et al.*, 1982; Compton *et al.*, 1983). In contrast, polypeptide hormones bind to specific plasma membrane receptors and thus require a signal transduction pathway to mediate gene transcription. Although stimulation of the classical 'second messenger' pathways resulting from the interaction of peptide ligands with their receptors has been shown to correlate with changes in gene expression, factors that mediate these effects at the level of gene transcription have not been identified. It is generally believed, however, that activation of the cyclic AMP and phosphoinositol pathways results in a cascade of interactions leading to the regulation of specific genes through phosphorylation of nuclear proteins by protein kinases A and C (cAMP-dependent and Ca^{2+}–phospholipid-dependent enzymes, respectively) (Murdoch *et al.*, 1982, 1985). Consensus DNA sequences located near cyclic AMP- and phorbol ester-responsive genes have been proposed (Deutsch *et al.*, 1987), suggesting a model akin to the prokaryotic transcriptional system which utilizes a cyclic AMP receptor protein in the regulation of catabolite repression (Ebright *et al.*, 1984).

Studies of proglucagon gene expression in a rat islet cell line (RIN-1056A cells) (Philippe *et al.*, 1987*a*) have shown a stimulatory effect by the phorbol ester, phorbol myristate acetate (PMA), and the analog of diacylglycerol, 1,2-dioctanoylglycerol (OAG) on proglucagon gene transcription as well as on steady state levels of mRNA encoding proglucagon (Philippe *et al.*, 1987*b*) (Fig. 4.8). No effects of either PMA or OAG were observed on the expression of the insulin, somatostatin, or angiotensinogen genes, which were also expressed at basal levels in the RIN-1056 cells. Nor were any effects of dibutyryl cAMP or cortisol seen on the expression of the glucagon gene. Inasmuch as PMA and OAG are believed to activate protein kinase C by mimicking the endogenous activator, diacylglycerol, it seems reasonable to conclude that at least one component of the intracellular transmission of stimuli for proglucagon gene transcription in the pancreas involves a phospholipid-dependent or protein kinase C pathway.

Recently reported observations of proglucagon gene expression in primary cultures of intestinal cells isolated from fetal rats suggest that cyclic AMP may stimulate both peptide secretion (Brubaker, 1988) and proglucagon mRNA levels (Drucker & Brubaker, 1988). No effects of phorbol esters were seen on glucagon gene expression in primary cultures of intestinal cells. Thus, it is possible that the metabolic signaling pathway

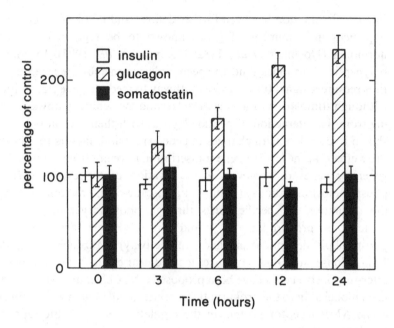

Fig. 4.8. Time course of glucagon, insulin, and somatostatin mRNA accumulation following stimulation by phorbol myristate acetate (PMA). PMA (100 nM) was added at time 0, and RNA was extracted from cells at 3, 6, 12, and 24 h. Quantification of each mRNA was done by densitometric scanning of the autoradiographs. Values are derived from five different experiments, means ± s.d., and represent ratios of glucagon-specific mRNA to internal control of β-actin or angiotensinogen-specific mRNAs.

responsible for regulating the expression of the proglucagon gene differs in pancreatic A- and intestinal L-cells; in pancreatic islet cells, proglucagon gene expression may be modulated by a protein kinase C pathway and in the intestinal cells the corresponding pathway of gene expression may be mediated by a cAMP-dependent protein kinase.

4.8. Direct analyses of potential DNA regulatory elements responsible for the cell-specific expression of the proglucagon gene

Analyses of the cell-specific expression of genes in the endocrine and exocrine pancreas indicate that control and regulation of gene transcription is mediated by specific *cis*-acting DNA sequences. *Cis*-acting sequences have been shown to promote transcription, hormonal regulation, and developmental and tissue-specific expression in a variety of genetic systems (Darnell, 1982; Staudt *et al.*, 1986; Treisman, 1986; Landolfi *et al.*, 1986). These experiments have utilized the technology of gene transfer to introduce and express DNA fragments encoding entire genes, or chimeric fusion genes linked to sensitive reporter functions, into mouse embryos or tissue culture cells *in vitro*.

To characterize the factors that mediate proglucagon gene transcription, the 5'-flanking sequences of the rat proglucagon gene were analyzed for the presence of *cis*-sequences that promote proglucagon gene expression (Drucker *et al.*, 1987*a*). To facilitate analyses of gene transcription, successively 5' shortened sequences of the glucagon gene 5'-flanking sequence were fused to the coding sequence of the reporter enzyme bacterial chloramphenicol acetyl transferase (CAT). By this approach the relative extents of transcription conferred by the different glucagon gene sequences were reflected directly by the amounts of bacterial enzyme synthesized after transfection of the DNA plasmids into various islet and other animal cell lines.

The studies show that 249 bases of glucagon gene 5'-flanking sequence, including the promoter and transcriptional start site, was sufficient to direct transcription of the CAT gene in both rat and hamster islet cell lines (Drucker *et al.*, 1987*b*) (Fig. 4.9). This fusion gene was expressed at much higher levels in islet cells that expressed the endogenous glucagon gene than in cell lines that did not express the glucagon gene; the expression of both the endogenous and experimentally introduced fusion genes were highly correlated. One possible explanation for the correlation between the endogenous and the fusion gene expression involves the existence of *trans*-acting factors capable of recognizing both endogenous and introduced glucagon gene regulatory sequences. Evidence for the importance of *cis*- and *trans*-acting factors in the regulation of glucagon gene transcription comes from the observation that rat glucagon gene sequences are transcriptionally active in hamster islet cell lines (Drucker *et al.*, 1987*a, b*). The high degree of expression of the rat glucagon–CAT fusion genes in hamster islet cell lines indicates that the *cis*- and *trans*-acting factors which promote glucagon gene transcription are likely to be highly conserved across species. Additional evidence that glucagon gene regulatory sequences are highly conserved comes from comparison of the human and rat glucagon gene 5'-flanking sequences, which show 88% homology within the proximal 130 base pairs of the 5'-flanking regions (White & Saunders, 1986).

The capability of rat glucagon gene 5'-flanking sequences to compete for the binding of *trans*-acting factors with the GLUCAT (-350)

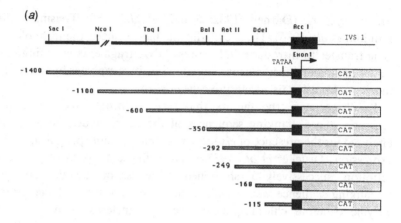

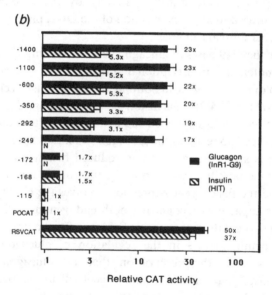

Fig. 4.9. Structure and relative transcriptional expression of
glucagon–CAT fusion genes in rat and hamster islet cell lines. The
structure of the rat glucagon gene 5'-flanking and 5'-untranslated
sequences (plasmid PrGluS-1) used to construct the GLUCAT
plasmids is shown diagrammatically in (a). Fusion genes contain the
indicated amounts of 5'-flanking sequence and 58 bp of exon 1. Values
for chloramphenicol acetyltransferase activity are normalized to the
chloramphenicol acetyltransferase activity obtained with PO–CAT,
the promoterless chloramphenicol acetyltransferase vector. For
analyses of chloramphenicol acetyltransferase activity after transient
transfection of hamster cell lines (b), $\frac{1}{100}$ and $\frac{1}{2}$ of the cell extract from a
60 mm plate of InR1-G9 and HIT cells respectively, was used.

reporter in a hamster cell line provides further evidence for the existence of *trans*-acting factors in the control of glucagon gene transcription (Drucker *et al.*, 1987*a, b*). Deletion analyses of the rat glucagon gene 5′-flanking region revealed that the fusion genes containing only 249 bp of 5′-flanking sequences are as transcriptionally active as those containing 1400 bp of glucagon gene 5′-flanking sequence. Further deletion to − 168 results in a marked decrease in transcriptional activity in glucagon-producing islet cell lines, whereas further deletion to − 115 completely abolishes all transcriptional activity. The sharp decline in transcriptional activity observed by deletion of sequences from − 249 to − 168 suggests that a *cis*-acting element(s) important for the augmentation of glucagon gene transcription resides within this region. Deletion of 5′-flanking sequence upstream of − 168 resulted in similar transcriptional activity in glucagon- and insulin-producing cell lines. One explanation for these results points to the potential existence of a sequence with glucagon cell-specific enhancer-like activity located upstream of − 168. Enhancers are generally viewed as *cis*-acting DNA elements capable of increasing transcription from a heterologous promoter in a manner that is independent of distance and orientation. DNA sequences which act in *cis* to enhance gene transcription may be 5′, within an intron, or 3′ to the transcribed sequences of the gene. Preliminary observations indicate that at least three sequences in the 5′-flanking region of the rat glucagon gene possess enhancer-like activity.

4.9. Summary

Diversification of the expression of the glucagon-related genes has occurred at two levels: (i) multiplication of the numbers of structurally-related genes within the mammalian genome; and (ii) alternative posttranslational processing of a prohormonal polyprotein to provide different spectra of bioactive peptides in pancreas compared to intestines. The pancreatic endocrine A-cells produce and secrete predominantly glucagon, a liver-active hormone that stimulates glycogenolysis and

Caption to Fig. 4.9 (*cont.*)

Hamster cell lines were cotransfected with the plasmid pXGH5, a plasmid containing growth hormone sequences under the control of the mouse metallothionein promoter. Chloramphenicol acetyltransferase activity in transient assays was normalized to the growth hormone secreted into the media. The values represent the means ±s.e. of three or more independent determinations; N, not determined. Note that the abscissae are logarithmic scales.

gluconeogenesis to raise blood glucose levels in countermeasure to the actions of insulin. The intestinal L-cells, however, do not produce glucagon but rather glicentin and the glucagon-like peptides. One of these peptides, glucagon-like peptide I (7–37), has potent insulinotropic actions on pancreatic B-cells and may be an incretin factor responsible for the augmentation of glucose-stimulated insulin release in response to oral nutrients.

The structure of the glucagon gene is notable in that the five introns precisely divide the gene into six separate exonic domains that have distinctive functions. Four of the exons individually encode the signal peptide, glucagon, glucagon-like peptide I and glucagon-like peptide II of preproglucagon.

The factors responsible for the regulation of glucagon gene transcription are poorly understood. It is likely that in both pancreatic A- and intestinal L-cells regulation is mediated by both the cAMP and phospholipid pathways of signal transduction. The extracellular factors and A- and L-cell receptors responsible for activating these pathways are largely unknown. It seems clear, however, that nutrients, glucose, amino acids, and probably catecholamines modulate the secretion of glucagon and the glucagon-like peptides. Studies of the DNA elements and *trans*-acting DNA-binding proteins responsible for the cell-specific expression of the glucagon gene have just begun. As has been found in similar studies of the expression of the proinsulin gene, it appears that multiple DNA control elements and DNA-binding proteins are involved in the cell-specific transcription of the proglucagon gene.

I am particularly indebted to Richard H. Goodman, Pauline K. Lund, Gerhard Heinrich, Jacques Philippe, Daniel J. Drucker and Svetlana Mojsov for working with me in the studies described in this chapter. I thank Janice Canniff for preparing the manuscript. The studies described were supported in part by U.S.P.H.S. Grant DK30834.

References

Andrews, P.C., Hawke, D.H., Lee, T.D., Legesse, K., Noe, B.D. & Shively, J.E. (1986). Isolation and structure of the principal products of preproglucagon processing, including an amidated glucagon-like peptide. *Journal of Biological Chemistry* 261, 8128–33.

Andrews, P.C. & Ronner, P. (1985). Isolation and structures of glucagon and glucagon-like peptide from catfish pancreas. *Journal of Biological Chemistry* 260, 3910–14.

Aoki, T.T., Müller, W.A., Brennan, M.F. & Cahill, G.F. Jr (1974). Effect of glucagon on amino acid and nitrogen metabolism in fasting man. *Metabolism* 23, 805–14.

Assan, R., Attali, J.R., Ballerio, G., Boillot, J. & Girard, J.R. (1977). Glucagon secretion induced by natural and artificial amino acids in perfused rat pancreas. *Diabetes* 26, 300–7.

Bell, G.I., Santerre, R.F. & Mullenbach, G.T. (1983). Hamster preproglucagon contains the sequence of glucagon and two related peptides. *Nature* 302, 716–18.

Böttger, I., Dobbs, R., Faloona, G.R. & Unger, R.H. (1973). The effects of triglyceride absorption upon glucagon, insulin, and gut glucagon-like immunoreactivity. *Journal of Clinical Investigation* **52**, 2532–41.

Brown, J.C. (1971). A gastric inhibitory polypeptide. I. The amino acid composition and the tryptic peptides. *Canadian Journal of Biochemistry* **49**, 255–61.

Brubaker, P.L. (1988). Control of glucagon-like immunoreactive peptide secretion from fetal rat intestinal cultures. *Endocrinology* **123**, 220–6.

Cherrington, A.D., Chiasson, J.L., Liljenquist, J.E., Jennings, A.S., Keller, U. & Lacy, W.W. (1976). The role of insulin and glucagon in the regulation of basal glucose production in the postabsorptive dog. *Journal of Clinical Investigation* **58**, 1407–18.

Compton, J.G., Schrader, W.T. & O'Malley, B.W. (1983). DNA sequence preference of the progesterone receptor. *Proceedings of the National Academy of Sciences of the U.S.A.* **80**, 16–20.

Conlon, J.M. (1980). The glucagon-like polypeptides – order out of chaos? *Diabetologia* **18**, 85–8.

Creutzfeldt, W. (1979). The incretin concept today. *Diabetologia* **16**, 75–85.

Darnell, J.E., Jr (1982). Variety in the level of gene control in eukaryotic cells. *Nature* **297**, 365–7.

Deutsch, P.J., Jameson, L.J. & Habener, J.F. (1987). Cyclic AMP responsiveness of human gonadotropin-α gene transcription is directed by a repeated 18-base pair enhancer. *Journal of Biological Chemistry* **262**, 12169–74.

Dobbs, R., Sasaki, H., Faloona, G., Valverde, I., Baetens, D., Orci, L. & Unger, R. (1975). Glucagon: role in the hyperglycemia of diabetes mellitus. *Science* **187**, 544–7.

Doehmer, J., Barinaga, M., Vale, W., Rosenfeld, M.G., Verma, I.M. & Evans, R.M. (1982). Introduction of rat growth hormone gene into mouse fibroblasts via a retroviral DNA vector: Expression and regulation. *Proceedings of the National Academy of Sciences of the U.S.A.* **79**, 2268–72.

Drucker, D.J. & Brubaker, P.L. (1988). Glucagon gene expression in rat intestine. *Endocrine Society Abstracts* **70**, 307.

Drucker, D.J., Mojsov, S. & Habener, J.F. (1986). Cell-specific post-translational processing of preproglucagon expressed from a metallothionein-glucagon fusion gene. *Journal of Biological Chemistry* **261**, 9637–43.

Drucker, D.J., Philippe, J., Jepeal, L. & Habener, J.F. (1987a). Cis-acting DNA sequence controls glucagon gene expression in pancreatic islet cells. *Transactions of the Association of American Physicians* **100**, 109–15.

Drucker, D.J., Philippe, J., Jepeal, L. & Habener, J.F. (1987b). Glucagon gene 5'-flanking sequences promote islet cell-specific gene transcription. *Journal of Biological Chemistry* **262**, 15659–65.

Drucker, D.J., Philippe, J., Mojsov, S., Chick, W.L. & Habener, J.F. (1987c). Glucagon-like peptide I stimulates insulin gene expression and increases cyclic AMP levels in a rat islet cell line. *Proceedings of the National Academy of Sciences of the U.S.A.* **84**, 3434–8.

Ebright, R.H., Cossart, P., Gicquel-Sanzey, B. & Beckwith, J. (1984). Mutations that alter the DNA sequence specificity of the catabolite gene activator protein of *E. coli*. *Nature* **311**, 232–5.

George, S.K., Uttenthal, L.O., Ghiglione, M. & Bloom, S.R. (1985). Molecular forms of glucagon-like peptides in man. *FEBS Letters* **192**, 275–8.

Gerich, J.E., Schneider, V., Dippe, S.E., Langlois, M., Noacco, C., Karam, J.H, & Forsham, P.H. (1974). Characterization of the glucagon response to hypoglycemia in man. *Journal of Clinical Endocrinology and Metabolism* **38**, 77–82.

Heinrich, G., Gros, P. & Habener, J.F. (1984a). Glucagon gene sequence: four of six exons encode separate functional domains of rat preproglucagon. *Journal of Biological Chemistry* **259**, 14082–7.

Heinrich, G., Gros, P., Lund, P.K., Bentley, R.C. & Habener, J.F. (1984b). Preproglucagon messenger ribonucleic acid: nucleotide and encoded amino acid sequences of the rat pancreatic complementary deoxyribonucleic acid. *Endocrinology* **115**, 2176–81.

Holst, J.J., Ørskov, C., Nielsen, O.V. & Schwartz, T.W. (1987). Truncated glucagon-like peptide I, an insulin-releasing hormone from the distal gut. *FEBS Letters* **211**, 169–74.

Hoshino, M., Yanaihara, C., Hong, Y.-M., Kishida, S., Katsumaru, Y., Vandermeers, A., Vandermeers-Piret, M.-C., Robberecht, P., Christophe, J. & Yanaihara, N. (1984). Primary structure of helodermin, a VIP-secretin-like peptide isolated from gila monster venom. *FEBS Letters* **178**, 233–9.

Kreymann, B., Williams, G., Ghatei, M.A. & Bloom, S.R. (1987). Glucagon-like peptide-I 7–36: A physiological incretin in man. *The Lancet* **ii**, 1300–4.

Landolfi, N.F., Capra, J.D. & Tucker, P.W. (1986). Interaction of cell-type-specific nuclear proteins with immunoglobulin V_H promoter region sequences. *Nature* **323**, 548–51.

Lawrence, A.M., Tan, S., Hojvat, S. & Kirsteins, L. (1977). Salivary gland hyperglycemic factor: an extrapancreatic source of glucagon-like material. *Science* **195**, 70–2.

Lopez, L.C., Frazier, M.L., Su, C.-J., Kumar, A. & Saunders, G.F. (1983). Mammalian pancreatic preproglucagon contains three glucagon-related peptides. *Proceedings of the National Academy of Sciences of the U.S.A.* **80**, 5485–9.

Lund, P.K., Goodman, R.H., Dee, P.C. & Habener, J.F. (1982). Pancreatic preproglucagon cDNA contains two glucagon-related coding sequences arranged in tandem. *Proceedings of the National Academy of Sciences of the U.S.A.* **79**, 345–9.

Lund, P.K., Goodman, R.H., Montminy, M.R., Dee, P.C. & Habener, J.F. (1983). Anglerfish islet pre-proglucagon II; nucleotide and corresponding amino acid sequence of the cDNA. *Journal of Biological Chemistry* **258**, 3280–4.

Matsuyama, T., Namba, M., Nonaka, K., Tarui, S., Tanaka, R. & Shima, K. (1980). Decrease in blood glucose and release of gut glucagon-like immunoreactive materials by bombesin infusion in the dog. *Endocrinologica Japonica* **1**, 115–19.

Mojsov, S., Heinrich, G., Wilson, I.B., Ravazzola, M., Orci, L. & Habener, J.F. (1986). Preproglucagon gene expression in pancreas and intestine diversifies at the level of post-translational processing. *Journal of Biological Chemistry* **261**, 11880–9.

Mojsov, S., Weir, G.C. & Habener, J.F. (1987). Insulinotropin: glucagon-like peptide I(7–37) co-encoded in the glucagon gene is a potent stimulator of insulin release in the perfused rat pancreas. *Journal of Clinical Investigation* **79**, 616–19.

Murdoch, G.H., Rosenfeld, M.G. & Evans, R.M. (1982). Eukaryotic transcriptional regulation and chromatin-associated protein phosphorylation by cyclic AMP. *Science* **218**, 1315–17.

Murdoch, G.H., Waterman, M., Evans, R.M. & Rosenfeld, M.G. (1985). Molecular mechanisms of phorbol ester, thyrotropin-releasing hormone, and growth factor stimulation of prolactin gene transcription. *Journal of Biological Chemistry* **260**, 11852–8.

Mutt, V., Jorpes, J.E. & Magnusson, S. (1970). Structure of porcine secretin. The amino acid sequence. *European Journal of Biochemistry* **15**, 513–19.

Mutt, V. & Said, S.I. (1974). Structure of the porcine vasoactive intestinal octacosapeptide. The amino-acid sequence. Use of kallikrein in its determination. *European Journal of Biochemistry* **42**, 581–9.

Novak, U., Wilks, A., Buell, G. & McEwen, S. (1987). Identical mRNA for preproglucagon in pancreas and gut. *European Journal of Biochemistry* **164**, 553–8.

Ørskov, C., Holst, J.J., Knuhtsen, S., Baldissera, F.G.A., Poulsen, S.S. & Nielsen, O.V. (1986). Glucagon-like peptides GLP-I and GLP-II, predicted products of the glucagon gene, are secreted separately from pig small intestine but not pancreas. *Endocrinology* **119**, 1467–75.

Parker, D.S., Raufman, J.-P., O'Donohue, T.L., Bledsoe, M., Yoshida, H. & Pisano, J.J.

(1984). Amino acid sequences of helospectins, new members of the glucagon superfamily, found in gila monster venom. *Journal of Biological Chemistry* **259**, 11751–5.

Philippe, J., Chick, W.L. & Habener, J.F. (1987*a*). Multipotential phenotypic expression of genes encoding peptide hormones in rat insulinoma cell lines. *Journal of Clinical Investigation* **79**, 351–8.

Philippe, J., Drucker, D.J. & Habener, J.F. (1987*b*). Glucagon gene transcription in an islet cell line is regulated via a protein kinase C-activated pathway. *Journal of Biological Chemistry* **262**, 1823–8.

Philippe, J., Mojsov, S., Drucker, D.J. & Habener, J.F. (1986). Proglucagon processing in a rat islet cell line resembles phenotype of intestine rather than pancreas. *Endocrinology* **119**, 2833–9.

Pipeleers, D.G., Schuit, F.C., Van Schravendijk, C.F.H. & Van De Winkel, M. (1985). Interplay of nutrients and hormones in the regulation of glucagon release. *Endocrinology* **117**, 817–23.

Sagor, G.R., Al-Mukhtar, M.Y.T., Ghatei, M.A., Wright, N.A. & Bloom, S.R. (1982). The effect of altered luminal nutrition on cellular proliferation and plasma concentrations of enteroglucagon and gastrin after small bowel resection in the rat. *British Journal of Surgery* **69**, 14–18.

Samols, E., Weir, G.C. & Bonner-Weir, S. (1983). Intra-islet insulin–glucagon–somatostatin relationships. In *Handbook of experimental pharmacology*, vol. 66, part II (ed. P.J. Lefebvre), pp. 133–62. New York: Springer-Verlag.

Seino, S., Welsh, M., Bell, G.I., Chan, S.J. & Steiner, D.F. (1986). Mutations in the guinea pig preproglucagon gene are restricted to a specific portion of the prohormone sequence. *FEBS Letters* **203**, 25–30.

Spiess, J., Rivier, J., Thorner, M. & Vale, W. (1982). Sequence analysis of a growth hormone releasing factor from a human pancreatic islet tumor. *Biochemistry* **21**, 6037–40.

Staudt, L.M., Singh, H., Sen, R., Wirth, T., Sharp, P.A. & Baltimore, D. (1986). A lymphoid-specific protein binding to the octamer motif of immunoglobulin genes. *Nature* **323**, 640–3.

Sutton, C.A. & Martin, T.F.J. (1982). Thyrotropin-releasing hormone (TRH) selectively and rapidly stimulates phosphatidylinositol turnover in GH pituitary cells: a possible second step of TRH action. *Endocrinology* **110**, 1273–80.

Tager, H.S., Hohenboken, M., Markese, J. & Dinerstein, R.J. (1980). Identification and localization of glucagon-related peptides in rat brain. *Proceedings of the National Academy of Sciences of the U.S.A.* **77**, 6229–33.

Tager, H.S. & Markese, J. (1979). Intestinal and pancreatic glucagon-like peptides. Evidence for identity of higher molecular weight forms. *Journal of Biological Chemistry* **254**, 2229–33.

Thim, L. & Moody, A.J. (1981). The primary structure of porcine glicentin (proglucagon). *Regulatory Peptides* **2**, 139–50.

Treisman, R. (1986). Identification of a protein-binding site that mediates transcriptional response of the c-*fos* gene to serum factors. *Cell* **46**, 567–74.

Unger, R.H. & Orci, L. (1981). Glucagon and the A cell: physiology and pathophysiology. *New England Journal of Medicine* **304**, 1518–24.

White, J.W. & Saunders, G.F. (1986). Structure of the human glucagon gene. *Nucleic Acids Research* **14**, 4719–30.

5

The structure and regulation of the somatostatin gene

JOEL F. HABENER

5.1. Introduction

Somatostatin is a tetradecapeptide that regulates the release of pituitary, pancreatic, and gastrointestinal hormones (Reichlin, 1983, 1987). Initially identified in the hypothalamus as an inhibitor of growth hormone secretion (Brazeau et al., 1973), somatostatin has subsequently been found in the extrahypothalamic brain, spinal cord, retina, gastrointestinal tract, pancreatic islets, and thyroid (Patel & Reichlin, 1978; Arimura et al., 1975; Rorstad et al., 1979; Hökfelt et al., 1975). In addition to inhibiting the secretion of a number of peptide hormones, somatostatin has been proposed to act as a neurotransmitter and to modulate gastrointestinal motility (Barker, 1976; Gerich & Patton, 1978).

The diverse functions and the widespread distribution of the tetradecapeptide somatostatin (somatostatin-14) have focused attention on the biosynthesis of the hormone. Several studies have shown that somatostatin-14 is synthesized as part of a larger precursor. A 28-amino acid form of the hormone (somatostatin-28) has been identified in extracts of porcine hypothalamus (Schally et al., 1980), gastrointestinal tract (Pradayrol et al., 1980) and bovine hypothalamus (Esch et al., 1980). The biological actions of somatostatin-28 are similar to those of somatostatin-14 but in addition the larger peptide may have functions distinct from those of the tetradecapeptide (Meyers et al., 1980; Mandarino et al., 1981; Brown et al., 1981).

The amino acid sequence of somatostatin-14 has been conserved remarkably well throughout evolution; the sequences of the fish and mammalian (human and rat) peptides are identical (Fig. 5.1). The variety of somatostatin-related peptides present in the animal kingdom, however, is quite diverse. The catfish endocrine pancreas contains both a

MAMMALS

```
                    ┌─────────────────┐
          A G C K N F F W K T F T S C    ST-14
S A N S N P A M A P R E R L A G C K N F F W K T F T S C    ST-28
    S N P A M A P R E R L A G C K N F F W K T F T S C    ST-25
```

LOW VERTEBRATES (FISH)

```
                    ┌─────────────────┐
          A G C K N F F W K T F T S C    ST-14 I
          A G C K N F [Y] W K [G] F T S C    ST-14 II
[D][N][T][V][T][S][K][P][L][N] C [M] N [Y] F W K [S][R] T [A] C    ST-22
          A G [T][A][D][C] F W K [Y][C][V]    URO II
                    └───────────┘
```

Fig. 5.1. Primary structures of some of the somatostatins and somatostatin-related peptides. The cysteine bridges are denoted by heavy lines. The boxed residues of the fish peptides differ from the highly conserved mammalian sequences. ST, somatostatin. The fish ST-14I and ST-14II peptides are encoded in separate prohormones of the anglerfish endocrine pancreas. The ST-22 peptide is from a prohormone in the catfish pancreas. URO II is urotensin II, a peptide isolated from the caudal neurosecretory system of fish.

somatostatin-14 and a 22-amino acid peptide that is structurally distinct but similar to somatostatin-14 (Oyama *et al.*, 1980a; Andrews *et al.*, 1984) and the anglerfish endocrine pancreas contains two different versions of somatostatins, one identical to mammalian somatostatin-14 and another peptide of 14 amino acids that differs in sequence from somatostatin-14 in two of the fourteen residues (Noe & Spiess, 1983) (Fig. 5.1). Somatostatin-25 has been isolated from extracts of hypothalami and intestines (Benoit *et al.*, 1982). The amounts of these N-terminally shortened derivatives of somatostatin-28, however, are small; it seems possible that these truncated peptides are proteolytic degradation products resulting from the isolation procedure, rather than authentic natural peptides (Benoit, 1987). Urotensin II is a peptide, found in the caudal neurosecretory system of fish, that is evolutionarily related to somatostatin-14 (Pearson *et al.*, 1980).

5.2. Identification of biosynthetic precursors to somatostatin-14

Pulse and pulse–chase labeling studies in hypothalamus (Zingg & Patel, 1982) and pancreatic islets (Patzelt *et al.*, 1980; Noe & Spiess, 1983), as well as analyses of the products of cell-free synthesis programmed by mRNAs from pancreas (Goodman *et al.*, 1980a; Oyama *et al.*, 1980b;

Shields, 1980*b*), hypothalamus (Joseph-Bravo *et al.*, 1980) and intestine (Goodman *et al.*, 1981) indicate that somatostatin-14 and somatostatin-28 are derived from larger precursors, prosomatostatins, of apparent molecular masses of 12–16 kDa. As described below the advent of recombinant DNA technology led directly to the complete elucidation of the structures of the prosomatostatins.

5.3. Recombinant cDNAs encoding preprosomatostatins

Nucleotide sequencing of cloned cDNAs has provided the amino acid sequences of several different preprosomatostatins. Molecular cloning of the cDNAs encoding the anglerfish islet preprosomatostatins indicated that there are at least two separate preprosomatostatins, preprosomatostatin I and preprosomatostatin II (Hobart *et al.*, 1980; Goodman *et al.*, 1980*b*). The nucleotide sequences and the deduced amino acid sequences encoded therein were determined (Goodman *et al.*, 1980*a*; Hobart *et al.*, 1980). Subsequently, cDNAs encoding the single rat preprosomatostatin were obtained by screening a cloned cDNA library prepared from a rat medullary carcinoma of the thyroid, a tumor that produces somatostatin peptides (Goodman *et al.*, 1982, 1983; Funckes *et al.*, 1983). In addition, cloned cDNAs generated from the catfish pancreas revealed two cDNAs coding for somatostatin-14 (Taylor *et al.*, 1981) and a different but structurally related peptide, somatostatin-22 (Magazin *et al.*, 1982). The nucleotide and corresponding amino acid sequence of a cDNA encoding human preprosomatostatin has also been elucidated (Shen *et al.*, 1982). Both the nucleotide and amino acid sequences of the human preprosomatostatin are strikingly homologous to those of the rat preprosomatostatin; 95% of the nucleotides are identical in the sequences and the encoded 116 amino acids differ in only four residues, all four of which represent conservative substitutions (Fig. 5.2).

There is a considerable nucleotide and encoded amino acid homology between the anglerfish preprosomatostatin I and the rat (and human) preprosomatostatin cDNAs (Fig. 5.3) (see also Argos *et al.*, 1983). This high degree of homology is attested to by the fact that it was possible to use the anglerfish preprosomatostatin I as a hybridization probe to select recombinant rat preprosomatostatin cDNAs prepared from a rat medullary thyroid carcinoma (Goodman *et al.*, 1982, 1983). In both the anglerfish I and II, catfish, human, and rat preprosomatostatins, the sequences of somatostatin-28 and somatostatin-14 reside at the carboxy-terminus of the precursor. The somatostatin-14 sequences of the anglerfish I and rat precursors are identical, and the somatostatin-28 sequences are very

```
                    CGGACCCUGCGUCUAGACUGACCACCGCGCCUCAAGCUCGGCGUCGUCUCGAGGCAGGGGAG
                    •  •    ••  •  •• •••••  •   •  •              •  •
Rat
Human  5' ACACAAGCCCGCUUUAGGAGGCGAGGGUUCGGCAGCCAUCCGUCCUGCCUGCUGCUCGAGUUGACCAGCACUCGCCAGCUCGGCGUCGGCCGCCGAG

Rat    Met Leu Ser Cys Arg Leu Gln Cys Ala Leu Ala Ala Leu [Cys] Ile Val Leu Ala Leu Gly Gly Val Thr Gly [Ala] Pro Ser
       AUG CUG UCC UGC CGU CUC CAG UGC GCG CUG GCG GCC CUG  UGC  AUC GUC CUG GCC CUG GGC GGU GUC ACC GGC  GCG  CCC UCG
       ••• ••• ••• ••• •• ••• ••• ••• •• ••• •• ••• ••• •••  •••  ••• ••• ••• ••• •• ••• •••  •  •• •••  ••   ••  •••
       AUG CUG UCC UCC CGC CUC CAG UGC GCC CUG GCU GCC CUG  UCC  AUC GUC CUG GCC CUG GGC UGU GUC ACC GGC  GCU  CCC UCG
Human  Met Leu Ser Ser Arg Leu Gln Cys Ala Leu Ala Ala Leu [Ser] Ile Val Leu Ala Leu Gly Cys Val Thr Gly [Ala] Pro Ser

Rat    Asp Pro Arg Leu Arg Gln Phe Leu Gln Lys Ser Leu Ala Ala Ala [Thr] Gly Lys Gln Glu Leu Ala Lys Tyr Phe Leu Ala
       GAC CCC AGA CUC CGC CAG UUU CUG CAG AAG UCU CUG GCG GCC ACC  ACC  GGG AAA CAG GAA CUG GCC AAG UUC UUC CUC GCA
       ••• ••• ••• ••• •• ••• ••• ••• ••• ••• •• ••• ••• •• •••        ••• ••• ••• ••• ••• ••• ••• •• ••• •• •••
       GAC CCC AGA CUC CGU CAG UUU CUG CAG AAG UCC CUG GCU GCU UCC  GCG  GGG AAG CAG GAG CUG GCC AAG UAC UUC UUG GCA
Human  Asp Pro Arg Leu Arg Gln Phe Leu Gln Lys Ser Leu Ala Ala [Val] Gly Lys Gln Glu Leu Ala Lys Tyr Phe Leu Ala

Rat    Glu Leu Leu Ser Glu Leu Pro Asn Gln Thr Glu Asn Asp Ala Leu Glu Pro Glu Asp Leu [Pro] Gln Ala Ala Glu Gln Asp Glu
       GAA CUG CUG UCU GAG CUG CCC AAC CAG ACA GAG AAC GAU GCU CUG GAG CCU GAG GAU CUG  CCC  CAG GCA GCU GAG CAG GAC GAG
       •• ••• ••• ••  •• •• •• ••• ••• •• ••• ••• ••• •• ••  ••• ••• ••• ••• •••       ••• ••• ••• ••• ••• •• •••
       GAG CUG CUG UCU GAA CUG CCU GAA GAU CUG GAG AAU CCU GAA GCA CUG GAG CCG GAU CUG  UCC  CAG GCA GCU GAG CAG GAU GAA
Human  Glu Leu Leu Ser Glu Leu Pro Glu Asp Leu Glu Asn Pro Glu Ala Leu Glu Pro Asp Leu [Ser] Gln Ala Ala Glu Gln Asp Glu

                                                                               ST-28
Rat    Met Arg Leu Glu Leu Gln Arg Ser Ala Asn Ser Asn Pro Ala Met Ala Pro Arg Glu Arg Lys Ala Gly Cys Lys Asn Phe
       AUG AGG CUG GAG CUG CAG AGG UCU GCC UCU AAC UCG GCU CCA AUG GCA CCC CGG GAG AGG AAA GCU GGC UGC AAG AAC UUC
       ••• ••• •• ••• •• •• •• •• ••  ••  ••  •  ••  ••• •••  •• ••• ••• ••• ••• ••• ••  •• ••• ••• •• ••• •••
       AUG AGG GAG CUG GAG AGG UCU GCC AAC UCA AAC CCG GCU AUG GCU CCG CGA GAA CGC AAG GCU GGC UGC AAG AAU UUC
Human  Met Arg Glu Leu Glu Arg Ser Ala Asn Ser Asn Pro Ala Met Ala Pro Arg Glu Arg Lys Ala Gly Cys Lys Asn Phe

             ST-14
Rat    Phe Trp Lys Thr Phe Asn Ser Cys Stop
       UUC UGG AAG ACA UCC UGU UAG  CUUU——AA-UAUUGUUGCUCAGCCAGGACCUCAGAUCCCUCAUCUCUUCCUUAACUCCCAGCCCCC-C-C
       ••• ••• ••• ••  •• •• •••     •••••  •• ••••• •  ••••
       UUC UGG AAG ACU UUC ACA UCC  UGU UAG  CUUUCUUAACUAGAUUGUCCAU-UCAGACCUGAUCCCUC—————————————GCCCCCACAC
Human  Phe Trp Lys Thr Phe Asn Ser Cys Stop

Rat    CCCAAUGCUC———————AACUAGACCCUGCG-UUUAGAAAUUGAAGACUGCAAAUACAAAUAAUAAAUUAUGGUGAAAUUAUG(A)n   3'
       •••   •• •••       •   • ••
Human  CCC-AU-CUCUCUUCCCUAAUCCUCCAAGACGCGAGACCCCUUGCAUUAGAAACUGAAACUGAAACUGAAACAAAUACAAAUAAAUUGUGAAAUAU——(A)n
```

Fig. 5.2. Comparison of the rat and human preprosomatostatin cDNAs and encoded protein precursors. The sequences designated ST-28 and ST-14 indicate the two bioactive somatostatin peptides cleaved from prosomatostatin during posttranslational processing (see Figs. 5.4, 5.5). The vertical arrow points to the glycyl–alanyl peptide bond cleaved during cotranslational removal of the amino-terminal signal sequence of 24 amino acids. The boxed residues are the four out of the 116 amino acids of the rat and human precursors that differ. The dots indicate identical nucleotides, emphasizing the marked similarities between the two mRNAs.

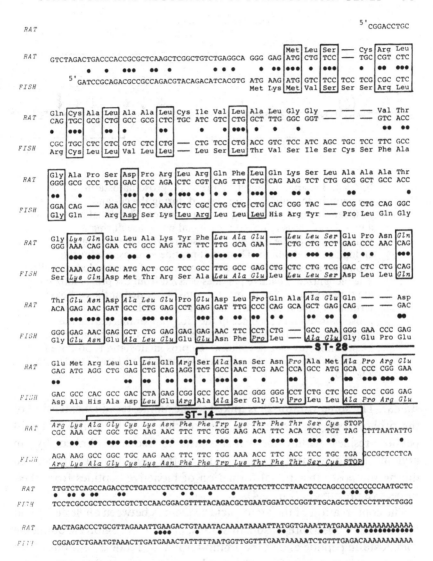

Fig. 5.3. Comparison of anglerfish I and rat preprosomatostatin cDNAs and encoded protein precursors. Sequences corresponding to somatostatin-14 (ST-14) and somatostatin-28 (ST-28) are indicated. Boxes denote amino acids strictly conserved between the fish and rat precursors. Dots refer to nucleotides conserved between the two cDNAs. To maintain homology between the NH$_2$-terminal positions of the precursors, it was necessary to delete three codons from the rat cDNA and one codon from the fish cDNA.

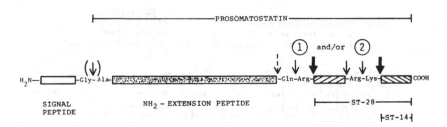

Fig. 5.4. Diagram of the structure of preprosomatostatin indicating
the peptide domains that are cleaved during the co- and
posttranslational processing of the precursor. The signal peptide at the
amino-terminus is removed by a signal peptidase during the
translocation of the nascent precursor polypeptide through the
endoplasmic reticulum. The signal peptide is necessary for the
prohormone to enter the secretory pathway of neuroendocrine cells.
The recognized bioactive peptides cleaved from the prosomatostatin
are somatostatin-28 and somatostatin-14, located at the carboxyl
terminus of the precursor. The processing of these peptides from
prosomatostatin takes place in the Golgi apparatus during the
sequestration of the peptides into secretory vesicles and granules.
Depending upon the tissue in which prosomatostatin is produced,
processing may occur by either routes designated 1 or 2 (circled) (see
Fig. 5.5).

similar. The somatostatin-28 is linked to the amino-terminal extension of
the prohormone by a single arginine residue and the somatostatin-14
peptide resides within the somatostatin-28 sequence linked by an
arginine–lysine residue, characteristic of the sites that are processed
during the posttranslational formation of bioactive peptides from their
respective precursors (Fig. 5.4).

The conservation of the amino acid sequences between anglerfish I and
the rat (and human) preprosomatostatins, both in the region encoding the
somatostatin peptide and in the amino-terminal peptide extensions of the
precursors, is even more remarkable when one considers that this conservation of sequences is much greater than that present between the
sequences of the two anglerfish preprosomatostatins. The somatostatin-14
and somatostatin-28 peptides of the anglerfish preprosomatostatin II
differ in two and eight amino acids, respectively, from the corresponding
somatostatin-14 and somatostatin-28 peptides of the anglerfish preprosomatostatin I (Hobart et al., 1980), and the two amino-terminal peptides
(including the signal sequence) show no discernible homology in amino
acid sequence (Hobart et al., 1980).

These observations of a high degree of homology between anglerfish I and rat preprosomatostatin and little or no homology between anglerfish II preprosomatostatin and anglerfish I and rat preprosomatostatins indicate that, during evolution, the rat and anglerfish I preprosomatostatins have been under a similar selective pressure to conserve their common sequences, and as a consequence, we can infer that the rat and anglerfish I precursors share similar biological functions that probably differ from the functions served by the anglerfish II and catfish-22 preprosomatostatins. Furthermore, the similarities in the sequences of the amino-terminal peptide of the anglerfish I and rat preprosomatostatins suggest that these peptides may have some intrinsic biological function apart from those of the somatostatin-14 and somatostatin-28 peptides.

5.4. Posttranslational processing of prosomatostatin

The different patterns of the posttranslational processing of the prosomatostatins appear to be quite complex and dependent upon the particular tissue in which the prosomatostatin is synthesized and upon the animal species. One question of interest has been whether or not somatostatin-28 is a precursor of somatostatin-14, or alternatively whether or not somatostatin-14 and somatostatin-28 are independently cleaved from prosomatostatin (Fig. 5.5). The consensus derived from the many pulse–chase studies that have been carried out in different tissues indicates that both circumstances occur. That is, in the hypothalamus (Zingg & Patel, 1982) and pancreatic islets (Patel et al., 1985) somatostatin-14 is cleaved directly from prosomatostatin and in the cells of the dorsal root ganglion somatostatin-28 is a precursor of somatostatin-14 (Harmar et al., 1982). It has been suggested that the precursor–product relationship as well as its importance may vary from one tissue to another or even at different stages of development (Benoit, 1987).

The remarkably high extent of conservation of the NH_2-terminal extension peptides of the prosomatostatins suggests that additional biologically relevant peptides may reside in these sequences. At least two additional cleavage sites have been identified in mammalian prosomatostatins resulting in the formation of peptides with as yet unknown biologic functions (Fig. 5.6). Cleavage at a Leu–Leu bond (position 56–57) of preprosomatostatin results in the formation of a 32-amino acid N-terminal fragment of prosomatostatin (Schmidt et al., 1985) and a larger (6 kDa) C-terminal fragment containing the carboxy-terminal somatostatin-28 (Benoit et al., 1982) (Fig. 5.6). The former peptide has been isolated from porcine intestine and structurally characterized (Schmidt

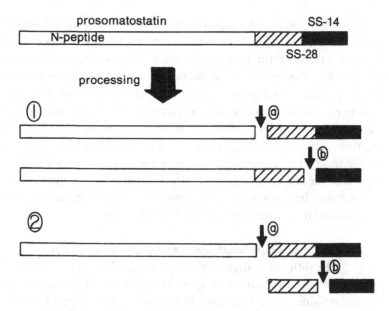

Fig. 5.5. Schematic depiction of alternate pathways for the posttranslational processing of prosomatostatin. Both pathways appear to exist depending upon the type of tissue as well as the animal species analyzed. For example, the intestinal cells appear to utilize pathway 1a, the pancreatic islet cells pathway 1b, and the hypothalamus pathways 2a and 2b.

et al., 1985). The 6 kDa peptide is present in hypothalamic extracts at relatively high concentrations during human fetal development. Levels of 6 kDa prosomatostatin are high at 14 weeks of fetal development and fall progressively thereafter to levels that are no longer detectable at 22 weeks of gestation (Ackland *et al.*, 1983). The 6 kDa peptide is also the predominant form of prosomatostatin formed by the posttranslational processing of the somatostatin precursor in various tissues of mice expressing a metallothionein–somatostatin fusion gene (Low *et al.*, 1985).

A second site of processing of prosomatostatin appears to be the Lys at position 34 of the preprosomatostatin resulting in the formation of a dodecapeptide consisting of the N-terminal 12 amino acids of prosomatostatin. This peptide, named 'antrin', is formed uniquely and in high amounts in the stomachs of mammals, where it accounts for up to one third of the total prosomatostatin-derived peptides (Benoit *et al.*, 1988).

The posttranslational processing of the two prosomatostatins of the anglerfish endocrine pancreas has been studied extensively (Noe *et al.*, 1987). As discussed earlier, the pancreas of the anglerfish contains two

RAT PREPROSOMATOSTATIN

1   10   20
MLSCRLQCALAALCIVLALG

         30   40
GVT**GA**PSDPRLRQFL**QK**SLA

         50   60
AATGKQELAKYFLAE**LL**SEP

         70   80
NQTENDALEPEDLPQAAEQD

         90   100
EMRLEL**QR**SANSNPAMAPRE

         110   116
RKAGCKNFFWKTFNSC

Fig. 5.6. Amino acid sequence of rat preprosomatostatin, depicting the residues that are cleaved either during cellular posttranslational processing or possibly during isolation of the peptides. Evidence for posttranslational processing of arginine 88 and arginine–lysine 101–102 to provide the bioactive peptides somatostatin-28 and somatostatin-14 is well substantiated. The biological significance of the cleavages at lysine 37 and leucine–leucine 56–57 is not yet understood. Cleavage at glycine–alanine 24–25 removes the amino-terminal signal sequence. Amino acid residues are given in the simplified, more convenient single letter code.

distinct but structurally related prosomatostatins encoded by two separate, non-allelic genes (Hobart *et al.*, 1980). By immunochemical studies it was determined that the two prosomatostatins are present in two different populations of islet cells (Noe *et al.*, 1987). Analyses of the structures of peptides isolated from anglerfish islets, as well as pulse and pulse–chase labeling studies, revealed that the two prosomatostatins are processed quite differently (Fig. 5.7). The prosomatostatin-I, which contains the somatostatin-14 identical in sequence to the mammalian tetradecapeptides, is processed to liberate somatostatin-14. In contrast, the processing of the prosomatostatin-II, which contains the somatostatin-14 sequence with two amino acid substitutions, results in the formation of a somatostatin-28 and no somatostatin-14 is produced. Additionally, several different peptides appear to be processed by cleavages at single basic residues located in the N-terminal regions of the prosomatostatins. This alternative pattern of posttranslational processing of these two closely related prosomatostatins strongly suggests that differences exist in the

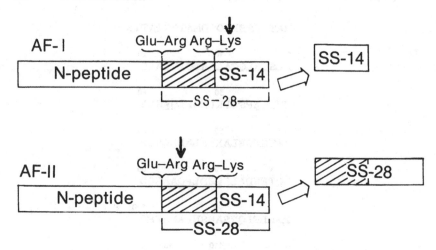

Fig. 5.7. Diagram showing the two different islet cell-specific pathways of the processing of the two anglerfish prosomatostatins encoded by two separate genes (Noe *et al.*, 1987). Different populations of islet cells express the two genes and the processing of the gene-I (AF-I) produces somatostatin-14 (SS-14), whereas other cells express gene II (AF-II) and produce somatostatin-28 (SS-28).

regulatory functions of the processed peptides synthesized in, and presumably secreted from, the two distinct islet cell phenotypes.

The posttranslational processing of prosomatostatin has been studied in different cell lines into which recombinant fusion genes have been integrated and expressed (Warren & Shields, 1984; Sevarino *et al.*, 1987). Processing of the prosomatostatin to the somatostatin-14 and somatostatin-28 peptides was observed to occur more efficiently in some cell types (insulinoma and pituitary corticotrophs) than in others (pituitary somatotrophs and adrenal medulla). It was suggested that prosomatostatin processing requires specific pathways which are present in some neuroendocrine cells, but not in others (Sevarino *et al.*, 1987).

5.5. Structures of the preprosomatostatin genes

The nucleotide sequences have been determined for the rat (Montminy *et al.*, 1984; Tavianini *et al.*, 1984) and human (Shen & Rutter, 1984) genes. The sequences of the two genes are remarkably similar, sharing approximately 85% identity in bases. The genes each consist of two exons separated by an intron (Figs. 5.8, 5.9). The introns of the two genes are located between the codons for Gln(+ 22) and Glu(+ 23) of preprosomatostatin. The introns of the rat and human genes consist of 630

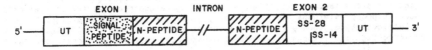

Fig. 5.8. Schematic diagram of the overall topography of the rat preprosomatostatin gene. A single intron interrupts the transcriptional region of the gene within the amino-terminal (N-peptide) sequence of preprosomatostatin. UT = the 5'- and 3'-untranslated sequences of the primary RNA transcript. SS-28 and SS-14 denote the peptides somatostatin-28 and somatostatin-14, respectively.

and 877 base pairs, respectively. The intron of the human gene contains several short directly repeated sequences and a long stretch of alternating purine and pyrimidine bases, thereby accounting for its greater length compared with the intron of the rat gene.

With just a single intron, the preprosomatostatin gene appears to be an unusually small and compact gene. Unlike the preproglucagon gene, which has six exons, each of which encodes a distinct functional domain (Mojsov *et al.*, 1986), the single intron of the preprosomatostatin gene resides in the middle of the N-terminal extension peptide of the prosomatostatin sequence and does not clearly separate the peptide domains, except as one might consider the N-terminal signal peptide as a functional domain separate from the C-terminal domain that contains somatostatin-14 and somatostatin-28 (Figs. 5.8, 5.9).

5.6. Regulation of somatostatin gene expression

Much more is known about the actions of somatostatin than is known about the factors that regulate the secretion and biosynthesis of this hormone (Reichlin, 1983, 1987). The widespread distribution of somatostatin-producing cells in many different tissues suggests the situation in which the regulation of somatostatin secretion and synthesis is complex. In the endocrine pancreas and the intestine, it is reasonable to conjecture that the regulation of somatostatin secretion is under the control of both systemic nutrients and local paracrine transmitters. In the central nervous system, regulation may be even more complex, involving an interplay of numerous neurotransmitters.

Although the precise repertoire of extracellular regulators of somatostatin production are as yet not defined, it is known that the transcription of the somatostatin gene is regulated through the cAMP-dependent signal transduction pathway. Several studies have demonstrated that analogs of cAMP and forskolin, an activator of adenylate cyclase, increase the

```
gaagtggaccagccgaatagctttaagcacccttgcacatacacacgaccgttaagcatgatggcaagtccagtaatc
tgagtacattgacaggtacccaactgtgtgtgctgatgtattgctggccaaggactgaaggatctcagtaattaatcatg
cacctatgtggcggaaatatgggatatgcatgtcgacactgagtgaaggcaagattatttggtctgtgtggcgtggagaa
tttcatgtgcctgtgtgggtgcaggctttcttttttcttcaaaaaaaaaaaaaataaaccactttagatcgtgtcgcctccc
ctcacttctttgattgattttgcgaggctaatggtgcgtaaaagcactggtgagatctgggggcgcctccttggctgacg
        (-31)                                      (+1)
tcagagagagagttttaaaaaggggagaccgtggagagctcgatAGCGGCTGAAGGAGACGCTACTGGAGTCGTCTCTGC
```
```
                                                                  -24
                                                       met leu ser cys
TGCCTGCGGACCTGCGTCTAGACTGACCCACCGCGCTCAAGCTCGGCTGTCTGAGGCAGGGGAGATG CTG TCC TGC
```
```
arg leu gln cys ala leu ala ala leu cys Ile val leu ala leu gly gly val thr gly
CGT CTC CAG TGC GCG CTG GCC GCG CTC TGC ATC GTC CTG GCT TTG GGC GGT GTC ACC GGG
+1                              +10                                        +20
ala pro ser asp pro arg leu arg gln phe leu gln lys ser leu ala ala ala tyr gly
GCG CCC TCG GAC CCC AGA CTC CGT CAG TTT CTG CAG AAG TCT CTG GCG GCT GCC ACC GGG
```
```
lys gln
AAA CAG gtaaggaaatggctgggactcgtcccctttgcgaattccccggccttcccttagtcttgctgtagcccctg
cgacaggtgttttagcgggcgcttctcagagtcgctcagcccctgagctcccagggaaactttttgaagtctagggtccgc
tcttactcgttccagaattgatcggcgctggtggtcacccttgcaggtaagttcccccttcgctttcaggaaaattccgaa
agcctgcaagagagcggggagagactgagctctatccctggtactggcacgagggttgctgacccaggtgctgaaaaaaa
atccggcaagaactcaggtccatggtccatttcgtgtctcataaaggaaaatggagctgctcaaactattggcatactat
atttacaaaacgacttcctatcatccatggtttctctgtgtttttaaggcatagcactttctgaaagacttgggtttgagg
aagcttttttccctgtgcataatctagtgaatatagcagccatccatattactgtggaaacttggtttgaatgattaaa
```
```
                                                          glu leu ala
tcttattttcaaaccgcatttctccctttctcccattcccccttttgctctcctccctgccctatccagGAA CTG GCC
         +30                                        +40
lys tyr phe leu ala glu leu leu ser glu pro asn gln thr glu asn asp ala leu glu
AAG TAC TTC TTG GCA GAA CTG CTG TCT GAG CCC AAC CAG ACA GAG AAC GAT GCC CTG GAG
         +50                                        +60
pro glu asp leu pro gln ala ala glu gln asp glu met arg leu glu leu gln arg|ser
CCT GAG GAT TTG CCC CAG GCA GCT GAG CAG GAC GAG ATG AGG CTG GAG CTG CAG AGG TCT
         +70                      SS-28                 +80       SS-14
ala asn ser asn pro ala met ala pro arg glu arg lys|ala gly cys lys asn phe phe
GCC AAC TCG AAC CCA GCC ATG GCA CCC CGG GAA CGC AAA|GCT GGC TGC AAG AAC TTC TTC
         +90
trp lys thr phe thr ser cys|stop
TGG AAG ACA TTC ACA TCC TGT|TAG CTTTAATATTGTTGTCTCAGCCAGACCTCTGATCCCTCTCCTGCAAA
```
```
TCCCATATCTCTTCCTTAACTCCCAGCCCCCCCCCCCCAATGCTCAACTAGACCCTGCGTTAGAAATTGAAGACTGTAAT
```
```
TACAAAATAAAATTATGGTGAAATTATGAA
```

Fig. 5.9. Nucleotide and deduced encoded protein sequence of the rat somatostatin gene. The mRNA sequence encoding preprosomatostatin is shown in large type. The single intron and 5'-flanking region of the gene sequence are shown in small type. Numbering of the encoded preprosomatostatin sequence begins at −24. The prosomatostatin, after removal of the signal sequence begins at +1 and spans 92 residues to +92. The entire preprosomatostatin consists of 116 amino acids. The bioactive peptides cleaved from prosomatostatin are somatostatin-14 (SS-14, boxed) and somatostatin-28 (SS-28, overlined). The variant, constitutive transcriptional promoter (TATA box) in the 5'-flanking region of the gene is underlined (data from Montminy et al., 1984).

transcriptional rate of the somatostatin gene (Montminy *et al.*, 1986; Andrisani *et al.*, 1987).

The cloning of the somatostatin genes has expedited analyses of the identification of the control regions of the gene. By fusing the 5'-flanking sequences of the somatostatin gene to convenient enzymatic reporters, such as bacterial chloramphenicol acetyl transferase, and transfecting the fusion genes into animal cells, it has been possible to demonstrate directly cAMP-mediated stimulation of somatostatin gene transcription. Once a DNA sequence responsible for cAMP-responsive gene transcription was detected in the 5'-flanking region of the somatostatin gene, a series of 5'-shortened sequences were tested for cAMP-stimulated transcription (Montminy *et al.*, 1986; Andrisani *et al.*, 1987). These analyses led to the identification of a short sequence of nucleotides, an octanucleotide, as necessary for the cAMP-mediated transcription (Fig. 5.10). This octanucleotide sequence, TGACGTCA, is a perfect palindrome with dyad symmetry about the CG axis. The identical or very similar 8 bp DNA element has been identified in a substantially large number of other cAMP-responsive genes (see Montminy *et al.*, 1986).

Further dissection of the control elements of the somatostatin gene by DNA deletional-expression analyses indicate that the full complement of sequence information responsible for both cell-specific and cAMP-responsive transcription resides in approximately 60 bp of DNA flanking the transcriptional start site of the gene (Fig. 5.10*b*). The transcriptional control of the somatostatin gene appears to come about by the binding of specific proteins to DNA sequence elements located within this relatively short (60 bp) of DNA flanking the bases involved in initiating the transcription of the gene (Montminy & Bilezikjian, 1987; Andrisani *et al.*, 1988). At least three different DNA-binding proteins have been implicated in the transcriptional activation of the somatostatin gene. One of these proteins is a specific cyclic AMP-response element DNA-binding protein that recognizes and binds to the cyclic AMP-responsive DNA sequence of the motif TGACGTCA and has been termed CREB for cyclic AMP-response element binding protein (Montminy & Bilezikjian, 1987). The other two proteins appear to augment somatostatin gene transcription by binding to regions of the DNA adjacent to the cyclic AMP response element (Andrisani *et al.*, 1988). It is tempting to speculate that the proteins that recognize these specific DNA sequences signal the activation of RNA polymerase II to begin the synthesis of an mRNA template from the somatostatin gene. Much experimental work remains to be done to finally elucidate the molecular workings of the expression of the

(a)

$\qquad$ CRE $\qquad$ TATA BOX

Transcription

gatctgggggcgcctccttggctTGACGTCAgagagagagTTTAAAaaggggagaccgtggagagctcgatA

-70   -48   -41   -31   +1

(b)

Deletions	% Activity in CA-77 cells
PBx△-750	100
PBx△-500	100
PBx△-420	70
PBx△-315	70
PBx△-250	94
PBx△-101	70
PBx△-70	94
PBx△-60	100
PBx△-43	4.8

Fig. 5.10. (a) Sequence of 5'-flanking region of the transcriptional unit of the rat preprosomatostatin gene. The two regions of the nucleotide sequence believed to be important for the transcription of the gene are highlighted in large type. The CRE is a palindromic sequence that appears to mediate cAMP-responsivity as well as basal activity for gene transcription. The TATA box element (TTTAAA) is a constitutive promoter located in most animal cell genes at a position 24 to 28 bp upstream from the site that signals the start of transcription (designated + 1). Notably, most if not all of the information for cell-specific and cAMP-regulated transcription of the somatostatin gene appear to reside in only 60 bp of DNA flanking the transcriptional site. (b) Relative transcriptional activity provided by DNA sequence of various lengths flanking the start of transcription of the rat preprosomatostatin gene. The table shows the activity of transcription of a bacterial chloramphenicol acetyl transferase reporter

somatostatin gene. It seems certain, however, that the somatostatin gene is a highly useful model gene with which to analyze the factors responsible for the transcription of an animal cell gene under the influence of the cAMP-mediated signal transduction pathway.

5.7. Summary

The structure of the somatostatin gene has been highly conserved throughout evolution with the encodement of the identical peptide somatostatin-14 at the carboxy-terminus of the prosomatostatin. Mammals appear to have a single gene encoding preprosomatostatin but for unexplained reasons the fishes (anglerfish and catfish) have two or more similar but distinct non-allelic genes. The complete structures of the human and rat preprosomatostatin genes have been determined. The two genes share a remarkably high extent of sequence similarity each with a single intron located at the identical position in the mRNA. The rat and human preprosomatostatins differ in only four conservative amino acid substitutions out of 116 amino acid residues. The posttranslational processing of the prosomatostatins appears to vary markedly depending upon the tissue in which it is produced and upon the particular animal species. The major recognized peptides processed from the prosomatostatins are somatostatin-14 and somatostatin-28. Both of these peptides share similar bioactivites in the suppression of the secretory activities of many different endocrine and exocrine glands. Additional peptides appear to be cleaved from other regions of the prosomatostatins but the biologic actions of these peptides, if any exist, remain unknown.

Analyses of the factors that control the expression of the preprosomatostatin gene have just begun. The gene is expressed in specific cells in the pancreatic islets and in a wide variety of neuroendocrine cells in the intestines, brain and thyroid gland. The important regulatory region of the gene appears to be unusually compact, consisting of only 60 bp of DNA flanking the transcriptional start site. Importantly, the regulatory sequence contains a cyclic AMP-response element of 8 bp that appears to mediate transcriptional responses of the gene through the cAMP-dependent pathway of signal transduction. Much of the work yet to be

Caption to Fig. 5.10 (*cont.*)
> gene after transfection into rat thyroid carcinoma (CA-77) cells in response to different lengths of DNA sequences from 750 to 43 bp. Note that full transcriptional activity remains in a DNA fragment of only 60 bp. (Data are from Andrisani *et al.*, 1987.)

done will involve the isolation and characterization of the DNA-binding proteins that interact with these DNA control elements to modulate cell-specific gene transcription.

I am grateful for the contributions of my co-workers, Richard H. Goodman, Pauline C. Lund, and Marc R. Montminy, and to Janice Canniff for typing the manuscript. Some of the studies described were supported by N.I.H. grant DK-30457.

References

Ackland, J., Ratter, R., Bourne, G.L. & Rees, L.H. (1983). Characterization of immunoreactive somatostatin in human fetal hypothalamic tissue. *Regulatory Peptides* 5, 95–101.

Andrews, P.C., Pubols, M.H., Hermodson, M.A., Sheares, B.T. & Dixon, J.E. (1984). Structure of the 22-residue somatostatin from catfish. An *o*-glycosylated peptide having multiple forms. *Journal of Biological Chemistry* 259, 13267–72.

Andrisani, O.M., Hayes, T.E., Roos, B. & Dixon, J.E. (1987). Identification of the promoter sequences involved in the cell specific expression of the rat somatostatin gene. *Nucleic Acids Research* 15, 5715–28.

Andrisani, O.M., Pot, D.A., Zhu, Z. & Dixon, J.E. (1988). Three sequence-specific DNA-protein complexes are formed with the same promoter element essential for expression of the rat somatostatin gene. *Molecular and Cellular Biology* 8, 1947–56.

Argos, P., Taylor, W.L., Minth, C.D. & Dixon, J.E. (1983). Nucleotide and amino acid sequence comparisons of preprosomatostatins. *Journal of Biological Chemistry* 258, 8788–93.

Arimura, A., Sato, H., Dupont, A., Nishi, N. & Schally, A.V. (1975). Somatostatin: abundance of immunoreactive hormone in rat stomach and pancreas. *Science* 189, 1007–9.

Barker, J.L. (1976). Peptides: Roles in neuronal excitability. *Physiological Reviews* 56, 435–52.

Benoit, R. (1987). Peptides derived from mammalian prosomatostatin. In *Somatostatin. Basic and Clinical Status* (ed. S. Reichlin), pp. 33–50. New York: Plenum Publishing Corporation.

Benoit, R., Ling, N., Alford, B. & Guillemin, R. (1982). Seven peptides derived from prosomatostatin in rat brain. *Biochemical and Biophysical Research Communications* 107, 944–50.

Benoit, R., Ravazzola, M., Bennett, H.P.J., Fried, G., Gravel, G. & Orci, L. (1988). Antrin is present in several mammals and secreted by endocrine cells of the stomach. *Abstracts of the Endocrinological Society (Endocrinology)* 122, 308 (suppl.).

Brazeau, P., Vale, W., Burgus, R., Ling, N., Butcher, M., Rivier, J. & Guillemin, R. (1973). Hypothalamic polypeptide that inhibits secretion of immunoreactive pituitary growth hormone. *Science* 179, 77–9.

Brown, M., Rivier, J. & Vale, W. (1981). Somatostatin 28: selective action on the pancreatic β-cell and brain. *Endocrinology* 108, 2391–3.

Esch, F., Böhlen, P., Ling, N., Benoit, R., Brazeau, P. & Guillemin, R. (1980). Primary structure of ovine hypothalamic somatostatin-28 and somatostatin-25. *Proceedings of the National Academy of Sciences of the U.S.A.* 77, 6827–31.

Funckes, C.L., Minth, C.D., Deschenes, R., Magazin, M., Tavianini, M.A., Sheets, M., Collier, K., Weith, H.L., Aron, D.C., Roos, B.A. & Dixon, J.E. (1983). Cloning and characterization of a mRNA-encoding rat preprosomatostatin. *Journal of Biological Chemistry* 258, 8781–7.

Gerich, J.E. & Patton, G.S. (1978). Somatostatin: physiology and clinical applications. *Medical Clinics of North America* 62, 375–92.

Goodman, R.H., Aron, D.C. & Roos, B.A. (1983). Rat pre-prosomatostatin. Structure and processing by microsomal membranes. *Journal of Biological Chemistry* **258**, 5570–3.

Goodman, R.H., Jacobs, J.W., Chin, W.W., Lund, P.K., Dee, P.C. & Habener, J.F. (1980a). Nucleotide sequence of a cloned structural gene coding for a precursor of pancreatic somatostatin. *Proceedings of the National Academy of Sciences of the U.S.A.* **77**, 5869–73.

Goodman, R.H., Jacobs, J.W., Dee, P.C. & Habener, J.F. (1982). Somatostatin-28 encoded in a cloned cDNA obtained from a rat medullary thyroid carcinoma. *Journal of Biological Chemistry* **257**, 1156–9.

Goodman, R.H., Lund, P.K., Barnett, F.H. & Habener, J.F. (1981). Intestinal pre-prosomatostatin. Identification of mRNA coding for a precursor by cell-free translation and hybridization with a cloned islet cDNA. *Journal of Biological Chemistry* **256**, 1499–1501.

Goodman, R.H., Lund, P.K., Jacobs, J.W. & Habener, J.F. (1980b). Pre-prosomatostatins. Products of cell-free translations of messenger RNAs from anglerfish islets. *Journal of Biological Chemistry* **255**, 6549–52.

Harmar, A., Ivell, R. & Keen, P. (1982). The *de novo* biosynthesis of somatostatin and a related peptide in isolated rat dorsal root ganglia. *Brain Research* **242**, 365–8.

Hobart, P., Crawford, R., Shen, L., Pictet, R. & Rutter, W.J. (1980). Cloning and sequence analysis of cDNAs encoding two distinct somatostatin precursors found in the endocrine pancreas of anglerfish. *Nature* **288**, 137–41.

Hökfelt, T., Efendic, S., Hellerström, C., Johansson, O., Luft, R. & Arimura, A. (1975). Cellular localization of somatostatin in endocrine-like cells and neurons of the rat with special references to the A_1-cells of the pancreatic islets and to the hypothalamus. *Acta Endocrinologica* **80** (suppl. 200), 1–41.

Joseph-Bravo, P., Charli, J.L., Sherman, T., Boyer, H., Bolivar, F. & McKelvy, J.F. (1980). Identification of a putative hypothalamic mRNA coding for somatostatin and of its product in cell-free translation. *Biochemical and Biophysical Research Communications* **94**, 1004–12.

Low, M.J., Hammer, R.E., Goodman, R.H., Habener, J.F., Palmiter, R.D. & Brinster, R.L. (1985). Tissue-specific post-translational processing of pre-prosomatostatin encoded by a metallothionein–somatostatin fusion gene in transgenic mice. *Cell* **41**, 211–19.

Magazin, M., Minth, C.D., Funckes, C.L., Deschenes, R., Tavianini, M.A. & Dixon, J.E. (1982). Sequence of a cDNA encoding pancreatic preprosomatostatin-22. *Proceedings of the National Academy of Sciences of the U.S.A.* **79**, 5152–6.

Mandarino, L., Stenner, D., Blanchard, W., Nissen, S., Gerich, J., Ling, N., Brazeau, P., Bohlen, P., Esch, F. & Guillemin, R. (1981). Selective effects on somatostatin-14, -25, and -28 on *in vitro* insulin and glucagon secretion. *Nature* **291**, 76–7.

Meyers, C.A., Murphy, W.A., Redding, T.W., Coy, D.H. & Schally, A.V. (1980). Synthesis and biological actions of prosomatostatin. *Proceedings of the National Academy of Sciences of the U.S.A.* **77**, 6171–4.

Mojsov, S., Heinrich, G., Wilson, I.B., Ravazzola, M., Orci, L. & Habener, J.F. (1986). Preproglucagon gene expression in pancreas and intestine diversifies at the level of post-translational processing. *Journal of Biological Chemistry* **261**, 11880–9.

Montminy, M.R. & Bilezikjian, L.M. (1987). Binding of a nuclear protein to the cyclic-AMP response element of the somatostatin gene. *Nature* **328**, 175–8.

Montminy, M.R., Goodman, R.E., Horovitch, S.J. & Habener, J.F. (1984). Primary structure of the gene encoding rat preprosomatostatin. *Proceedings of the National Academy of Sciences of the U.S.A.* **81**, 3337–40.

Montminy, M.R., Sevarino, K.A., Wagner, J.A., Mandel, G. & Goodman, R.H. (1986). Identification of a cyclic-AMP-responsive element within the rat somatostatin gene. *Proceedings of the National Academy of Sciences of the U.S.A.* **83**, 6682–6.

Noe, B.D., Mackin, R.B., McDonald, J.K., Andrews, P.C., Dixon, J.E. & Spiess, J. (1987). Cotranslational and posttranslational proteolyte processing of preprosomatostatin-I and preprosomatostatin-II in intact islet tissue. In *Somatostatin. Basic and clinical status* (ed. S. Reichlin), pp. 59–70. New York: Plenum Publishing Corporation.

Noe, B.D. & Spiess, J. (1983). Evidence for biosynthesis and differential post-translational proteolytic processing of different (pre)prosomatostatins in pancreatic islets. *Journal of Biological Chemistry* **258**, 1121–8.

Oyama, H., Bradshaw, R.A., Bates, O.J. & Permutt, A. (1980*a*). Amino acid sequence of catfish pancreatic somatostatin I. *Journal of Biological Chemistry* **255**, 2251–4.

Oyama, H., O'Connell, K. & Permutt, A. (1980*b*). Cell-free synthesis of somatostatin. *Endocrinology* **107**, 845–7.

Patel, Y.C. & Reichlin, S. (1978). Somatostatin in hypothalamus, extrahypothalamic brain and peripheral tissues of the rat. *Endocrinology* **102**, 523–30.

Patel, Y.C., Zingg, H.H. & Srikant, C.B. (1985). Somatostatin-14 like immunoreactive forms in the rat: characterization, distribution and biosynthesis. In *Somatostatin* (ed. Y.C. Patel & G.S. Tannenbaum), pp. 71–87. New York: Plenum Publishing Corporation.

Patzelt, C., Tager, H.S., Carroll, R.J. & Steiner, D.F. (1980). Identification of prosomatostatin in pancreatic islets. *Proceedings of the National Academy of Sciences of the U.S.A.* **77**, 2410–14.

Pearson, D., Shively, J.E., Clark, B.R., Geschwind, I.I., Barkley, M., Nishioka, R.S. & Bern, H.A. (1980). Urotensin II: A somatostatin-like peptide in the caudal neurosecretory system of fishes. *Proceedings of the National Academy of Sciences of the U.S.A.* **77**, 5021–4.

Pradayrol, L., Jornvall, H., Mutt, V. & Ribet, A. (1980). N-terminally extended somatostatin: the primary structure of somatostatin-28. *FEBS Letters* **109**, 55–8.

Reichlin, S. (1983). Somatostatin. *New England Journal of Medicine* **309**, 1495–501.

Reichlin, S. (ed.) (1987). *Somatostatin. Basic and clinical status.* New York: Plenum Publishing Corporation.

Rorstad, O.P., Epelbaum, J., Brazeau, P. & Martin, J.B. (1979). Chromatographic and biological properties of immunoreactive somatostatin in hypothalamic and extrahypothalamic brain regions of the rat. *Endocrinology* **105**, 1083–90.

Schally, A.V., Huang, W.-Y., Chang, R.C.C., Arimura, A., Redding, T.W., Millar, R.P., Hunkapiller, M.W. & Hood, L.E. (1980). Isolation and structure of prosomatostatin: a putative somatostatin precursor from pig hypothalamus. *Proceedings of the National Academy of Sciences of the U.S.A.* **77**, 4489–93.

Schmidt, W.E., Mutt, V., Kratzin, H., Carlquist, M., Conlon, J.M. & Creutzfeldt, W. (1985). Isolation and characterization of proSS(1–32), a peptide derived from the N-terminal region of porcine preprosomatostatin. *FEBS Letters* **192**, 141–6.

Sevarino, K.A., Felix, R., Banks, C.M., Low, M.J., Montminy, M.R., Mandel, G. & Goodman, R.H. (1987). Cell-specific processing of preprosomatostatin in cultured neuroendocrine cells. *Journal of Biological Chemistry* **262**, 4987–93.

Shen, L.-P., Pictet, R.L. & Rutter, W.J. (1982). Human somatostatin I: Sequence of the cDNA. *Proceedings of the National Academy of Sciences of the U.S.A.* **79**, 4575–9.

Shen, L.-P. & Rutter, W.J. (1984). Sequence of the human somatostatin I gene. *Science* **224**, 168–71.

Shields, D. (1980*a*). *In vitro* biosynthesis of fish islet preprosomatostatin: Evidence of processing and segregation of a high molecular weight precursor. *Proceedings of the National Academy of Sciences of the U.S.A.* **77**, 4074–8.

Shields, D. (1980*b*). *In vitro* biosynthesis of somatostatin. *Journal of Biological Chemistry* **255**, 11625–8.

Tavianini, M.A., Hayes, T.E., Magazin, M.D., Minth, C.D. & Dixon, J.E. (1984). Isolation, characterization, and DNA sequence of the rat somatostatin gene. *Journal of Biological Chemistry* **259**, 11798–803.

Taylor, W.L., Collier, K.J., Deschenes, R.J., Weith, H.L. & Dixon, J.E. (1981). Sequence analysis of a cDNA coding for a pancreatic precursor to somatostatin. *Proceedings of the National Academy of Sciences of the U.S.A.* **78**, 6694–8.

Warren, T.G. & Shields, D. (1984). Expression of preprosomatostatin in heterologous cells: biosynthesis, post-translational processing and secretion of mature somatostatin. *Cell* **39**, 547–55.

Zingg, H.H. & Patel, Y.C. (1982). Biosynthesis of immunoreactive somatostatin by hypothalamic neurons in culture. *Journal of Clinical Investigation* **70**, 1101–9.

6

The mosaic evolution of the pancreatic polypeptide gene

HIROSHI YAMAMOTO, HIDETO YONEKURA AND KOJI NATA

6.1. Introduction

Pancreatic polypeptide, a 36 amino acid carboxyamidated peptide hormone, is synthesized in the PP cells of the islets of Langerhans (Solcia *et al.*, 1985). The polypeptide has been reported to modulate insulin and somatostatin secretion (Murphy *et al.*, 1981; Trimble *et al.*, 1982) and to inhibit pancreatic exocrine and gastric secretion (Adrian *et al.*, 1981; Hazelwood, 1981). In canine islet cells, the pancreatic polypeptide precursor was demonstrated to be posttranslationally processed to produce pancreatic polypeptide and an icosapeptide, a second stable product derived from the carboxy-terminal region of the precursor (Schwartz & Tager, 1981). Icosapeptide-like peptides have also been detected in human, sheep and cat pancreases (Schwartz *et al.*, 1984; Nielsen *et al.*, 1986). However, in 1986 Yamamoto *et al.* demonstrated that rat pancreatic polypeptide precursor deduced from the nucleotide sequence of the pancreatic polypeptide precursor mRNA did not contain any sequence similar to the icosapeptide (Yamamoto *et al.*, 1986). Recent characterizations of several mammalian pancreatic polypeptide mRNAs (Boel *et al.*, 1984; Toothman & Paquette, 1987; Yonekura *et al.*, 1988; Blackstone *et al.*, 1988) have shown that the pancreatic polypeptide domain of the precursor is well conserved, whereas there is a high degree of divergence in the carboxy-terminal domain; the mouse and guinea pig precursors also contained no sequence similar to the icosapeptide. As the structural analyses were extended to the gene level (Leiter *et al.*, 1985; Yonekura *et al.*, 1988), it became evident that the blend of structural conservation and divergence in the pancreatic polypeptide precursor may have been generated through the 'mosaic evolution' of the pancreatic polypeptide gene (Yamamoto *et al.*, 1986; Yonekura *et al.*, 1988); the exons coding for the

```
Rat              TCAGTCCAGCAGAAGGTAGGTGTCTGGGTCCAATACCCACTCGCTCAGGACACAGG
Mouse                          ::::::::C::::C::T::::A:::::C:::::::
Guinea-Pig                               GGCTC.CTCA:GTT::::
Canine                                 T:CGC:C::T:::::T:G::
Human            ATCCGG:CTG::GCA:GT:C:CG::T::::T:G:G::::T:TA:::T:::T:C::
                 -50       -40       -30      `-20       -10       -1
```

```
                                              Putative Signal Peptide
                 1                       10                            20
Human            :   :  Ala  :  Arg Leu  :   :   :   :  Leu  :   :   :   :  Cys  :   :   :
Canine           :  Pro Ala  :  Cys Arg  :   :  Phe  :  Leu  :   :  Ala Cys  :   :
Guinea-Pig       :  Thr Ala Thr Arg Cys  :   :  Trp  :  Leu  :   :  Gly  :  Cys Met  :   :   :
Mouse            :   :   :   :   :  Cys  :   :   :   :   :  Val  :   :   :   :   :   :   :   :
Rat
                 Met Ala Val Ala Tyr Tyr Cys Leu Ser Leu Phe Leu Leu Ser Thr Trp Val Ala Leu Leu

                 ATG GCC GTC GCA TAC TAC TGC CTC TCC CTG TTT CTC CTA TCC ACT TGG GTG GCT CTG CTG
Mouse            ::: ::: ::: ::: ::: :G: ::: ::: ::: ::: G:: ::: ::: ::: ::: ::: ::: ::: ::: :::
Guinea-Pig       ::: A:: :C: A:C CG: :G: ::: T:G :GG ::: C:G ::: ::G GG: ::: ::C A:: ::C ::: :::
Canine           ::: C:T :C: ::C :G: CG: ::: ::: :T: ::: C:G ::: ::G ::A G:C ::T ::: ::: ::: T::
Human            ::: ::T :C: ::: CG: CT: ::: ::: ::: ::: C:G ::: ::G ::: ::C ::C ::: ::: ::: T:A
                 1         10          20          30          40          50          60
```

```
                                                   30                               40
Human            :   :   :   :  Leu  :   :  Gln  :   :   :   :   :  Val  :   :   :   :  Asn
Canine           :   :   :  Pro Leu  :  Thr Arg  :   :   :   :   :  Val  :   :   :   :  Asp
Guinea-Pig       :  Pro  *   *   *  Glu  :   :   :   :   :   :   :  Val  :   :   :   :  Asp
Mouse            :   :   :   :   :  Thr  :   :   :   :   :   :   :   :   :   :   :   :   :
Rat
                 Leu Gln Pro Leu Gln Gly Ala Trp Gly Ala Pro Leu Glu Pro Met Tyr Pro Gly Asp Tyr

                 CTA CAG CCC CTG CAG GGT GCC TGG GGA GCC CCA CTG GAG CCA ATG TAC CCG GGG GAC TAT
Mouse            ::G ::: ::: ::: ::: ::G A:: ::: ::: ::: ::: ::: ::: ::: ::: ::A ::C ::: :::
Guinea-Pig       ::T :C: *** *** *** :AA ::: ::: ::: ::C ::: ::: ::T G:: ::C ::: ::: G:C
Canine           ::G ::: ::A :CA :T: ::: A:: C:: ::G ::: ::G ::: ::: ::: G:: ::T ::: ::: A::
Human            ::: ::: ::A :T: ::: ::: CA: ::: ::: ::G ::: ::: ::: ::: G:: ::: ::A ::: A::
                           70          80          90          100         110         120
```

```
                                                   Pancreatic Polypeptide
                                         50                                  60
Human            :   :  Pro  :   :  Met  :   :   :  Ala Ala Asp  :   :   :   :   :   :  Met  :
Canine           :   :  Pro  :   :  Met  :   :   :  Ala Ala Glu  :   :   :   :   :  Met  :
Guinea-Pig       :   :  Pro Gln  :  Met  :   :   :  Ala Ala Glu Met  :   :   :   :   :  Met  :
Mouse            :   :  Pro  :   :  Met  :   :   :   :   :   :   :   :   :   :   :   :   :   :
Rat
                 Ala Thr His Glu Gln Arg Ala Gln Tyr Glu Thr Gln Leu Arg Arg Tyr Ile Asn Thr Leu

                 GCT ACG CAT GAA CAG AGG GCT CAA TAC GAA ACT CAG CTC CGA AGA TAC ATC AAC ACA CTG
Mouse            ::G ::A :C: ::G ::: :T: ::A ::: ::T ::: ::: ::: ::C ::: ::: ::: ::: ::: :::
Guinea-Pig       ::G ::: :CC C:G ::: :T: ::: ::C G:C G:: A:G ::C ::: ::C ::: ::: ::: ::: :TG :::
Canine           ::C ::A :CA ::G ::: :T: ::C ::G ::: :CG G:: G:: ::: ::C ::: ::: ::: ::: :TG :::
Human            ::C ::A :CA ::G ::: :T: ::C ::G ::T :C: G:: G:T ::: ::T ::: ::: ::: ::: :TG :::
                           130         140         150         160         170         180
```

```
                                                   Icosapeptide
                                         70                                  80
Human            :   :   :   :   :   :   :   :  His Lys Glu Asp Thr Leu Ala Phe Ser Glu Trp Gly
Canine           :   :   :   :   :   :   :   :   :  Arg Gly Glu Met Arg Asp Ile Leu Glu Trp Gly
Guinea-Pig       :   :   :   :   :   :   :  Ser Ala  :  Glu Asp  :  Leu Gly Leu Pro Val Trp Arg
Mouse            :   :   :   :   :   :   :   :  Ala  :  Glu Glu Asn Thr Gly Gly Leu Pro Gly Val
Rat
                 Thr Arg Pro Arg Tyr Gly Lys Arg Asp Glu Asp Thr Ala Gly Leu Pro Gly Arg Gln Leu

                 ACC AGG CCT AGG TAT GGG AAG AGA GAT GAG GAC ACA GCT GGG CTT CCT GGA AGG CAG CTC
Mouse            ::: ::: ::: ::: ::: ::: ::: ::: :CC ::: ::G GAG AAC ACA GG: GGA CTT CCT GGA G:G
Guinea-Pig       ::T C:: ::C ::: ::C ::: ::: ::C :CC ::: ::G GAT ::G CT: GGC TTG CCG GT: TG: :G:
Canine           ::: ::: ::C ::: ::: ::: ::: ::A ::: ::C AGA :GA GA: ATG C:: GAC ATC CTG GAA TG: GG:
Human            ::: ::: ::: ::: ::: ::: ::: ::A ::: C:C A:A ::G GAC A:G CT: GCC TTC TCG GA: TG: GGG
                           190         200         210         220         230         240
```

```
                                                    90                                              100
Human        Ser  :  His Ala Ala Val Pro Arg Glu  :  Ser  :  Leu Asp Leu
Canine       Ser  :  His Ala Ala Ala Pro Arg Glu  :      Asp Glu
Guinea-Pig   Gln Ser His Ala Ala Ala Pro Gly  :  Ser His Arg His Pro Pro  :  Gly Leu Pro Ala
Mouse        Gln Leu Ser Pro Cys Thr Ser Pro Pro Val Gly Leu Ile Pro Cys Ser Ala Pro Trp Ser
Rat          Pro Pro Cys Thr Ser Leu Leu Val Gly Leu Met Pro Cys Ala Ala Ala Arg Ser

             CCA CCA TGC ACC AGC CTT CTG GTG GGT TTG ATG CCC TGC GCT GCT GCC CGG AGC TGA AGG
Mouse        :AG :TC :C: C:: T:: ACC AGC CCC CCA G:T GGC TTG ATT C:C TGC T:T GC: CC: ::G ::C
Guinea-Pig   :AG T:C CA: G:G GCT GCC :C: :GT ::G :CC CAT :G: CA: C:A C:: ::: G:: CT: CCT GCA
Canine       T:C ::C CAT G:A GC: GCC :CC AG:|:AG C:: ::: GA: GAG TAA TGC CA: :TC CAA GT: :T:
Human        T:C ::G CAT G:T GCT G:C :CC AG:|:AG C:C :GC ::G CTG :AC TTA TAA T:C CA: CTT CT:
                 250             260             270             280             290             300


Guinea-Pig   Ala Lys Gly Gly Thr Gly Val Ser Gly Ser Pro Pro Lys Pro Trp Asp Cys Leu Pro Cys

Rat          CAG GGA TCT CAG CCC AAT GGA TTT GTA GCC TCC CTT CTG TCT CTC TCC TAC CCA TGG CTG
Mouse        TGA A:G CAG GGA G:T CTG CCC AA: :G: CTT G:A GCC TCT CT: G:: :T: AG: ::G ::: :::
Guinea-Pig   GCT AA: GGG GGA A:A GGG :TG :C: :GC AG: C:T :CG AA: C:A TGG GA: :GT :TC CCC TGC
Canine       :CA CCT CTG :CT :T: :GG CC: A:G CC: ::: :A: ::C TCC C:: ::G CA: CC: TGG CCA AA:
Human        TCT CCT A:G ACT ::A TGA :C: GCG CC: ::: CAG ::C TCC C:: ::G CA: CCT TGG CTC TG:
                 310             320             330             340             350             360


Guinea-Pig   Arg Ala His Ser Leu Pro Ser Gln Ser

Rat          TAG GAC GGT ACA GGC CCA GCT CTT GCA ACA ACTGCTCCTTGCAAACTAAATAAAGCAAAGGAAAAATTAT
Mouse        ::: ::G A:: ::: ::T ::: ::: ::: T:C T:T G:A:TCTTGCAAC:::CTGC:CCCTTGC:AACT::::A:A
Guinea-Pig   CG: :C: CAC T:T CT: ::C T:G :AG AGC TG: GTCCG:GGA::AGC:GCGCCACGTA:C:GCCCT::CC:CC
Canine       CTT :CT CCC TGC TCT :AC A:A :AG A:T :A: TAAAGCAAG:CA::GTC polyA
Human        CCA A:G CT: G:T CC: TGC T:C :AC A:: GGC T:AATAAAGCAAGTCAA:GCC polyA
                 370             380             390             400             410             420             430


Rat          polyA
Mouse        GCAAATTAAAATCAT polyA
Guinea-Pig   CCCCCCCCCCGCGCCTCTGCCCCCTTCCGCTCCCACAATAAAGCAAGCGGACCCG polyA
                 440             450             460             470             480
```

Fig. 6.1. Nucleotide and deduced amino acid sequences of rat, mouse, guinea pig, canine and human prepropancreatic polypeptide cDNAs. Nucleotide residues are numbered in the 5′ to 3′ direction, beginning with the first residue of the ATG triplet encoding the initiator methionine; the nucleotides on the 5′ side of residue 1 are indicated by negative numbers. Amino acid residues are numbered beginning with the initiator methionine. Identity of the nucleotide and amino acid sequences with the rat sequences is indicated by colons. Sequences of the putative signal peptide and the polyadenylation signal are underlined. Sequences of pancreatic polypeptides and human and canine icosapeptides are boxed. An asterisk represents an amino acid or nucleotide deletion.

pancreatic polypeptide domain and those coding for the carboxy-terminal domain have evolved at different rates because they were faced with different selective constraints.

6.2. Pancreatic polypeptide precursors

6.2.1. Pancreatic polypeptide precursor with icosapeptide
Like other polypeptide hormones, the pancreatic polypeptide can be assumed to be synthesized as part of a large precursor protein. Radioimmunoassay of extracts from human islet cell tumors led to the detection of a larger form of pancreatic polypeptide (Adrian *et al.*, 1978), which has been shown with the use of isolated canine pancreatic islets to be a biosynthetic precursor of pancreatic polypeptide. Gel filtration analysis allowed an estimation of the molecular mass of the larger immunoreactive species as being between 7.2 and 9 kDa, and the larger peptide was converted to a smaller peptide having a molecular mass similar to that of pancreatic polypeptide (4.3 kDa) (Schwartz *et al.*, 1980; Paquette *et al.*, 1981). A second stable product of the pancreatic polypeptide precursor was also identified, designated 'icosapeptide' because it consisted of 20 amino acids (Schwartz & Tager, 1981).

A cDNA for pancreatic polypeptide precursor was first isolated by Boel *et al.* from a human pancreas cDNA library (Boel *et al.*, 1984). The library was constructed from total human pancreas poly(A)$^+$ RNA and screened with a 17-base oligodeoxyribonucleotide mixture corresponding to Glu^{15}–Gln^{16}–Met^{17}–Ala^{18}–Gln^{19}–Tyr^{20} of human pancreatic polypeptide. One clone hybridized to the probe mixture and was found to carry an insert of about 530 base pairs. The nucleotide sequence of the insert was in total agreement with the known amino acid sequence of human pancreatic polypeptide (Chance *et al.*, 1979). Leiter *et al.* (1984) and Takeuchi & Yamada (1985) also isolated cDNA clones encoding pancreatic polypeptide precursor from human pancreatic islet cell tumors. The pancreatic polypeptide precursor cDNA (Fig. 6.1) encoded a protein of 95 amino acids. The 36 amino acid sequence of pancreatic polypeptide was flanked by a 29 amino acid putative signal sequence at the amino terminus and Gly–Lys–Arg followed by a 27 amino acid peptide at the carboxyl terminus. The glycine residue seems to serve as an amino donor to the carboxyl terminus of the pancreatic polypeptide (Bradbury *et al.*, 1982). The pair of basic amino acid residues are considered to constitute a processing site for a dibasic specific endopeptidase (Steiner *et al.*, 1980). The sequence of the first 20 amino acids in the carboxy-terminal heptacosapeptide was identical to the structure of human icosapeptide (Schwartz et al., 1984). This showed that the human pancreatic polypeptide is synthesized as a 95 amino acid precursor of molecular mass 10.433 kDa ('prepropancreatic polypeptide'). Following the cotranslational removal of a 29 amino acid signal sequence from the prepropancreatic polypeptide, a 66 amino acid prohormonal form of molecular mass 7.509 kDa ('propancreatic polypeptide') is generated; it contains at its amino terminus the sequence of pancreatic polypeptide. The propancreatic polypeptide is further cleaved to produce the 36 amino acid pancreatic polypeptide and the carboxy-terminal 27 amino acid peptide. The 27 amino acid peptide is probably cleaved at an arginyl residue to produce the icosapeptide and an additional carboxy-terminal heptapeptide. A proline-directed arginyl cleavage mechanism has been suggested for the generation of the icosapeptide (Schwartz, 1986, 1987).

Toothman & Paquette (1987) isolated a canine prepropancreatic polypeptide cDNA from cultured pseudoislets of dogs. A cDNA library was constructed from endocrine cell cultures, which had been prepared and pooled from the uncinate processes of canine pancreases for pancreatic polypeptide-secreting cell enrichment, and screened with 17-base and 11-base oligodeoxyribonucleotides corresponding to $Glu^4-Pro^5-Val^6-Tyr^7-Pro^8-Gly^9$ and $Gln^{16}-Met^{17}-Ala^{18}-Gln^{19}$ of canine pancreatic polypeptide (Glover et al., 1984). The cloned cDNA contained the entire coding sequence of canine prepropancreatic polypeptide, consisting of 93 amino acids (molecular mass 10.427 kDa) (Fig. 6.1). The sequence of canine pancreatic polypeptide (Glover et al., 1984) occurred in the middle of the precursor and was flanked at the amino terminus by a 29 amino acid signal sequence and at the carboxyl terminus by a 25 amino acid sequence which was identical to a peptide previously isolated from canine islets (Boel et al., 1984). The 25 amino acid peptide is probably cleaved at an arginyl residue to produce the icosapeptide and an additional carboxy-terminal pentapeptide (Schwartz & Tager, 1981).

6.2.2. Pancreatic polypeptide precursor without icosapeptide

cDNA clones encoding rat pancreatic polypeptide precursor were isolated from rat islets of Langerhans (Yamamoto et al., 1986). The islet cDNA library was screened with a 60 base oligodeoxyribonucleotide probe that was designed on the basis of the amino acid sequence of rat pancreatic polypeptide (Kimmel et al., 1984) and the nucleotide sequence of human prepropancreatic polypeptide (Boel et al., 1984). Approximately 200 out of 240 000 transformants of the rat islet cDNA library hybridized with the oligodeoxyribonucleotide, suggesting that the prepro-pancreatic polypeptide mRNA content in rat islet poly(A)$^+$ RNA was about 0.08%. The longest cDNA inserts (about 660 nucleotides) were isolated from the hybridization-positive clones and their nucleotide sequences were determined. As shown in Fig. 6.1, the amino acid sequence (residues 30–65) deduced from nucleotide residues 88–195 was in total agreement with the rat pancreatic polypeptide amino acid sequence that was determined with the purified peptide from rat pancreas (Kimmel *et al.*, 1984). Rat pancreatic polypeptide was found to be synthesized as a 98 amino acid precursor, which began at Met (residue 1) and ended at Ser (residue 98). Its molecular mass was estimated to be 10.975 kDa. The amino acid sequence (residues 1–29) deduced from nucleotide residues 1–87 had the characteristics of a typical signal peptide (Kreil, 1981); this sequence was regarded as a putative signal peptide in rat prepropancreatic polypeptide. Glycine and a pair of basic amino acids (residues 66–68, Gly–Lys–Arg) lay adjacent to the carboxyl terminus of rat pancreatic polypeptide. Thus, rat pancreatic polypeptide was considered to be liberated from the precursor by proteolytic cleavage at residues 29–30 and 66–67 and to be amidated on its carboxyl terminus. Furthermore, rat cDNA encoded a carboxy-terminal peptide consisting of 30 amino acids (residues 69–98). However, the rat carboxy-terminal 30 amino acid peptide lacked the monobasic processing site (Pro–Arg) and did not contain any sequence similar to that of the icosapeptide.

   The cDNA encoding mouse pancreatic polypeptide precursor was isolated and sequenced by Nata *et al.* (Nata, 1988; Yonekura *et al.*, 1988). A cDNA library was constructed from mouse islet poly(A)$^+$ RNA, and 60 000 clones were screened with a synthetic 60 base oligodeoxyribonu-cleotide, corresponding to the 20 amino acid amino terminal of rat pancreatic polypeptide (Yamamoto *et al.*, 1986). Sixty positive clones were identified, suggesting, because the hybridization was conducted in stringent conditions, that there is a high degree of homology between mouse and rat pancreatic polypeptides. The relative abundance of prepro-pancreatic polypeptide mRNA in mouse islet poly(A)$^+$ RNA was esti-mated to be 0.1%, comparable with that in rat islet poly(A)$^+$ RNA (Yamamoto *et al.*, 1986). The nucleotide sequence of the cDNA was found to code for a 100 amino acid protein containing a sequence at amino acid residues 30–65 that was quite similar to rat pancreatic polypeptide (Fig. 6.1). The 36 amino acid sequence differs from that of rat pancreatic polypeptide at only two residues, and from human pancreatic polypeptide at six residues. Therefore, this amino acid sequence (residues 30–65,

Fig. 6.1) can reasonably be identified as the previously un-isolated mouse pancreatic polypeptide. In the mouse prepropancreatic polypeptide, pancreatic polypeptide was flanked on the amino terminus by a putative signal peptide and on the carboxyl terminus by Gly–Lys–Arg followed by a 32 amino acid peptide. Like the rat precursor, the mouse carboxy-terminal region contained no sequence similar to the icosapeptide.

More recently, Blackstone *et al.* (1988) identified a guinea pig prepropancreatic polypeptide cDNA in a guinea pig pancreas cDNA library by hybridization using a human prepropancreatic polypeptide cDNA (Takeuchi & Yamada, 1985) under conditions of low stringency. The 126 residue precursor sequence was shown to include a 26 residue amino-terminal signal peptide followed by the 36 amino acid pancreatic polypeptide sequence and a large carboxy-terminal extension (Fig. 6.1). The predicted sequence of guinea pig pancreatic polypeptide was in agreement with the amino acid sequence determined by Eng *et al.* (1987). The carboxy-terminal extension lacked the monobasic processing site (Pro–Arg) and contained no sequence similar to the icosapeptide. It is perhaps significant that, at the dibasic amino acid processing site on the carboxy-terminal side of the pancreatic polypeptide domain in guinea pig prepropancreatic polypeptide, serine is substituted for arginine. The occurrence at this Lys–Ser site of efficient proteolytic cleavage has been demonstrated by radioimmunoassay of protein fractions gel-filtered from guinea pig pancreas extract; this assay also indicated the normal carboxy-amidation of mature guinea pig pancreatic polypeptide (Blackstone *et al.*, 1988).

6.2.3. *Structural conservation and divergence in prepropancreatic polypeptides*

The nucleotide and deduced amino acid sequences of the rat, mouse, guinea pig, canine and human prepropancreatic polypeptide cDNAs are aligned in Fig. 6.1. Inspection of the sequences in Fig. 6.1 reveals two distinct regions in the precursor: one is highly conserved and the other is extremely divergent among these species.

The region of conservation extends from amino acid 1 through 68, and comprises the signal peptide (1–29), the pancreatic polypeptide (30–65) and Gly–Lys–Arg (66–68). The conservation in this region indicates that these segments play significant roles in the generation of the peptide hormone and the exertion of its biological activity, which may have imposed strong mutational constraints upon the corresponding genomic region. It should be noted that the carboxy-terminal hexapeptide of pancreatic polypeptide and the following glycine residue exhibit complete identity in the amino acid sequence in all the five mammalian species (amino acid residues 60–66 in Fig. 6.1). The sequence identity is consistent with analyses of the bioactivity of synthetic fragments of pancreatic polypeptide, in which it was shown that the carboxy-terminal hexapeptide of pancreatic polypeptide mimicked the actions of the whole molecule on gastric and pancreatic secretion and that the des-carboxy-terminal–Tyr–NH_2 pancreatic polypeptide was devoid of both gastric and pancreatic actions (Lin, 1980).

The region of divergence composes the carboxy-terminal segment of prepropancreatic polypeptide (amino acid residues 69–98, 69–100, 69–126, 69–93 and 69–95 in the rat, mouse, guinea pig, canine and human precursors, respectively) (see Fig. 6.1). These carboxy-terminal extensions differ greatly in sequence and size. The most striking mark of the divergence in this region is the presence or absence of the icosapeptide structure. The human and canine possess, but rat, mouse and guinea pig prepropancreatic polypeptides lack, the icosapeptide sequence at the carboxyl terminus. Proline-directed arginyl cleavage sites that have been suggested as being involved in the liberation of the icosapeptide from the precursor (Schwartz, 1986, 1987) are also missing in the rodent carboxy-terminal regions. Although the presence of the icosapeptide is also known in ovine and feline pancreases (Schwartz *et al.*, 1984; Nielsen *et al.*, 1986), larger species variation is observed among the icosapeptide segments of the human, canine, ovine and feline precursors than among the pancreatic polypeptide segments. The extreme divergence in the carboxy-terminal region of the mammalian prepropancreatic polypeptide suggests that the primary structure of the carboxy-terminal peptide extension is relatively unimportant and that its presence may be necessary merely to provide a minimum critical length for translocation of the nascent peptide across the endoplasmic reticulum membrane (Steiner, 1984) or to ensure that pancreatic polypeptide precursor proteins are in appropriate conformation for proteolytic processing.

Yamamoto *et al.* (1986) compared amino acid and nucleotide sequences in rat and human representative prohormones: as shown in Fig. 6.2, there are considerable homologies in all the functionally divided domains of preproinsulin, preproglucagon, preprosomatostatin, preprovasoactive intestinal peptide, and preproopiomelanocortin. The sequence homology in the rat and human carboxy-terminal regions of prepropancreatic polypeptide is much lower than in any of the domains of other prohormones, including preproinsulin C-peptide which has so far been con-

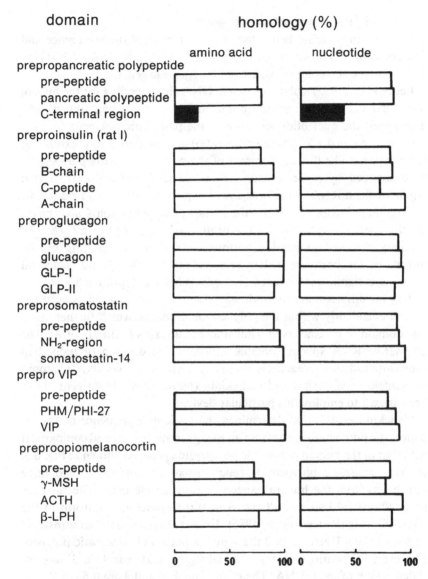

Fig. 6.2. Sequence homology in functionally divided domains of rat and human peptide hormone precursors and corresponding mRNAs (Yamamoto *et al.*, 1986).

sidered to be the most variable (Kimura, 1979). Thus, prepropancreatic polypeptide seems to be a unique prohormone in that the amino-terminal two thirds are highly conserved whereas the carboxy-terminal one third exhibits a high degree of species divergence.

6.3. Pancreatic polypeptide genes

To understand better the molecular basis of the divergence and conservation seen in pancreatic polypeptide precursors, it was desirable to determine the structure of pancreatic polypeptide genes.

Leiter *et al.* (1985) isolated genomic DNA clones coding for the human prepropancreatic polypeptide from a human liver genomic library, and determined the nucleotide sequence of the pancreatic polypeptide gene. The gene spanned 1.7 kilobase pairs (kbp) and contained four exons and three introns. The distinct domains of the prepropancreatic polypeptide were encoded by separate exons; exon 1 encoded the 5'-untranslated region of the mRNA, exon 2 the signal sequence and the sequence of the pancreatic polypeptide, exon 3 the icosapeptide, and exon 4 a carboxy-terminal heptapeptide and the 3'-untranslated region of the mRNA. The diversity achieved in peptide hormones, through the splitting by the introns of the hormone coding regions into functional and structural domains, is significant for several reasons. When a polyprotein has regions coding for different structural domains separated, changes may accumulate independently within the polypeptide domains, which further allow one domain to be conserved while another diverges (Gilbert, 1978). The alternative RNA splicing obtained through the division of individual exons into functional regions is also likely to enhance diversity. Calcitonin (Rosenfeld *et al.*, 1984) and tachykinin (Nawa *et al.*, 1984) genes have been shown to employ this particular device.

Yonekura *et al.*(1988) determined the nucleotide sequence of the rat pancreatic polypeptide gene from an islet genomic library and compared it with that of the human pancreatic polypeptide gene. As shown in Fig. 6.3, the rat pancreatic polypeptide gene consists of four exons and three introns, as does the human pancreatic polypeptide gene. The rat gene spans about 1.4 kbp and all the exon–intron junctions conform to the 'GT–AG' rule (Padgett *et al.*, 1986). Exon 1 encodes the 5'-untranslated region of the mRNA, exon 2 the signal peptide and pancreatic polypeptide, exon 3 the entire carboxy-terminal region, and exon 4 the 3'-untranslated region of the mRNA. The rat introns 1, 2 and 3 are 0.53, 0.19 and 0.09 kbp long, whereas the human introns 1, 2 and 3 are 0.76, 0.24 and 0.19 kbp long, respectively. Rat exons 1 and 2, which are 56 and 191 bp long, respectively, are exactly the same in length as human exons 1 and 2. Exon 2, encoding the signal peptide and pancreatic polypeptide, displays 81% homology to the human sequence. On the other hand, exons 3 and 4 are very different in length and quite heterologous in nucleotide sequence in rat and human genes. Rat exon 3, encoding the carboxy-terminal

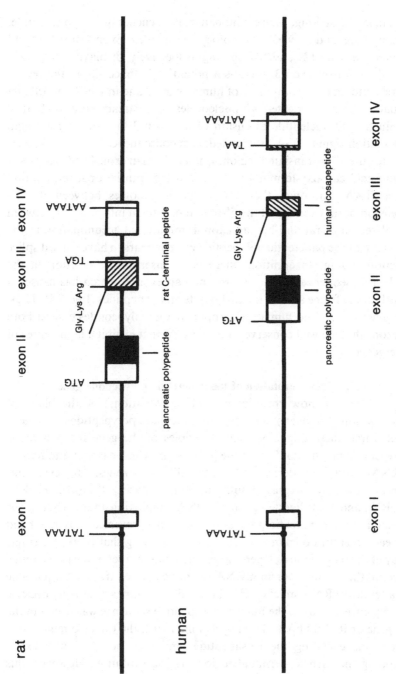

Fig. 6.3. Organization of human and rat pancreatic polypeptide genes.

region, is 112 bp long, whereas human exon 3, encoding the icosapeptide, is 72 bp long, and only 51% homology is displayed. Exon 4 of the rat and human genes are 129 and 147 bp long, respectively, displaying only 39% homology. In rat exon 3, there is a potential 5'-splice site at the region corresponding to the boundary of human exon 3 and intron 3, but the site actually spliced is located 42 nucleotides downstream (Fig. 6.4). It is within this 42 nucleotide extension of rat exon 3 that the translational termination signal, TGA, is encoded; as a consequence, rat exon 4 codes only for the 3'-untranslated region, whereas human exon 4 codes for the 7 amino acid carboxy-terminal peptide and the translational termination signal TAA. Although there are some homologies between the 3'-extension of rat exon 3 and the 5'-region of human intron 3 and between the 5'-region of rat and human exon 4, mutational accumulation in the 3'-region of the pancreatic polypeptide gene appears to have caused splice junction sliding. In addition, there is an insertion or deletion in the nucleotide sequences of the rat and human exon 3, which has caused a translational frame shift of the carboxy-terminal peptides (Fig. 6.4). Thus, the genes for rat and human prepropancreatic polypeptides consist both of exons that are well conserved and of exons that exhibit a high degree of divergence.

6.4. Mosaic evolution of the pancreatic polypeptide gene

We can now consider possible explanations for the blend of conservation and divergence in prepropancreatic polypeptide at the gene level. First, there might be multiple copies of the gene for pancreatic polypeptide in a haploid set of the genome, in which case rat and human mRNAs would be derived from two different genes. However, this appears unlikely because Southern blot analysis of *Dra*I-, *Eco*RI-, *Eco*RV-, and *Sca*I-digested genomic DNA, employing a rat cDNA probe containing the pancreatic polypeptide sequence, revealed only one band for each digest (Yonekura *et al.*, 1988), suggesting that there is a single copy of the pancreatic polypeptide gene in a haploid set of the rat genome. Second, the rat and human mRNA might be derived from a unique gene by alternative RNA splicing. However, this also appears unlikely because no sequence similar to the human icosapeptide sequence was found in the rat gene or its 1.8 kbp 3'-flanking region. Third, the rat gene might have lost exons encoding the icosapeptide, and then acquired new exons encoding the carboxy-terminal region, during evolution. However, this also appears unlikely because some parts of rat exon 3 and a part of rat exon 4 have a certain degree of homology to human sequences, as shown exon 3

```
rat          gTyrGlyLysArgAsp    GluAspThrAlaGlyLeuProGlyArgGlnLeuProProCysT
      - - ccagGTATGGGAAGAGAGAT...GAGGACACAGCTGGGCTTCCTGGAAGGCAGCTCCACCATGCA
           **********      ***  *      *** *  *  *  * *****
human - - ccagGTATGGGAAAGACAACAAAGAGAGACAC.GCTGGCCTTCTCGAGTGGGGTCCCCGCATGCT
           gTyrGlyLysArgHisLysGluAspTh  rLeuAlaPheSerGluTrpGlySerProHisAla
```

```
rat       hrSerLeuLeuValGlyLeuMetProCysAlaAlaAlaAlaArgSerEnd
      CCAGCCTTCTGGTGGTTTGATGCCTGCTGCCCCGGAGCTGAAGGCAGgtctgg - - - -
      * *  * *****  *******  ***    *    * *   ***  *
human GCTGTCCCCAGgtgagtttgactccctgccctgtctgtccaggctccctggggctgaaa - - - -
      AlaValProAr
```

exon 4

```
rat   - - acagGGATCTCAGCCCAATGGATTTGTAGCCTCCCTTCTGTCTCCTACCATGCT - - - -
           ******* **********  **** *  * * *  * **  *
human - - acagGGAGCTCAGCCCGCTGGACTTATAATGCCACCTTCTGTCTCCTACGACTCCATGAG - - - -
           gGluLeuSerProLeuAspLeuEnd
```

Fig. 6.4. Splice junction sliding and translational frame shift in the 3' region of the pancreatic polypeptide gene (Yonekura et al., 1988). Exon 3, exon 3 – intron 3 boundary, and the 5' part of exon 4 of rat and human pancreatic polypeptide genes were aligned according to maximum homology of their nucleotide sequences. Capital letters indicate exons, and lower case letters are used for introns. Numbering (+ 1) begins at the cap site. Underlined letters indicate nucleotides that agree with the consensus sequence of the 5' splice site. Asterisked nucleotides indicate identical nucleotides in rat and human sequences. Dots indicate gaps in rat and human sequences in the maximal matched alignment. Closed arrows indicate the 5' splice site of rat and human introns 3; the open arrow indicates the deletion or insertion which has resulted in translational frame shift.

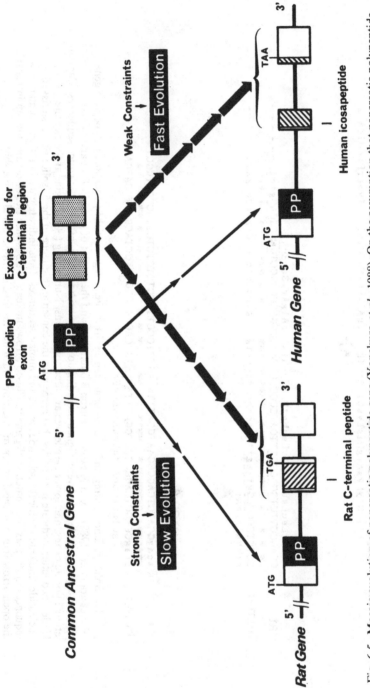

Fig. 6.5. Mosaic evolution of pancreatic polypeptide gene (Yonekura *et al.*, 1988). On the assumption that pancreatic polypeptide genes have evolved from a common ancestral gene, exons coding for the amino-terminal region and those for the carboxy-terminal region may have evolved at different rates because they were faced with different selective constraints. The regions coding for the carboxy-terminal regions of rat and human prepropancreatic polypeptide are indicated as cross-hatched bars; PP, regions coding for pancreatic polypeptide.

Table 6.1. *Evolutionary rates of domains of prepropancreatic polypeptide and of corresponding genomic regions.*

domains	
signal peptide	1.73
pancreatic polypeptide	1.59
carboxy-terminal region[a]	9.63
exon 2[b] 1st base of codon	1.00
2nd base of codon	0.98
3rd base of codon	2.22
exon 3[c] 1st base of codon	2.68
2nd base of codon	2.64
3rd base of codon	3.08

Rates are expressed in amino acid replacements per 10^9 years and in nucleotide substitutions per 10^9 years (Kimura, 1983). The rat and the human sequences were aligned to maximize homology: [a] amino acid residues 69–98 and 69–95; [b] nucleotide residues 589–779 and 815–1005; [c] nucleotide residues 970–1075 and 1256–1363, respectively (Yonekura *et al.*, 1988; Leiter *et al.*, 1985).

in Fig. 6.4. Therefore, as shown in Fig. 6.5, it is reasonable to assume that the structural conservation and divergence in prepropancreatic polypeptide have been generated through the 'mosaic evolution' of the pancreatic polypeptide gene: the exons coding for the amino-terminal portion and those for the carboxy-terminal portion may have evolved at different rates because they were faced with different selective constraints. The evolutionary rate estimated for the carboxy-terminal domain is about six times faster than those for the amino-terminal domains (Table 6.1). The rates of base substitutions are different in the first two bases and the third base of codons in exon 2, whereas the rates are almost equal in the first, second and third bases of codons in exon 3 (Table 6.1). This implies that the carboxy-terminal domain has been faced with few selective pressures to eliminate base substitutions that might result in amino acid replacements. The splice junction sliding and the translational frame shift (Fig. 6.4) help to account for the divergence in the amino acid sequences of the carboxy-terminal regions. The icosapeptide or carboxy-terminal segment *per se* may not have an essential biological function, and more base substitutions could have been tolerated by the corresponding genomic region than by the region that codes for the pancreatic polypeptide segment.

The term mosaic evolution was coined by DeBeer for the quasi-independent evolution of different anatomical characteristics (DeBeer,

1954). Modern knowledge has extended the principle of mosaic evolution from vertebrates to nearly all kinds of animals and plants, and from the gross morphological level to the molecular level (Stebbins, 1983). For example, the evolutionary rates differ markedly among protein species, and non-coding DNA evolves much more rapidly than coding, genic DNA. The prepropancreatic polypeptide and its gene may represent an example of 'intramolecular' or 'intragenic' mosaic evolution.

Peptide YY (Tatemoto & Mutt, 1980) and neuropeptide Y (Tatemoto *et al.*, 1982), 36 amino acid peptides with an amino-terminal tyrosine and a carboxy-terminal tyrosine amide, exhibit similarities in amino acid sequence to pancreatic polypeptide. Pancreatic polypeptide, peptide YY and neuropeptide Y have therefore been regarded as constituting a new family of structurally related peptides. The peptide YY precursor cDNA from rat intestine (Leiter *et al.*, 1987) codes for a 98 residue protein (molecular mass 11.121 kDa) in which the peptide YY is immediately preceded by a 28 amino acid signal sequence and followed by Gly–Lys–Arg plus 31 additional amino acids. The neuropeptide Y cDNAs from human pheochromocytoma (Minth *et al.*, 1984) and from rat brain (Higuchi *et al.*, 1988) encode 97 and 98 residue proteins (molecular mass 10.839 and 11.032 kDa), with neuropeptide Y being flanked by a signal peptide and by Gly–Lys–Arg plus a 30 amino acid carboxy-terminal peptide. The carboxy-terminal region of prepropancreatic polypeptide differs extensively in amino acid sequence from the equivalent regions of preproneuropeptide Y and prepropeptide YY, as well as showing extreme species divergence. On the other hand, the carboxy-terminal region of preproneuropeptide Y resembles that of prepropeptide YY and exhibits a high degree of interspecies conservation, which suggests that it may be functionally important (Larhammar *et al.*, 1987). Therefore, mosaic evolution (Fig. 6.5) is a phenomenon unique to the pancreatic polypeptide gene, and presumably occurred after the pancreatic polypeptide gene was segregated from the peptide YY and neuropeptide Y genes.

References

Adrian, T.E., Bloom, S.R., Besterman, H.S. & Bryant, M.G. (1978). PP–physiology and pathology. In *Gut hormones* (ed. S.R. Bloom), pp. 254–60. Edinburgh: Churchill Livingstone.

Adrian, T.E., Greenberg, G.R. & Bloom, S.R. (1981). Actions of pancreatic polypeptide in man. In *Gut hormones* 2nd edn (ed. S.R. Bloom & J.M. Polack), pp. 206–12. Edinburgh: Churchill Livingstone.

Blackstone, C.D., Seino, S., Takeuchi, T., Yamada, T. & Steiner, D.F. (1988). Novel organization and processing of the guinea pig pancreatic polypeptide precursor. *Journal of Biological Chemistry* **263**, 2911–16.

Boel, E., Schwartz, T.W., Norris, K.E. & Fill, N.P. (1984). A cDNA encoding a small common precursor for human pancreatic polypeptide and pancreatic icosapeptide. *EMBO Journal* **3**, 909–12.

Bradbury, A.F., Finnie, M.D.A. & Smyth, D.G. (1982). Mechanism of C-terminal amide formation by pituitary enzymes. *Nature* **298**, 686–8.

Chance, R.E., Moon, N.E. & Johnson, M.G. (1979). Human pancreatic polypeptide (HPP) and bovine pancreatic polypeptide (BPP). In *Methods of hormone radioimmunoassay* 2nd edn (ed. B.M. Jaffe & H.R. Behrman), pp. 657–72. New York: Academic Press.

DeBeer, G.R. (1954). *Archaeopteryx* and evolution. *Advancement of Science* **11**, 160–70.

Eng, J., Huang, C.-G., Pan, Y.-C.E., Hulmes, J.D. & Yalow, R.S. (1987). Guinea pig pancreatic polypeptide: structure and pancreatic content. *Peptides* **8**, 165–8.

Gilbert, W. (1978). Why genes in pieces? *Nature* **271**, 501.

Glover, I.D., Barlow, D.J., Pitts, J.E., Wood, S.P., Tickle, I.J., Blundell, T.L., Tatemoto, K., Kimmel, J.R., Wollmer, A., Strassburger, W. & Zhang, Y.-S. (1985). Conformational studies on the pancreatic polypeptide hormone family. *European Journal of Biochemistry* **142**, 379–85.

Hazelwood, R.L. (1981). Synthesis, storage, secretion, and significance of pancreatic polypeptide in vertebrates. In *The islets of Langerhans* (ed. S.J. Cooperstein & D. Watkins), pp. 275–318. New York: Academic Press.

Higuchi, H., Yang, H.-Y.T. & Sabol, S.L. (1988). Rat neuropeptide Y precursor gene expression: mRNA structure, tissue distribution, and regulation by glucocorticoids, cyclic AMP, and phorbol ester. *Journal of Biological Chemistry* **263**, 6288–95.

Kimmel, J.R., Pollock, H.G., Chance, R.E., Johnson, M.G., Reeve, J.R. Jr, Taylor, I.L., Miller, C. & Shively, J.E. (1984). Pancreatic polypeptide from rat pancreas. *Endocrinology* **114**, 1725–31.

Kimura, M. (1979). The neutral theory of molecular evolution. *Scientific American* **241**, 94–104.

Kimura, M. (1983). *The neutral theory of molecular evolution.* Cambridge University Press.

Kreil, G. (1981). Transfer of proteins across membranes. *Annual Review of Biochemistry* **50**, 317–48.

Larhammar, D., Ericsson, A. & Persson, H. (1987). Structure and expression of the rat neuropeptide Y gene. *Proceedings of the National Academy of Sciences of the U.S.A.* **84**, 2068–72.

Leiter, A.B., Keutmann, H.T. & Goodman, R.H. (1984). Structure of a precursor to human pancreatic polypeptide. *Journal of Biological Chemistry* **259**, 14702–5.

Leiter, A.B., Montminy, M.R., Jamieson, E. & Goodman, R.H. (1985). Exons of the human pancreatic polypeptide gene define functional domains of the precursor. *Journal of Biological Chemistry* **260**, 13013–17.

Leiter, A.B., Toder, A., Wolfe, H.J., Taylor, I.L., Cooperman, S., Mandel, G. & Goodman, R.H. (1987). Peptide YY: structure of the precursor and expression in exocrine pancreas. *Journal of Biological Chemistry* **262**, 12984–8.

Lin, T.-M. (1980). Pancreatic polypeptide: isolation, chemistry, and biological function. In *Gastrointestinal hormones* (ed. G.B.J. Glass), pp. 275–306. New York: Raven Press.

Minth, C.D., Bloom, S.R., Polak, J.M. & Dixon, J.E. (1984). Cloning, characterization and DNA sequence of a human cDNA encoding neuropeptide tyrosine. *Proceedings of the National Academy of Sciences of the U.S.A.* **81**, 4577–81.

Murphy, W.A., Fries, J.L., Meyers, C.A. & Coy, D.H. (1981). Human pancreatic polypeptide inhibits insulin release in the rat. *Biochemical and Biophysical Research Communications* **101**, 189–93.

Nata, K. (1988). Mosaic evolution of prepropancreatic polypeptide: molecular cloning of mouse prepropancreatic polypeptide cDNA and comparison with corresponding sequences of other mammalian species. Tohoku University Medical Dissertation no. 982.

124   H. YAMAMOTO et al.

Nawa, H., Kotani, H. & Nakanishi, S. (1984). Tissue-specific generation of two preprotachy-kinin mRNAs from one gene by alternative RNA splicing. *Nature* **312**, 729–34.

Nielsen, H.V., Gether, U. & Schwartz, T.W. (1986). Cat pancreatic eicosapeptide and its biosynthetic intermediate: conservation of a monobasic processing site. *Biochemical Journal* **240**, 69–74.

Padgett, R.A., Grabowski, P.J., Konarska, M.M., Seiler, S. & Sharp, P.A. (1986). Splicing of messenger RNA precursors. *Annual Review of Biochemistry* **55**, 1119–50.

Paquette, T.L., Gingerich, R. & Scharp, D. (1981). Altered amidation of pancreatic polypeptide in cultured dog islet tissue. *Biochemistry* **20**, 7403–8.

Rosenfeld, M.G., Amara, S.G. & Evans, R.M. (1984). Alternative RNA processing: determining neuronal phenotype. *Science* **225**, 1315–20.

Schwartz, T.W. (1986). The processing of peptide precursors: 'Proline-directed arginyl cleavage' and other monobasic processing mechanisms. *FEBS Letters* **200**, 1–10.

Schwartz, T.W. (1987). Cellular peptide processing after a single arginyl residue: studies on the common precursor for pancreatic polypeptide and pancreatic icosapeptide. *Journal of Biological Chemistry* **262**, 5093–9.

Schwartz, T.W., Gingerich, R.L. & Tager, H.S. (1980). Biosynthesis of pancreatic polypep-tide. Identification of a precursor and a co-synthesized product. *Journal of Biological Chemistry* **255**, 11494–8.

Schwartz, T.W., Hansen, H.F., Håkanson, R., Sundler, F. & Tager, H.S. (1984). Human pancreatic icosapeptide: Isolation, sequence, and immunocytochemical localization of the COOH-terminal fragment of the pancreatic polypeptide precursor. *Proceedings of the National Academy of Sciences of the U.S.A.* **81**, 708–12.

Schwartz, T.W. & Tager, H.S. (1981). Isolation and biogenesis of a new peptide from pancreatic islets. *Nature* **294**, 589–91.

Solcia, E., Capella, C., Usellini, L., Fiocca, R. & Sessa, F. (1985). The PP cell. In *The diabetic pancreas* (ed. B.W. Volk & E.R. Arquilla), pp. 107–15. New York: Plenum Medical Book Company.

Stebbins, G.L. (1983). Mosaic evolution: an integrating principle for the modern synthesis. *Experientia* **39**, 823–34.

Steiner, D.F. (1984). The biosynthesis of insulin: genetic, evolutionary, and pathophysiolo-gic aspects. *The Harvey Lectures* **78**, 191–228.

Steiner, D.F., Quinn, P.S., Chan, S.J., Marsh, J. & Tager, H.S. (1980). Processing mechanisms in the biosynthesis of proteins. *Annals of the New York Academy of Sciences* **343**, 1–16.

Takeuchi, T. & Yamada, T. (1985). Isolation of a cDNA clone encoding pancreatic polypeptide. *Proceedings of the National Academy of Sciences of the U.S.A.* **82**, 1536–9.

Tatemoto, K., Carlquist, M. & Mutt, V. (1982). Neuropeptide Y: a novel brain peptide with structural similarities to peptide YY and pancreatic polypeptide. *Nature* **296**, 659–60.

Tatemoto, K. & Mutt, V. (1980). Isolation of two novel candidate hormones using a chemical method for finding naturally occurring polypeptides. *Nature* **285**, 417–18.

Toothman, P. & Paquette, T.L. (1987). Canine pancreatic polypeptide complementary deoxyribonucleic acid sequence: pancreatic polypeptide and insulin messenger ribonucleic acid distribution in the lobes of the pancreas. *Molecular Endocrinology* **1**, 413–19.

Trimble, E.R., Gerber, P.G. & Renold, A.E. (1982). Effects of synthetic human pancreatic polypeptide, synthetic bovine pancreatic polypeptide, and the C-terminal hexapeptide on pancreatic somatostatin and glucagon secretion in the rat. *Diabetes* **31**, 178–81.

Yamamoto, H., Nata, K. & Okamoto, H. (1986). Mosaic evolution of prepropancreatic polypeptide. *Journal of Biological Chemistry* **261**, 6156–9.

Yonekura, H., Nata, K., Watanabe, T., Kurashina, Y., Yamamoto, H. & Okamoto, H. (1988). Mosaic evolution of prepropancreatic polypeptide: II. Structural conservation and divergence in pancreatic polypeptide gene. *Journal of Biological Chemistry* **263**, 2990–7.

7

The structure and expression of genes of vasoactive intestinal peptide and related peptides

TAKASHI YAMAGAMI

7.1. Introduction

Said & Mutt (1972) first isolated a vasodilator peptide of 28 amino acids from porcine duodenum and designated it vasoactive intestinal peptide (VIP). Thereafter, VIP was found to be present not only in the duodenum but also in a variety of tissues, including central and peripheral nervous systems, small and large intestines, pancreas, lung and urogenital tract, functioning mainly as a neurotransmitter or neuromodulator (Håkanson *et al.*, 1982; Hökfelt *et al.*, 1982; Marley & Emson, 1982; Fahrenkrug, 1982). Islets of Langerhans are innervated by peptidergic as well as cholinergic and adrenergic nerve fibers: the peptidergic nerve fibers contain VIP immunoreactivity (Miller, 1981) and VIP has been shown to stimulate glucagon and insulin secretion from islets (Schebalin *et al.*, 1976). VIPomas, VIP-producing tumors, often arise in the islets of Langerhans, and many patients with VIPomas associated with the watery diarrhea, hypokalemia and achlorhydria (WDHA) syndrome have been reported (Verner & Morrison, 1958; Bloom *et al.*, 1973; Mekhjian & O'Dorisio, 1987).

In 1983, Itoh *et al.* first characterized the mRNA coding for human VIP precursor (Itoh *et al.*, 1983). The entire amino acid sequence of the precursor, deduced from the nucleotide sequence, indicated that the precursor protein contains not only VIP but also a novel PHI-like peptide of 27 amino acids, PHM-27. This article reviews the current knowledge of the structure and expression of the VIP/PHM-27 gene. In particular, it focusses on the organization of the VIP/PHM-27 gene and on the transcriptional regulation by cAMP and phorbol esters of VIP/PHM-27 gene expression.

1 **Signal Peptide**

```
                              1                                      10                                           40
                              Met Asp Thr Arg Asn Lys Ala Gln Leu Leu Val Leu Leu Thr   Leu Gly Asp Arg Ile Pro Phe Glu
GGGAGCACCACUGGCGGAGAGGCACAGAA AUG GAC ACC AGA AAU AAG GCC CAG CUC CUU GUG CUC CUG ACU   CUU ... GAC AGA AUA CCC UUU GAG
-30       -20       -10       -1 1                                 30
```

```
|Leu Leu Ser Val Leu Phe Ser|                                                                    60                      70
 Leu Leu Ser Val Leu Phe Ser   Gln Thr Ser Ala Trp Pro Leu Tyr Arg Ala Pro Ser Ala        Gly Ala Asn Glu Pro Asp   Thr Pro Tyr Tyr Asp
 CUU CUC AGU GUG CUU UUC UCA   CAG ACU UCA GCA UGG CCU CUU UAC AGG GCA CCU UCU GCA        GGA GCA AAU GAA CCU GAU   ACA CCC UAU UAU GAU
       20                          50                                                           150
```

PHM-27

```
                                       |His Ala Asp Gly Val Phe Thr Ser Asp|                                       100                    300
 Val Ser Arg Asn Ala Arg                His Ala Asp Gly Val Phe Thr Ser Asp   Phe Ser Lys Leu Leu Gly Gln Leu Ser Ala Lys Lys Tyr Leu Glu
 GUA UCC AGA AAU GCC AGG                CAU GCU GAU GGA GUC UUC ACC AGU GAU   UUC AGU AAG CUU CUG GGU CAA CUU UCG GCC AAA AAG UAC CUU GAG
       80                                    250                                   90                                              130
```

V I P

```
                                               |His Ser Asp Ala Val Phe Thr Asp Asn Tyr|                                  160
 Ser Leu Met   Gly Lys Arg                       His Ser Asp Ala Val Phe Thr Asp Asn Tyr   Gly Lys Ser Ser Glu Gly Glu Ser Pro Asp Phe
 UCU CUU AUG   GGA AAA CGU                       CAC UCA GAU GCU GUC UUC ACU GAC AAC UAU   GGA AAG AGC AGU GAG GGA GAA UCU CCC GAC UUU
       110        120                                 350                                    450                           400
```

```
|Thr Arg Leu Arg Lys Gln Met Ala Val Lys|  |Lys Tyr Leu Asn Ser Ile Leu Asn|
 ACC CGC CUU AGA AAA CAA AUG GCU GUA AAG    AAA UAU UUG AAC UCA AUU CUG AAU
       140                                       150
```

```
 Pro Glu Glu Leu Lys End End
 CCA GAA GAG UUA AAA UGA UGA   AAAAGACCUUUGGAGCAAAGCUGAUGACAACUCCCAGUGAAGGAAUGAUACGCAACAUAAUUAAAUUUAGAUUCU
       170                                                550                                                      600
       500
```

ACAUAAGUAAUUCAAGAAAACAACUUCAAUAUCCAAACCAAUAAAAAAUAUUGUGUGUUGUGAAUGUUGUGAUGUAUUCUAGCUAAUGUAAUGUAAUGUUACAUUGUAAAUAGUAU
650   700

UUGAGAGUUCUAAAUUUUGUCUUUAACUCAUAAAAAAGCCUGCAAUUUCAUAUGCUGUAUAUCCUUUCUAACAAAAAAAUAUAUUUUAUUGAAUAAGUAAUGCUAGGUUAAUCCAAUUAUA
750   800

UGAGACGUUUUUGAAGAGUAGUAAUUGUAAUAUGAGCAAAAUUGUGUGUUUAUUUAUAGAGUGUACUUAACUAUUCAGGGAGCAGAUAAUCAGUGUGUCUAAAAUUGAAUGUUAAGCA
850   900   950

GAUGGAAUGCGUGUGUU<u>AAUAAA</u>CCUCAAAAAUGUCUAAGAUAGUAACAUUGAAGUAAAAAGACAUUCUUCCAAAAAGAUUUUCAGAAAAUAUUAUGUGUUUCCAUAUUUAUAUAGGCAA
1000   1050

CCUUUAUUUUUUAAUGGCGUUUUAAAAAAUCUCAAAUUUGGAUUGCUAAUCACCAAAGCGUCUCUCCUGAUAGUCUUUCAGUUAAGGAGGACGACCCCUGCUUCUGACACUGAAACUUCC
1100   1150

CUUUCUCCUUGCUGUGUAAGUAUGUGUAAAAUGUGAAUGAAGUAAACACUCA**GUUCAAUAUAAAUAUUUUUGCCAUAA**
1200   1250   1279

Fig. 7.1. Primary structure of human preproVIP/PHM-27 and its mRNA (Itoh et al., 1983). The signal peptide, PHM-27 and VIP sequences are underlined. Polyadenylation signals are boxed.

7.2. Structure of the VIP precursor

Obata *et al.* (1981) identified the VIP precursors in a human neuroblastoma cell line (NB-1). The neuroblastoma cells were cultured with [^{35}S] methionine, and the cell extract was immunoprecipitated with antiserum to porcine VIP and analysed by SDS–polyacrylamide gel electrophoresis. Two products, one at the VIP position and one of estimated molecular mass 17.5 kDa, reacted specifically with the antiserum. In a chase experiment, the labeled 17.5 kDa product decreased, whereas the labeled product at the VIP position increased. Therefore, the 17.5 kDa product was regarded as a precursor of VIP (proVIP). In addition, by the translation of the neuroblastoma poly(A)-containing RNA in a reticulocyte lysate cell-free system, a primary translation product of VIP mRNA (preproVIP) of molecular mass about 20 kDa was identified.

The structure of human preproVIP mRNA was elucidated by Itoh *et al.* (1983). Poly(A)-containing RNA was isolated from human neuroblastoma cells, and was fractionated by sucrose gradient centrifugation. Double-stranded cDNA was synthesized using the partly purified mRNA as a template and was inserted into the *Pst*I cleavage site of the plasmid pBR322 by the oligo(dG)–oligo(dC) tailing method. A human VIP cDNA clone was isolated by a hybridization–translation assay. Fig. 7.1 shows the nucleotide sequence of the preproVIP mRNA, which encodes a protein of 170 amino acids, assuming that AUG at nucleotides 1–3 is the start codon and UGA at nucleotides 511–513 is the stop codon. The amino acid sequence (residues 1–21) deduced from nucleotide residues 1–63 possibly corresponds to the signal peptide, which generally contains a region rich in hydrophobic amino acid residues with large side chains and terminates in a residue having a small neutral side chain (Kreil, 1981). The amino acid sequence (residues 125–152) deduced from nucleotide residues 373–456 corresponds precisely to the amino acid sequences of porcine and bovine VIP, and thus to that of previously unidentified human VIP. The amino terminus of human VIP is preceded by a pair of basic amino acid residues (Lys–Arg) known to be frequent sites for the posttranslational processing of hormone precursors (Steiner *et al.*, 1980). The carboxyl terminus of human VIP is followed by a glycine and a pair of basic amino acid residues (Lys–Arg). The glycine residue seems to serve as an amino donor to the carboxyl terminus of human VIP, because both porcine and bovine VIP contain a carboxy-terminal Asn–NH$_2$ residue (Mutt & Said, 1974; Carlquist *et al.*, 1979). An interesting finding is that the deduced amino acid sequence (residues 81–107), which shows a remarkable similarity to that of

VIP, is almost identical to that of porcine PHI-27. PHI-27 is a peptide which was isolated from porcine upper intestinal tissue (Tatemoto & Mutt, 1981) and has amino-terminal histidine, carboxy-terminal isoleucine and 27 amino acid residues. The amino acid sequence (residues 81–107) differs from that of porcine PHI-27 only at residue 92 (Arg→Lys) and 107 (Ile→Met). Accordingly, Itoh *et al.* (1983) designated this peptide PHM-27. In fact, PHM-27 was later isolated from human colonic extracts (Tatemoto *et al.*, 1984) and the coexistence of PHM-27 and VIP was shown in a single population of neurons (Anand *et al.*, 1984). The carboxy-terminal methionine of PHM-27 is followed by a glycine residue and a pair of basic amino acid residues (Lys–Arg) just as with VIP, but the amino terminus of PHM-27 is preceded by one basic amino acid residue (Arg) (Fig. 7.1). Thus, VIP and PHM-27 are considered to be liberated from the common precursor, preproVIP/PHM-27, by proteolytic cleavage at the dibasic and/or monobasic cleavage sites, and then amidated. Both VIP and PHM-27 contain a pair of internal basic amino acid residues (Lys–Lys and Arg–Lys). This raises the possibility that tissue-specific posttranslational processing yields certain cryptic peptides. The nucleotide and deduced amino acid sequences of the human VIP/PHM-27 mRNA were confirmed by Bloom *et al.* (1983). Nishizawa *et al.* (1985) determined the nucleotide sequence of the rat preproVIP mRNA and found that the preproVIP contained both rat VIP and PHI-27; the deduced amino acid sequence of rat VIP was identical to that of human VIP, and rat PHI-27 differed by 4 amino acids from human PHM-27.

7.3. Structure of the VIP/PHM-27 gene

The nucleotide sequence of a part of human VIP/PHM-27 gene was determined by Bodner *et al.* (1985), Tsukada *et al.* (1985) and Linder *et al.* (1987). In 1988, Yamagami *et al.* isolated the VIP/PHM-27 gene from a human fetal liver genomic library and determined the complete nucleotide sequence of the gene (Yamagami *et al.*, 1988). As shown in Fig. 7.2, the VIP/PHM-27 gene spans 8837 base pairs (bp) and consists of seven exons and six introns. Exon I comprises 165 bp and encodes a part of the 5'-untranslated region of the VIP/PHM-27 mRNA. Exon II comprises 117 bp and encodes the rest of the 5'-untranslated region, the signal peptide of the preproVIP/PHM-27 and a part of the amino-terminal region. Exon III comprises 123 bp and encodes most of the amino-terminal region. Exon IV comprises 105 bp and encodes the PHM-27 sequence and a part of the intervening peptide. Exon V comprises 132 bp and encodes the rest of the intervening peptide, the sequence of VIP,

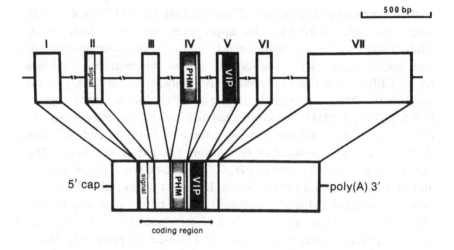

Fig. 7.2. Organization of exons and introns of the human
VIP/PHM-27 gene (Ohsawa *et al.*, 1987). The upper portion represents
the human VIP/PHM-27 gene and the lower portion the preproVIP
mRNA.

and a small part of the carboxy-terminal region. Exon VI comprises 89 bp
and encodes most of the carboxy-terminal region and a part of the
3'-untranslated region of the mRNA. Exon VII comprises 724 bp and
encodes the rest of the 3'-untranslated region. PHM-27 and VIP are
structurally related peptides, having 44% homology in amino acid
sequence and 52% homology in nucleotide sequence, but are encoded in
different exons, IV and V. This suggests that PHM-27-encoding and
VIP-encoding exons have been duplicated from an ancestral exon as in the
case of the glucagon/glucagon-like peptide gene (Bell *et al.*, 1983) and the
calcitonin/calcitonin-gene related peptide gene (Jonas *et al.*, 1985). In
addition, the 5' splice sites of introns IV and V downstream of PHM-27-
encoding exon IV and VIP-encoding exon V contain an identical stretch of
nine nucleotides (AGGTAAAGA). The 3' splice sites of these introns also
contain an identical stretch of six nucleotides (AGCAGT). Furthermore,
both introns IV and V fall after the second nucleotide of the serine codon,
AGC. The sequence conservation around the boundaries between the
PHM-27-encoding and VIP-encoding exons and their adjacent introns
suggests that the duplication has occurred over a broad area containing
the ancestral exon and its adjacent introns. The human VIP/PHM-27 gene
is located on chromosome 6 (Gozes *et al.*, 1987; Gotoh *et al.*, 1988).

7.4. Regulation of VIP/PHM-27 gene expression

7.4.1. Induction of VIP/PHM-27 gene expression by cyclic AMP and phorbol esters

The synthesis of proVIP has been shown to be induced by cyclic AMP (cAMP) in human neuroblastoma cells (Yanaihara et al., 1981; Obata et al., 1981). In order to clarify the mechanism of the induction of proVIP synthesis by cAMP, Hayakawa et al. determined the amount of VIP/PHM-27 mRNA in the neuroblastoma cells (Hayakawa et al., 1984; Okamoto et al., 1984). When neuroblastoma cells were grown in the presence of 1 mM dibutyryl cAMP, the synthesis of proVIP was stimulated 11-fold. The amount of VIP/PHM-27 mRNA determined both by hybridization with cloned VIP/PHM-27 cDNA and by a cell-free translation assay was also increased 11-fold in the dibutyryl cAMP-induced cells. Transcription of VIP/PHM-27 mRNA was observed in nuclei isolated from induced cells, but not in nuclei isolated from uninduced cells. These results indicate that cAMP-induced proVIP synthesis is achieved by an enhancement of the transcription rate of VIP/PHM-27 mRNA.

Ohsawa et al. (1985) found that tumor-promoting phorbol esters also increased the VIP/PHM-27 mRNA level in the human neuroblastoma cells. Sixteen nM 12-O-tetra-decanoylphorbol-13-acetate (TPA) increased the VIP/PHM-27 mRNA level about 5-fold. In the presence of both 1 mM dibutyryl cAMP and 16 nM TPA, the mRNA level increased as much as 30-fold. This increase was two times greater than the sum of those induced by the same concentration of each agent tested separately, indicating that the effects of dibutyryl cAMP and TPA are synergistic. Similar results were obtained with another potent tumor promoter, phorbol-12,13-didecanoate and with the combination of 3-isobutyl-1-methylxanthine (an inhibitor of cyclic nucleotide phospho-diesterase) and phorbol esters. The VIP/PHM-27 gene dosage was not changed by cAMP, TPA or both together. The intracellular cAMP level was essentially unaffected by phorbol esters. Phorbol esters appeared to increase the transcription of the VIP/PHM-27 mRNA independently on the cAMP-mediated pathway. In eukaryotic cells, cAMP is thought to exert its biochemical action through activating cAMP-dependent protein kinase (Krebs & Beavo, 1979) and phorbol esters are regarded as an activator of protein kinase C (Nishizuka, 1984). The two kinase systems may synergistically affect the expression of the VIP/PHM-27 gene.

```
gaattcaggataaatttgactgcagtactctttctttatgtttatttttctggaagcatccaaagttaacgctggctacaggacctctggtttacaatt    -1830
catttcctgaagaagaagaaataaaacacaattctataaagtctgtgattaaaacttgacctcagatggtctatgattccatttccaaattg          -1730
ttttgctttctagtttaatttcatagagaggaaatgcagttgtttggttgagtttactttattcagaggtaaaaaattgaatatttcttttattta      -1630
tcttattattattatatagagacaattgtctttgtattgggaaaagctactatagttactctttgtaaatgctaagtgcttgctagccaaatatg        -1530
atcagcttacatctttatgtttaaatttttctttctttatcattcagcagtcatttgtctgtgaccctcaattaagttgcttgctttctgctctta       -1430
aataaaattgttcactaagaatggggtgatgaagagctaagataaaaaattccatatgagaactgttcaacctctagtttgtaataatcaggtaaa      -1330
aaagatttcctggattaagccacaggaactctgcctaagtcagagctgtcaactgggaaacaaattccagacatttgaaacttaatcattaatt        -1230
tttctggtaactggattagaaaaatgattaagcatagcaggatatcttttttactggatcagtctgacttgaacgagggaaatagatactacaatcttaa  -1130
agtagaagaaagagactgtaaagccatctgtctgtctatgaacttagaaacaaatgtaaaataatacacatggcagagcctttaaataaaggaaaat    -1030
ggcatttcctctgttgttgggaaatttcatggcttaaactctgattgaatgttgctgactacctttagttacaaatccaca               - 930
atagaagtgaagtatttatcaagtacta TATAAAT gctttcaactatttgaagccacctgtcaagtgattgaatcagaaagttcaaaactggttt       - 832
gagtatgtgtgttggggatgagggtagaaaactaattctacatattgtcaaagact TATAAAA gtgataggaaaatcttaaaagaaagaaca         - 734
attaaattcctggagtatatagcctgacaacaattattttttggatcattgacacagtgtatgtaaagtgcattgggaattggccattttataca        - 634
aatatgaaatagcactactgtactgctttgtacaacagcgtgcatatcttactatcttagagactttgaaacacttatcttggattagtttcaaca       - 534
aagacgaaatgtaataatagaaggtgaaattatctctagagactttaaagttaaagtaggataaattcaagaaagattaaatctgaactatatgaatagca  - 434
taaaattaaaaatattaagagaattgatacttgtaataaagtataattaatacttttataagcgtaatgaacataagtgttgttgcttttatgcatttg    - 334
ttctcacacatgaagtctttattttctgaaggaatcccaatgccctggctataacactacaatcatttttctatttttacatttaattaaaggttaaaac  - 234
atttcacattgagcttgctacttgtttgtttaagaaaaagaaataaatttcctttccctttctttatcttttgaagttgctcactataaacta TATAAAT tag  - 136
```

cAMP/Phorbol Ester-Responsive Element

```
gcataatgacaattagtagaaacttcagccccattcatcc catggccgtcatactgtga cgtctttcagagcaacttttgtgattgctcagtcctaag    - 38
                                         -1 1  Exon I
tataaaccc TATAAAA tgatgggcttgaaatgctgg TCAGGGTAGAGTGAGAAGGCACCAGCACCAGGCAGCGAGTAACAGCCAACCCTTAGCCATTGCTAAG   59
GGCAGAGAACTGGTGCAGCGCTTTCTTACTCCCCAGGACTTCAGCCACCTAAGACCAGCTCCAAAACAAACCAGAACAGTCAGCTTCGCGGGGGAGCCACGACTG  159
GGCCAG gtaagtgaaaacttaccttctctcttttccctcctgccctcggaacttttcacaaataatgaatttctctttcttgaaagaatt              258
```

Fig. 7.3. Nucleotide sequence of the 5'-flanking region of the human VIP/PHM-27 gene (Yamagami *et al.*, 1988). Capital letters indicate exon I, and lower case letters are used for the 5'-flanking sequence and intron I. The four TATA boxes are boxed with a solid line. The 'cAMP/phorbol ester-responsive element' is underlined.

7.4.2. Identification of the inducible promoter of VIP/PHM-27 gene

Yamagami *et al.* (1988) determined the nucleotide sequence of 1.9 kilobase pairs of the 5'-flanking region of the human VIP/PHM-27 gene. In this region, four TATA box sequences were found at 28, 145, 772 and 900 bp upstream from the cap site, but there were no CAAT or GC box sequences (Fig. 7.3). The presence of several TATA box sequences and the different uses of the individual sequences have been reported in several genes, such as the mouse salivary and liver α-amylase gene (Young *et al.*, 1981), mouse myosin light chain gene (Robert *et al.*, 1984) and rat insulin-like growth factor II gene (Frunzio *et al.*, 1986). To see which TATA box sequence in the 5'-flanking region of the VIP/PHM-27 gene is the promoter inducible by dibutyryl cAMP and TPA, exon mapping analysis was done using the anti-sense RNA synthesized from the *Eco*RI fragment spanning − 1926 to 356 (Yamagami *et al.*, 1988). Only one RNase-resistant fragment of about 170 bases was detected with RNA from TPA- or dibutyryl cAMP-treated NB-1 cells. The amount of the fragment was also significantly increased with RNA from NB-1 cells which had been treated with both dibutyryl cAMP and TPA. The result suggested that there were no exons other than the 170 bp exon I in the region upstream from exon II and that among the four TATA box sequences only the TATA box sequence 28 bp upstream from the cap site was the promoter inducible by dibutyryl cAMP and TPA. This was confirmed by primer extension and mung-bean nuclease mapping analyses (Yamagami *et al.*, 1988).

7.4.3. The cis-*acting element and* trans-*acting factor which mediate regulation by cAMP and phorbol ester*

To identify the nucleotide sequence in the human VIP/PHM-27 gene responsible for regulation by cAMP and phorbol esters, chimeric genes were constructed in which different portions of the 5'-flanking region of the VIP/PHM-27 gene were fused to the bacterial chloramphenicol acetyltransferase (CAT) gene. The transcriptional activities of the fusion genes introduced into human neuroblastoma cells (NB-1 cells) were assayed by measuring CAT activity in the cell lysates. As shown in Fig. 7.4, cells transfected with a chimeric plasmid, pVIP(− 1372), containing the 1372 bp of the 5'-flanking sequence and a part of exon I, exhibited a slight increase in CAT activity after having been treated with TPA, and an approximately 5-fold increase when treated with dibutyryl cAMP. Simultaneous addition of both TPA and dibutyryl cAMP produced a

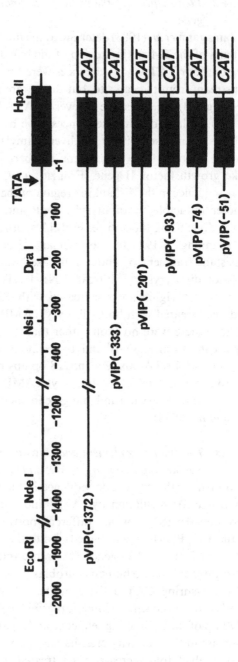

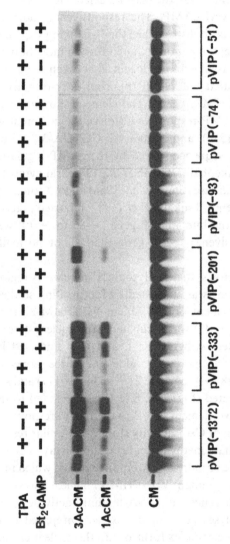

Fig. 7.4. Delimitation of cAMP/phorbol ester-responsive element by chloramphenicol acetyltransferase (CAT) assay (K. Ohsawa et al., unpublished data). Thin lines represent the 5'-flanking sequences of the human VIP/PHM-27 gene. The solid box indicates exon I of the gene. Numbers in parentheses indicate distances in base pairs from the transcription start point of the gene. Autoradiogram at bottom shows CAT activities in NB-1 cells that were transfected with the corresponding deletion mutants and then incubated with or without dibutyryl cAMP and TPA for 9 h. Positions of chloramphenicol (CM), 1-acetyl chloramphenicol (1AcCM) and 3-acetyl chloramphenicol (3AcCM) are shown on the left. The degree of TPA-induced increase in VIP/PHM-27 gene promoter activity was apparently smaller than that of the TPA-induced increase in VIP/PHM-27 mRNA level (Ohsawa et al., 1985). This may reflect an additional function of phorbol esters in the regulation of the VIP/PHM-27 gene expression; phorbol esters may in part stabilize the VIP/PHM-27 mRNA.

10-fold increase in CAT activity, demonstrating a synergistic effect. A deletion eliminating sequences upstream from -333 had little effect on the basal and induced levels of CAT activity. A deletion to -201 markedly reduced the basal level of expression, but did not affect the inducibility of the plasmid. pVIP(-93), the plasmid deleted to -93, still responded to TPA and dibutyryl cAMP. The responsiveness both to dibutyryl cAMP and to TPA was abolished when the 5′-flanking sequence was deleted to -74. Therefore, the region required for the induction by cAMP and phorbol esters was assumed to be localized between -93 and -75. This region was designated the 'cAMP/phorbol ester-responsive element' of the human VIP/PHM-27 gene. The element shows a high degree of homology with the cAMP-responsive region of the rat somatostatin gene (Montminy et al., 1986), and contains 5′-CGTCA-3′, a 5-bp motif within the 5′-regulatory sequence shared by the cAMP-regulated genes such as human preproenkephalin (Comb et al., 1986), rat cytosolic phosphoenolpyruvate carboxykinase (Short et al., 1986) and rat somatostatin (Montminy et al., 1986) genes. Tsukada et al. (1987) reported that the cAMP-responsive element of the human VIP gene (-86 to -70) was able to confer cAMP-responsiveness on a gene that is not normally regulated by cAMP.

The presence of a nuclear factor(s) that binds to the 5′-regulatory region of the human VIP/PHM-27 gene was suggested by footprinting studies. When a labeled promoter fragment of the human VIP/PHM-27 gene (nucleotide residues -337 to -49) was incubated with nuclear extract from NB-1 cells, lightly treated with DNase I, and then analysed by electrophoresis on a denaturing polyacrylamide gel, three areas of nuclease protection, site 1 (-88 to -69), site 2 (-153 to -117) and site 3 (-210 to -170), were detected (Fig. 7.5). Site 1 corresponded to the cAMP/phorbol ester-responsive element, site 3 to the region responsible for the basal level of expression of CAT activity (Fig. 7.4). Similar results were observed by dimethyl sulfate methylation protection assay. Thus, it became likely that there are one or more DNA-binding factors which may mediate the regulation of the human VIP/PHM-27 gene expression. Although no significant difference in DNA-binding activities was observed by footprinting analyses between the extracts prepared from induced and uninduced cells, the complex formed with the nuclear extract from induced cells was found to have migrated faster than that formed with the extract from uninduced cells by gel mobility shift assay (K. Ohsawa, T. Yamagami, H. Yamamoto & H. Okamoto, unpublished data). The induction of the VIP/PHM-27 gene expression may be

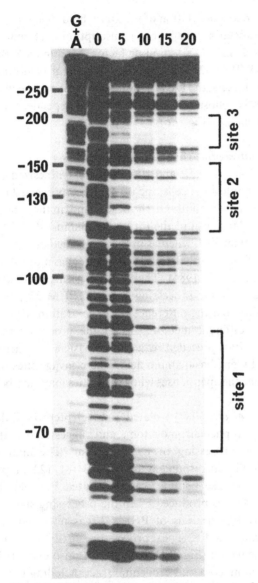

Fig. 7.5. DNase I footprinting of the promoter fragment of the human VIP/PHM-27 gene (K. Ohsawa *et al.*, unpublished data). A DNA fragment (− 337 to − 49, Fig. 7.3) was labeled at nucleotide residue − 49 of the non-coding strand, incubated with nuclear extracts from cAMP- and TPA-treated NB-1 cells, subsequently digested with DNase I, and analyzed on an 8% polyacrylamide gel. The amounts of extracts (micrograms of protein) are indicated above the lanes. Lane 0, no added extract; lane G + A, a chemical degradation sequencing ladder used for size markers; brackets, protected areas.

achieved through the phosphorylation of the DNA-binding factor(s) as in the case of the somatostatin gene whose transcription is stimulated by phosphorylation of a 43 kDa DNA-binding factor (Montminy & Bilezikjian, 1987). As the VIP/PHM-27 gene expression is synergistically induced by cAMP and phorbol esters, the DNA-binding factor(s) may be independently or interactively phosphorylated by cAMP-dependent protein kinase and protein kinase C (Fig. 7.6).

7.4.4. Alternative processing of the VIP precursor

As described above, VIP and PHM-27 are synthesized from a common precursor. In intestinal cells, the precursor is processed to yield VIP and PHM-27 in equal amounts. On the other hand, in the suprachiasmatic and periventricular nuclei of the rat hypothalamus, PHI-27 is 3–5 times more abundant than VIP, whereas in the ventromedial nucleus and the posterior region of the hypothalamus, VIP is 3–4 times more abundant than PHI-27 (Beinfeld et al., 1984). Therefore, it seems that there are tissue-specific mechanisms for processing the VIP/PHM-27 precursor, although the possibility that the difference in concentrations of VIP and PHM-27 reflects the different metabolisms of the processed peptides cannot be excluded. VIP is preceded by a pair of basic amino acids, but PHM-27 or PHI-27 by one basic amino acid. The cleavage sites may be recognized by different endopeptidases whose activities may vary between tissues.

Recently, Yiangou et al. (1987) isolated a novel preproVIP-derived peptide from an adrenal pheochromocytoma and suggested that there is another tissue-specific processing of the precursor. The amino acid sequence of the peptide corresponded to residues 81–122 of prepro-VIP/PHM-27 (Fig. 7.1); the peptide was designated PHV-42. It was concluded that PHV-42 was produced when the processing signal (Gly–Lys–Arg) in the carboxy-terminus of PHM-27 was not used and the peptide was further extended to the next processing site (Lys–Arg) located just before the VIP sequence (Fig. 7.1). PHV-42 was also extracted from human stomach, nasal mucosa and urogenital tract including the uterus. In contrast, PHV-42 could not be demonstrated in the large intestine or the central nervous system where PHM-27 was found in high concentrations. PHV-42 is much more potent than PHM-27 in reducing the force and frequency of uterine contraction and in relaxing the smooth muscles of the stomach, indicating the physiological role of the peptide in these organs (Yiangou et al., 1987).

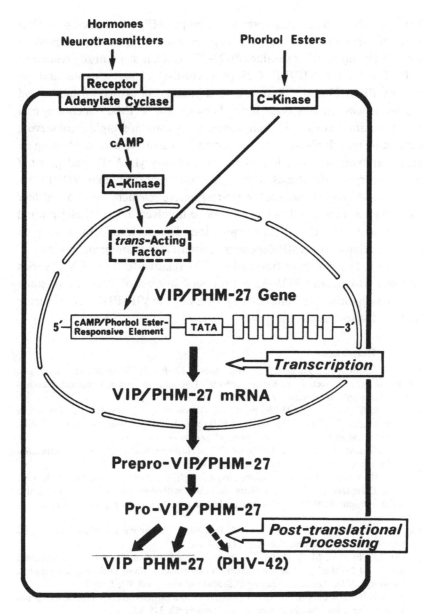

Fig. 7.6. VIP/PHM-27 gene expression and its regulation. A-kinase, cAMP-dependent protein kinase; C-kinase, protein kinase C.

7.5. Summary

Islets of Langerhans are innervated by VIP-containing nerve fibers; VIP-producing tumors often arise in the islets. As summarized in Fig. 7.6,

VIP is synthesized as a large precursor, preproVIP, from which proVIP is generated by removal of the signal peptide. Further proteolysis of proVIP yields VIP and another peptide, PHM-27, which is structurally related to VIP. The human VIP/PHM-27 gene consists of seven exons and six introns. PHM-27 and VIP are encoded in adjacent exons, IV and V, and the sequences around boundaries between the PHM-27-encoding and VIP-encoding exons and their adjacent introns are highly conserved, suggesting that duplication has occurred over a broad area containing an ancestral exon and its adjacent introns. Dibutyryl cAMP and phorbol esters synergistically stimulate the transcription of the human VIP/PHM-27 gene. A cAMP/phorbol ester-responsive element and a nuclear factor(s) that binds to the element may be involved in the transcriptional control of VIP/PHM-27 gene expression. The nuclear factor(s) may be phosphorylated by cAMP-dependent protein kinase and protein kinase C (Fig. 7.6). The concentration of PHM-27 relative to that of VIP varies between tissues, and PHV-42 is produced instead of PHM-27 in some tissues, suggesting mechanisms for processing the VIP/PHM-27 precursor that are tissue-specific.

References

Anand, P., Gibson, S.J., Yiangou, Y., Christofides, N.D., Polak, J.M. & Bloom, S.R. (1984). PHI-like immunoreactivity co-locates with the VIP-containing system in human lumbosacral spinal cord. *Neuroscience Letters* 46, 191–6.

Beinfeld, M.C., Korchak, D.M., Roth, B.L. & O'Donohue, T.L. (1984). The distribution and chromatographic characterization of PHI (peptide histidine isoleucine amide)-27-like peptides in rat and porcine brain. *Journal of Neuroscience* 4, 2681–8.

Bell, G.I., Sanchez-Pescador, R., Laybourn, P.J. & Najarian, R.C. (1983). Exon duplication and divergence in the human preproglucagon gene. *Nature* 304, 368–71.

Bloom, S.R., Christofides, N.D., Delamarter, J., Buell, G., Kawashima, E. & Polak, J.M. (1983). Diarrhoea in VIPoma patients associated with cosecretion of a second active peptide (peptide histidine isoleucine) explained by single coding gene. *The Lancet* ii, 1163–5.

Bloom, S.R., Polak, J.M. & Pearse, A.G.E. (1973). Vasoactive intestinal peptide and watery-diarrhoea syndrome. *The Lancet* ii, 14–16.

Bodner, M., Fridkin, M. & Gozes, I. (1985). Coding sequences for vasoactive intestinal peptide and PHM-27 peptide are located on two adjacent exons in the human genome. *Proceedings of the National Academy of Sciences of the U.S.A.* 82, 3548–51.

Breathnach, R. & Chambon, P. (1981). Organization and expression of eucaryotic split genes coding for proteins. *Annual Review of Biochemistry* 50, 349–83.

Carlquist, M., Mutt, V. & Jörnvall, H. (1979). Isolation and characterization of bovine vasoactive intestinal peptide (VIP). *FEBS Letters* 108, 457–60.

Comb, M., Birnberg, N.C., Seasholtz, A., Herbert, E. & Goodman, H.M. (1986). A cyclic AMP- and phorbol ester-inducible DNA element. *Nature* 323, 353–6.

Fahrenkrug, J. (1982). VIP as a neurotransmitter in the peripheral nervous system. In *Vasoactive intestinal peptide: advances in peptide hormone research series* (ed. S.I. Said), pp. 361–72. New York: Raven Press.

Frunzio, R., Chiariotti, L., Brown, A.L., Graham, D.E., Rechler, M.M. & Bruni, C.B. (1986). Structure and expression of the rat insulin-like growth factor II (rIGF-II) gene. *Journal of Biological Chemistry* **36**, 17138–49.

Gotoh, E., Yamagami, T., Yamamoto, H. & Okamoto, H. (1988). Chromosomal assignment of human VIP/PHM-27 gene to 6q26→q27 region by spot blot hybridization and *in situ* hybridization. *Biochemistry International* **17**, 555–62.

Gozes, I., Nakai, H., Byers, M., Avidor, R., Weinstein, Y., Shani, Y. & Shows, T.B. (1987). Sequential expression in the nervous system of c-*myb* and *VIP* genes, located in human chromosomal region 6q24. *Somatic Cell and Molecular Genetics* **13**, 305–13.

Håkanson, R., Sundler, F. & Uddman, R. (1982). Distribution and topography of peripheral VIP nerve fibers: functional implications. In *Vasoactive intestinal peptide: advances in peptide hormone research series* (ed. S.I. Said), pp. 121–44. New York: Raven Press.

Hayakawa, Y., Obata, K., Itoh, N., Yanaihara, N. & Okamoto, H. (1984). Cyclic AMP regulation of pro-vasoactive intestinal polypeptide/PHM-27 synthesis in human neuroblastoma cells. *Journal of Biological Chemistry* **259**, 9207–11.

Hökfelt, T., Schlutzberg, M., Lundberg, J.M., Fuxe, K., Mutt, V., Fahrenkrug, J. & Said, S.I. (1982). Distribution of vasoactive intestinal polypeptide in the central and peripheral nervous systems as revealed by immunocytochemistry. In *Vasoactive intestinal peptide: advances in peptide hormone research series* (ed. S.I. Said), pp. 65–90. New York: Raven Press.

Itoh, N., Obata, K., Yanaihara, N. & Okamoto, H. (1983). Human preprovasoactive intestinal polypeptide contains a novel PHI-27-like peptide, PHM-27. *Nature* **304**, 547–9.

Jonas, V., Lin, C.R., Kawashima, E., Semon, D., Swanson, L.W., Mermod, J-J., Evans, R.M. & Rosenfeld, M.G. (1985). Alternative RNA processing events in human calcitonin/-calcitonin-gene-related peptide gene expression. *Proceedings of the National Academy of Sciences of the U.S.A.* **82**, 1994–8.

Krebs, E.G. & Beavo, J.A. (1979). Phosphorylation-dephosphorylation of enzymes. *Annual Review of Biochemistry* **48**, 923–59.

Kreil, G. (1981). Transfer of proteins across membranes. *Annual Review of Biochemistry* **50**, 317–48.

Linder, S., Barkhem, T., Norberg, A., Persson, H., Schalling, M., Hökfelt, T. & Magnusson, G. (1987). Structure and expression of the gene encoding the vasoactive intestinal peptide precursor. *Proceedings of the National Academy of Sciences of the U.S.A.* **84**, 605–9.

Marley, P. & Emson, P. (1982). VIP as a neurotransmitter in the central nervous system. In *Vasoactive intestinal peptide: advances in peptide hormone research series* (ed. S.I. Said), pp. 341–60. New York: Raven Press.

Mekhjian, H.P. & O'Dorisio, T.M. (1987). VIPoma syndrome. *Seminars in Oncology* **14**, 282–91.

Miller, R.E. (1981). Pancreatic neuroendocrinology: peripheral neural mechanisms in the regulation of the islets of Langerhans. *Endocrine Reviews* **2**, 471–94.

Montminy, M.R. & Bilezikjian, L.M. (1987). Binding of a nuclear protein to the cyclic-AMP response element of the somatostatin gene. *Nature* **328**, 175–8.

Montminy, M.R., Sevarino, K.A., Wagner, J.A., Mandel, G. & Goodman, R.H. (1986). Identification of a cyclic-AMP-responsive element in the rat somatostatin gene. *Proceedings of the National Academy of Sciences of the U.S.A.* **83**, 6682–6.

Mutt, V. & Said, S.I. (1974). Structure of the porcine vasoactive intestinal octacosapeptide. *European Journal of Biochemistry* **42**, 581–9.

Nishizawa, M., Hayakawa, Y., Yanaihara, N. & Okamoto, H. (1985). Nucleotide sequence divergence and functional constraint in VIP precursor mRNA evolution between human and rat. *FEBS Letters* **183**, 55–9.

Nishizuka, Y. (1984). The role of protein kinase C in cell surface signal transduction and tumour promotion. *Nature* **308**, 693–8.

Obata, K., Itoh, N., Okamoto, H., Yanaihara, C., Yanaihara, N. & Suzuki, T. (1981). Identification and processing of biosynthetic precursors to vasoactive intestinal polypeptide in human neuroblastoma cells. *FEBS Letters* **136**, 123–6.

Ohsawa, K., Hayakawa, Y., Nishizawa, M., Yamagami, T., Yamamoto, H., Yanaihara, N. & Okamoto, H. (1985). Synergistic stimulation of VIP/PHM-27 gene expression by cyclic AMP and phorbol esters in human neuroblastoma cells. *Biochemical and Biophysical Research Communications* **132**, 885–91.

Ohsawa, K., Yamagami, T., Nishizawa, M., Inoue, C., Gotoh, E., Yamamoto, H. & Okamoto, H. (1987). Structure and expression of human vasoactive intestinal peptide/PHM-27 gene. In *Gunma symposia on endocrinology 24* (ed. Institute of Endocrinology, Gunma University), pp. 169–81. Tokyo: Center for Academic Publications; Utrecht: VNU Science Press BV.

Okamoto, H., Yanaihara, N., Hayakawa, Y., Itoh, N. & Obata, K. (1984). Biosynthesis and processing of human VIP/PHM-27 precursor. In *Endocrinology: international congress series 655* (ed. F. Labrie & L. Proulx), pp. 622–5. Amsterdam: Elsevier.

Robert, B., Daubas, P., Akimenko, M., Cohen, A., Garner, I., Guenet, J. & Buckingham, M. (1984). A single locus in the mouse encodes both myosin light chains 1 and 3, a second locus corresponds to a related pseudogene. *Cell* **39**, 129–40.

Said, S.I. & Mutt, V. (1972). Isolation from porcine intestinal wall of a vasoactive octacosapeptide related to secretin and glucagon. *European Journal of Biochemistry* **28**, 199–204.

Schebalin, M., Said, S.I. & Makhlouf, G.M. (1976). Stimulation of insulin and glucagon secretion by vasoactive intestinal peptide. *American Journal of Physiology* **232**, E197–200.

Short, J.M., Wynshaw-Boris, A., Short, H.P. & Hanson, R.W. (1986). Characterization of the phosphoenolpyruvate carboxykinase (GTP) promoter-regulatory region. *Journal of Biological Chemistry* **261**, 9721–6.

Steiner, D.F., Quinn, P.S., Chan, S.I., Marsh, J. & Tager, H.S. (1980). Processing mechanisms in the biosynthesis of proteins. *Annals of the New York Academy of Sciences* **343**, 1–16.

Tatemoto, K., Jörnvall, H., McDonald, T.J., Carlquist, M., Go, V.L.W., Johansson, C. & Mutt, V. (1984). Isolation and primary structure of human PHI (peptide HI). *FEBS Letters* **174**, 258–61.

Tatemoto, K. & Mutt, V. (1981). Isolation and characterization of the intestinal peptide porcine PHI (PHI-27), a new member of the glucagon-secretin family. *Proceedings of the National Academy of Sciences of the U.S.A.* **78**, 6603–7.

Tsukada, T., Fink, J.S., Mandel, G. & Goodman, R.H. (1987). Identification of a region in the human vasoactive intestinal polypeptide gene responsible for regulation by cyclic AMP. *Journal of Biological Chemistry* **262**, 8743–7.

Tsukada, T., Horovitch, S.J., Montminy, M.R., Mandel, G. & Goodman, R.H. (1985). Structure of the human vasoactive intestinal polypeptide gene. *DNA* **4**, 293–300.

Verner, J.V. & Morrison, A.B. (1958). Islet cell tumor and a syndrome of refractory watery diarrhea and hypokalemia. *American Journal of Medicine* **25**, 374–80.

White, J.W. & Saunders, G.F. (1986). Structure of the human glucagon gene. *Nucleic Acids Research* **14**, 4719–30.

Yamagami, T., Ohsawa, K., Nishizawa, M., Inoue, C., Gotoh, E., Yanaihara, N., Yamamoto, H. & Okamoto, H. (1988). Complete nucleotide sequence of human vasoactive intestinal peptide/PHM-27 gene and its inducible promoter. *Annals of the New York Academy of Sciences* **527**, 87–102.

Yanaihara, N., Suzuki, T., Sato, H., Hoshino, M., Okaru, Y. & Yanaihara, C. (1981). Dibutyryl cAMP stimulation of production and release of VIP-like immunoreactivity in a human neuroblastoma cell line. *Biomedical Research* **2**, 728–34.

Yiangou, Y., Marzo, V.D., Spokes, R.A., Panico, M., Morris, H.R. & Bloom, S.R. (1987).

Isolation, characterization, and pharmacological actions of peptide histidine valine 42, a novel preprovasoactive intestinal peptide-derived peptide. *Journal of Biological Chemistry* **262**, 14010–13.

Young, R.A., Hagenbüchle, O. & Schibler, U. (1981). A single mouse α-amylase gene specifies two different tissue-specific mRNAs. *Cell* **23**, 451–8.

8

Pancreastatin: a novel pancreatic hormone

KAZUHIKO TATEMOTO

8.1. Introduction

While most peptide hormones have been isolated on the basis of their specific biological activities, pancreastatin was discovered in pancreatic extracts by the presence of its unique C-terminal structure, a glycine amide (Tatemoto *et al.*, 1986). Since such a structure occurs only in peptide hormones and neuropeptides (Tatemoto & Mutt, 1980), it was anticipated that the peptide would be biologically active. Indeed, this 49 amino acid peptide was later named pancreastatin because it was found to inhibit insulin secretion (Tatemoto *et al.*, 1986). Subsequently, pancreastatin was found to inhibit exocrine pancreatic secretion as well (Ishizuka *et al.*, 1987; Funakoshi *et al.*, 1988). Pancreastatin is located in the islets of Langerhans (Shimizu *et al.*, 1987; Ravazzola *et al.*, 1988) and chromogranin A may serve as its precursor (Eiden, 1987; Huttner & Benedum, 1987). This novel peptide may therefore play important roles in regulating various pancreatic functions.

8.2. Isolation and primary structure

Pancreastatin was originally isolated from porcine pancreatic extracts after extraction with 0.5 M acetic acid, adsorption onto alginic acid, ethanol fractionation, gel filtration (Sephadex G-25), ion exchange chromatography (CM cellulose) and reverse-phase HPLC (Tatemoto *et al.*, 1986). The pancreastatin-containing fractions were identified by using a chemical method which measures the amounts of glycine amide released from tryptic digests of samples (Tatemoto & Mutt, 1978).

Porcine pancreastatin consists of 49 amino acid residues and has an N-terminal glycine and a C-terminal glycine amide. The primary structure of pancreastatin is different from that of all other known peptide

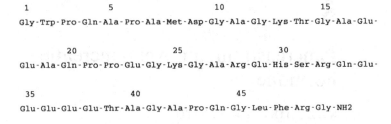

```
 1           5              10             15
Gly-Trp-Pro-Gln-Ala-Pro-Ala-Met-Asp-Gly-Ala-Gly-Lys-Thr-Gly-Ala-Glu-

       20             25             30
Glu-Ala-Gln-Pro-Pro-Glu-Gly-Lys-Gly-Ala-Arg-Glu-His-Ser-Arg-Gln-Glu-

 35             40             45
Glu-Glu-Glu-Glu-Thr-Ala-Gly-Ala-Pro-Gln-Gly-Leu-Phe-Arg-Gly-NH2
```

Fig. 8.1. Primary structure of porcine pancreastatin.

hormones and neuropeptides (Fig. 8.1). Synthetic pancreastatin, prepared by using solid-phase synthetic techniques, was found to be identical to the native peptide and to be equipotent in inhibiting insulin secretion. The C-terminal fragments, pancreastatin 14–49 and 33–49, also inhibit insulin secretion, suggesting that the C-terminal portion of the molecule is important for biological activity (Tatemoto *et al.*, 1986).

Recently, pancreastatin from human carcinoid tumors (Schmidt *et al.*, 1988) and bovine pancreas (Tatemoto *et al.*, 1988) were isolated and their structures were determined. Human and bovine pancreastatin consist of 53 and 47 amino acid residues, respectively. Comparison of their primary structures with that of porcine pancreastatin revealed a high degree of species variation, even between the C-terminal regions of the molecules that are thought to be responsible for biological activity (Fig. 8.2). This high degree of species variation in the structure of pancreastatin may be characteristic of this peptide family. All known structures of pancreastatin, however, contain a C-terminal glycine amide, the chemical feature used to discover this peptide in the pancreas.

```
pPS    GWPQAPAMDGAGKTGAEEAQPPEGKGAREHSRQ.EEEEETAGAPQGLFRG*

bPS    AAPGWPEDGAGKMGAEEAKPPEGKGEWAHSRQ..EEEEMARAPQVLFRG*

hPS    GESRSEALAVDGAGKPGAEEAQDPEGKGEQEHSQQKEEEEEMAVVPQGLFRG*
```

Fig. 8.2. Comparison of the primary structures of porcine, bovine, and human pancreastatin. PS, pancreastatin; asterisks, amidated. Identities to the porcine sequence are underlined. Spaces resulting from sequence alignment are indicated by dots.

8.3. Biological actions

8.3.1. Effects on pancreatic endocrine secretion

Pancreastatin seems to play important roles in regulating the endocrine functions of pancreatic islets. Pancreastatin inhibits glucose-stimulated insulin release from isolated perfused rat pancreas (Tatemoto *et al.*, 1986; Efendic *et al.*, 1987) and from isolated rat islets (Efendic *et al.*, 1987; Ishizuka *et al.*, 1988). Its effect on early-phase insulin release is more pronounced than on late-phase insulin release. It also augments arginine-induced glucagon release from the isolated pancreas (Efendic *et al.*, 1987). *In vivo*, pancreastatin lowers baseline plasma insulin levels, inhibits glucose-induced insulin secretion, stimulates glucagon secretion, and induces hyperglycemia in mice (Ahren *et al.*, 1988).

8.3.2. Effects on pancreatic exocrine secretion

In addition to affecting endocrine functions, pancreastatin also regulates exocrine functions of the pancreas. Pancreastatin inhibits cholecystokinin (CCK)-stimulated release of amylase from acinar cells (Ishizuka *et al.*, 1987) and inhibits pancreatic protein and fluid secretion stimulated by a diversion of bile–pancreatic juice from the intestine (Funakoshi *et al.*, 1988). These results indicate that pancreastatin inhibits CCK-stimulated pancreatic secretion. Pancreastatin also inhibits meal-stimulated pancreatic secretion (Miyasaka *et al.*, 1988), suggesting that the peptide is involved in the regulation of postprandial pancreatic exocrine secretion.

8.3.3. Other biological actions

Pancreastatin is a potent inhibitor of CCK-stimulated gallbladder contraction *in vivo* (Hashimoto *et al.*, 1988). It also enhances meal-stimulated gastric acid secretion (Hashimoto *et al.*, 1987). Interestingly, systemic administration of pancreastatin enhances memory retention in mice (Flood *et al.*, 1988). These results suggest that pancreastatin directly or indirectly affects a wide range of biological functions.

8.4. Localization

Immunohistochemical studies revealed that pancreastatin-immunoreactive cells were located in the islets of Langerhans, particularly in the secretory granules of insulin and somatostatin-containing cells (Ravazzola *et al.*, 1988). Immunoreactive pancreastatin was found not only in the pancreas but also in many other endocrine cells. Secretory granules of gut endocrine cells, adrenal chromaffin cells, and anterior pituitary cells likewise show intense pancreastatin immunofluorescence (Shimizu *et al.*, 1987; Ravazzola *et al.*, 1988). Tissue distribution of pancreastatin, studied using radioimmunoassay techniques, shows the presence of high concentrations of pancreastatin immunoreactivities in the pancreas, pituitary, gut, thyroid, and adrenal gland (Bretherton-Watt *et al.*, 1988). Many human endocrine tumors, such as gastrinoma and glucagonomas, contain high concentrations of pancreastatin immunoreactivities (Bishop *et al.*, 1988). Since localization of pancreastatin precisely coincides with that of chromogranin A, it is likely that pancreastatin is closely associated with chromogranin A.

Pancreastatin is also distributed in the central nervous system, including hypothalamus, cortex, thalamus, hippocampus, spinal cord, and dorsal ganglia. This distribution suggests that pancreastatin may function not only as an endocrine hormone but also as a neuromodulator of neuroendocrine, autonomic and sensory functions (Gibson *et al.*, 1987).

8.5. Chromogranin A serves as the precursor of pancreastatin

Shortly after the primary structure of porcine pancreastatin was reported in 1986, it was noted that the peptide has a striking sequence homology to a region of the bovine chromogranin A molecule (Eiden, 1987; Huttner & Benedum, 1987), the structure of which was deduced from its cDNA sequence during the same year (Iacangelo *et al.*, 1986; Benedum *et al.*, 1986). This observation suggests that chromogranin A serves as the precursor of pancreastatin. Until recently, however, this relationship remained unclear because of the large structural differences between the porcine and bovine sequences.

Chromogranin A is an acidic protein that is abundant in secretory granules of a variety of endocrine cells and neurons (Blaschko *et al.*, 1967; Winkler *et al.*, 1986). Despite its abundance in the endocrine and neuroendocrine systems, its biological functions have remained unknown for the past 20 years. The primary structures of chromogranin A were recently deduced from cDNA sequences of bovine (Iacangelo *et al.*, 1986; Benedum *et al.*, 1986; Ahn *et al.*, 1987), human (Konecki *et al.*, 1987), rat (Iacangelo *et al.*, 1988*b*) and porcine (Iacangelo *et al.*, 1988*a*) origins. These results, together with structural studies of pancreastatin, have greatly facilitated our understanding of the biological functions of chromogranin A.

Recent structural studies have shown that chromogranin A is the precursor of pancreastatin. The structures of porcine, human and bovine

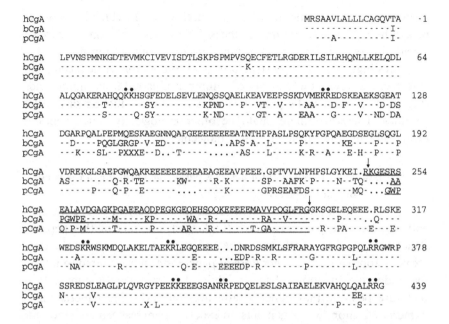

Fig. 8.3. A comparison of the primary structures of human, bovine, and porcine chromogranin A. The structures were deduced from the corresponding cDNA sequences: hCgA, human chromogranin A; bCgA, bovine chromogranin A; and pCgA, porcine chromogranin A. The regions encoding pancreastatin sequences are underlined. Identities to the human chromogranin A sequence are indicated by dashes. Spaces resulting from sequence alignment are indicated by dots. The processing sites for the formation of pancreastatin are indicated by arrows. Potential dibasic processing sites are shown by large dots.

pancreastatin were found to be identical to the corresponding regions of their chromogranin A molecules (Fig. 8.3). Although the structure of rat pancreastatin has not been determined, a region of the rat chromogranin A sequence also shows a high degree of sequence homology with the known pancreastatin sequences (Iacangelo *et al.*, 1988*b*). Furthermore, all chromogranin A molecules have a Gly–Gly–Lys structure, which is converted to a glycine amide during posttranslational processing. Indeed, the C-terminal glycine amide occurs at the corresponding position in human, bovine and porcine pancreastatin (Fig. 8.3). In all cases, the N-terminal pancreastatin sequence is preceded by a single arginine, a site where processing of many hormonal peptides occurs. These results confirm that chromogranin A is the precursor of pancreastatin. Thus, the determination of chromogranin A structures led to the elucidation of the biosynthesis of pancreastatin, and the isolation of pancreastatin led to the elucidation of possible biological functions of chromogranin A.

Interestingly, the primary structures of chromogranin A deduced from cDNA clones reveal the presence of a number of paired basic amino acids (Fig. 8.3). These may be processing sites for the production of smaller peptides. Therefore, there might be other biologically active peptides produced from the chromogranin A molecule.

8.6. Does pancreastatin act as a paracrine hormone in the islets of Langerhans?

Although pancreastatin is found in a variety of endocrine cells and neurons, the pancreas stores the largest amounts of this peptide. Pancreastatin affects both endocrine and exocrine functions of the pancreas. It appears that a higher dose of pancreastatin is required to inhibit the endocrine functions than to inhibit the exocrine functions. It is therefore possible that pancreastatin acts on endocrine pancreas in a paracrine fashion, whereas it acts on exocrine pancreas via an endocrine pathway.

Pancreastatin is present in high concentrations in secretory granules of the islets of Langerhans, particularly in those of B-cells. The peptide is processed from its precursor, chromogranin A, a major soluble protein released from the secretory granules of various endocrine cells. The secretory granules of B-cells also contain β-granin, an N-terminal fragment of chromogranin A (Hutton et al., 1985, 1988). The fact that β-granin is released from B-cells suggests that pancreastatin is co-secreted with β-granin from the same cells. Since pancreastatin modulates insulin secretion, it is of interest to know whether pancreastatin regulates insulin secretion in an autocrine fashion. If so, pancreastatin might be important in the pathogenesis and treatment of diabetes mellitus.

This work was supported by grants from NIH (DK-39188 and MH-23861).

References

Ahn, T.G., Cohn, D.V., Gorr, S.U., Ornstein, D.L., Kashdan, M.A. & Levine, M.A. (1987). Primary structure of bovine pituitary secretory protein I (chromogranin A) deduced from the cDNA sequence. *Proceedings of the National Academy of Sciences of the U.S.A.* **84**, 5043–7.

Ahren, B., Lindskog, S., Tatemoto, K. & Efendic, S. (1988). Pancreastatin inhibits insulin secretion and stimulates glucagon secretion in mice. *Diabetes* **37**, 281–5.

Benedum, U.M., Baeuerle, P.A., Konecki, D.S., Frank, R., Powell, J., Mallet, J. & Huttner,

W.B. (1986). The primary structure of bovine chromogranin A: a representative of a class of acidic secretory proteins common to a variety of peptidergic cells. *EMBO Journal* **5**, 1495–502.

Bishop, A., Bretherton-Watt, D., Hamid, Q.A., Fahey, M., Shepherd, N., Valentino, K., Tatemoto, K., Ghatei, M.A., Bloom, S.R. & Polak, J.M. (1988). The occurrence of pancreastatin in tumours of the diffuse neuroendocrine system. *Molecular and Cellular Probes* **2**, 225–35.

Blaschko, H., Comline, R.S., Schneider, F.H., Silver, M. & Smith, A.D. (1967). Secretion of a chromaffin granule protein, chromogranin from the adrenal gland after splanchnic stimulation. *Nature* **215**, 58–9.

Bretherton-Watt, D.G., Ghatei, M.A., Bishop, A.E., Facer, P., Valentino, K.L., Tatemoto, K., Roth, K., Polak, J.M. & Bloom, S.R. (1987). Pancreastatin-like immunoreactivity in porcine peripheral tissues. *Regulatory Peptides* **18**, 364.

Efendic, S., Tatemoto, K., Mutt, V., Quan, C., Chang, D. & Östenson, C. (1987). Pancreastatin and islet hormone release. *Proceedings of the National Academy of Sciences of the U.S.A.* **84**, 7257–60.

Eiden, L. (1987). Is chromogranin a prohormone? *Nature* **325**, 301.

Flood, J.F., Morley, J.E. & Tatemoto, K. (1988). Effects of systemic pancreastatin on memory retention. *Peptides* **9**, 1077–80.

Funakoshi, A., Miyasaka, K., Nakamura, R., Kitani, K., Shimizu, F., Nakano, I. & Tatemoto, K. (1988). Inhibitory effect of pancreastatin on pancreatic exocrine secretion in the conscious rat. *Gastroenterology* **94**, A138.

Gibson, S.J., Kar, S., Ballesta, J., Steel, J., Bretherton-Watt, D.G., Ghatei, M.A., Valentino, K.L., Tatemoto, K., Bloom, S.R. & Polak, J.M. (1987). Pancreastatin, a novel neuropeptide, is widely distributed throughout porcine brain, pituitary, spinal cord and dorsal root ganglia. *Regulatory Peptides* **18**, 376.

Hashimoto, T., Luis, F., Ishizuka, J., Gomez, G., Tatemoto, K., Greeley, G.H. Jr & Thompson, J.C. (1987). Pancreastatin can enhance meal-stimulated gastric acid secretion in the conscious dog. *Gastroenterology* **92**, A1428.

Hashimoto, T., Poston, G.J., Gomez, G., Tatemoto, K., Greeley, G.H. Jr & Thompson, J.C. (1988). Effect of pancreastatin on cholecystokinin-stimulated gallbladder contraction in vivo and in vitro. *Gastroenterology* **94**, A176.

Huttner, W. & Benedum, U.M. (1987). Chromogranin A and pancreastatin. *Nature* **325**, 305.

Hutton, J.C., Hansen, F. & Peshavaria, M. (1985). β-Granins: 21 kDa cosecreted peptides of the insulin granule closely related to adrenal medullary chromogranin A. *FEBS Letters* **188**, 336–40.

Hutton, J.C., Peshavaria, M., Johnston, C.F., Ravazzola, M. & Orci, L. (1988). Immunolocalization of betagranin: A chromogranin A-related protein of the pancreatic B-cell. *Endocrinology* **122**, 1014–20.

Iacangelo, A., Affolter, H.U., Eiden, L., Herbert, E. & Grimes, M. (1986). Bovine chromogranin A sequence and distribution of its messenger RNA in endocrine tissues. *Nature* **323**, 82–6.

Iacangelo, A.L., Fischer-Colbrie, R., Koller, K.J., Brownstein, M.J. & Eiden, L.E. (1988*a*). The sequence of porcine chromogranin A messenger RNA demonstrates chromogranin A can serve as the precursor for the biologically active hormone, pancreastatin. *Endocrinology* **122**, 2339–41.

Iacangelo, A., Okayama, H. & Eiden, L. (1988*b*). Primary structure of rat chromogranin A and distribution of its mRNA. *FEBS Letters* **227**, 115–21.

Ishizuka, J., Asada, I., Poston, G.J., Lluis, F., Tatemoto, K., Greeley, G.H. Jr & Thompson, J.C. (1987). A new pancreatic peptide, pancreastatin, regulates exocrine and endocrine secretions from the pancreas. *Gastroenterology* **92**, A1448.

Ishizuka, J., Singh, P., Greeley, G.H. Jr, Townsend, C.M. Jr, Cooper, C.W., Tatemoto, K. &

Thompson, J.C. (1988). A comparison of the insulinotropic and insulin-inhibitory actions of gut peptides on newborn and adult rat islet cells. *Pancreas* 3, 77–82.

Konecki, D.S., Benedum, U.M., Gerdes, H.H. & Huttner, W.B. (1987). The primary structure of human chromogranin A and pancreastatin. *Journal of Biological Chemistry* 262, 17026–30.

Miyasaka, K., Funakoshi, A., Nakamura, R., Kitani, K., Shimizu, F. & Tatemoto, K. (1988). Pancreastatin inhibits postprandial pancreatic exocrine secretion but not endocrine functions in the conscious rat. *Gastroenterology* 94, A306.

Ravazzola, M., Efendic, S., Östenson, C.G., Tatemoto, K., Hutton, J.C. & Orci, L. (1988). Localization of pancreastatin immunoreactivity in porcine endocrine cells. *Endocrinology* 123, 227–9.

Schmidt, W.E., Siegel, E.G., Kratzin, H. & Creutzfeldt, W. (1988). Isolation and primary structure of human pancreastatin. *Gastroenterology* 94, A409.

Shimizu, F., Ikei, N., Iwanaga, T. & Fujita, T. (1987). An immunochemical and immunohistochemical study on pancreastatin-like immunoreactivity using synthetic peptides. *Biomedical Research* 8, 457–62.

Tatemoto, K., Efendic, S., Mutt, V., Makk, G., Feistner, G.J. & Barchas, J.D. (1986). Pancreastatin, a novel pancreatic peptide that inhibits insulin secretion. *Nature* 324, 476–8.

Tatemoto, K. & Mutt, V. (1978). Chemical determination of polypeptide hormones. *Proceedings of the National Academy of Sciences of the U.S.A.* 75, 4115–19.

Tatemoto, K. & Mutt, V. (1980). Isolation of two novel candidate hormones using a chemical method for finding naturally occurring polypeptides. *Nature* 285, 417–18.

Tatemoto, K., Nakano, I., Ishida, K., Makk, G., Angwin, P., Chang, D. & Funakoshi, A. (1988). Chromogranin A is the precursor of pancreastatin, a novel hormone that inhibits insulin secretion. *Biomedical Research* suppl. 1, 10.

Winkler, H., Apps, D.K. & Fischer-Colbrie, R. (1986). The molecular function of adrenal chromaffin granules: established facts and unresolved topics. *Neuroscience* 18, 261–90.

9

The processing of peptide precursors

THUE W. SCHWARTZ

9.1. Introduction

The islet cell hormones are synthesized initially as parts of larger precursors, the genes for which have been described in Chapters 1, 4, 5, and 6. The primary translation products contain signals for biochemical modifications, proteolysis and derivatization, and signals for proper cellular handling. Based on this, the peptide is excised from the precursor and modified to the biologically active form. At the same time, the precursor is targeted to vesicles in the regulated export route of the cell, from which the activated peptide is eventually released. Previously, most attention was directed towards the biochemical part of the peptide activation, mainly the proteolytic cleavage of the precursor. However, the biochemical processing is intimately connected to cell biological events, as the different steps in the precursor modification occur in specific compartments of the cell with distinct biochemical compositions. Thus, the precursor will only be properly processed and derivatized if it is translocated and further targeted correctly in the secretory pathway of the cell.

The whole field of peptide precursor processing was founded in 1967 when Steiner discovered that insulin was synthesized as part of a larger precursor (Steiner & Oyer, 1967; Steiner et al., 1967). He subsequently delineated a processing mechanism (Nolan et al., 1971; Steiner et al., 1974) which since has proved to be of truly general importance not only for the activation of peptide hormones but also for neuropeptides, paracrine peptide messengers, pheromones, growth factors, etc. Five to ten years ago the structure of the precursors for the islet hormones had been characterized by deduction from the cloned cDNA. At the same time metabolic labeling studies had clarified most of the processing patterns for the individual precursors. However, although the structure of the primary translation products for the islet hormones have been known for many years, the signal epitopes [1] are only partly understood. Likewise, only a few of the processing enzymes, which have been pursued for nearly two decades, have been characterized; and the key enzymes that perform the primary endoproteolytic attack on the precursor are still elusive.

In the present chapter, the structure and processing pattern of the precursors for the individual islet cell hormones are presented, followed by a combined description of the biochemical and cell biological events in the precursor processing. The complex composition of the secretory vesicle, the organelle in which the final activation of the peptide takes place, is briefly discussed. The last part of the chapter deals with the processing of islet cell precursors in heterologous systems, i.e. in transfected cell lines, yeast and prokaryotes.

9.2. The precursors for the peptide hormones of the islets

Many peptides have over the years been described in the islets of Langerhans, mainly by immunohistochemical techniques. However, only four peptides are today established as true hormones of the islets: insulin, glucagon, pancreatic polypeptide (PP), and somatostatin. Only the precursors of these four peptides are described and discussed in this chapter.

The peptide precursors of the islets fall into two groups: (i) the precursors for insulin and PP, which are rather simple in their composition and which are almost exclusively expressed in the pancreas; and (ii) the precursors for glucagon and somatostatin, which hold the sequences of more than one biologically active peptide, and which are synthesized in several places in the body besides the pancreas. These last precursors are particularly interesting since they display a high degree of tissue-specific processing, the basis for which is still undetermined.

9.2.1. The insulin precursor

The insulin precursor (Fig. 9.1) is well known since it is depicted in almost all biochemistry textbooks. The amino acid sequence of proinsulin[2] was identified by Chance and coworkers at Eli Lilly, in fact before Steiner published his work on the biosynthetic insulin precursor, but owing to potential commercial interest the structure was first published later on (Chance et al., 1968). The preproinsulin was partly characterized by cell-free translation of purified mRNA from B-cells (Yip et al., 1975;

---

[1] Signal epitopes is in this chapter used as a designation for three-dimensional structural elements of a precursor that ensures the correct cellular handling and processing of the protein.
[2] The whole primary translation product is called prepro(peptide). The designation pro(peptide) is in general used only for the peptide which results from the primary translation product when the signal peptide is removed.

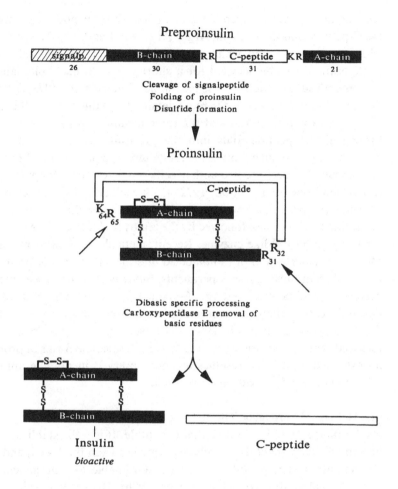

Fig. 9.1. Scheme for the organization and processing of the insulin precursor. The number of amino acid residues is indicated below the different peptide components in the preproinsulin. The single-letter code is used for the amino acid residues of the conversion sites, Arg^{31}–Arg^{32} (RR) and Lys^{64}–Arg^{65} (KR). In proinsulin, arrows point to the dibasic conversion sites.

Chan *et al.*, 1976) and the total sequence was later deduced from the cloned cDNA structure (Bell *et al.*, 1979).

(i) Precursor structure and processing pattern The two-chained insulin molecule is made from a single chained precursor consisting of an amino-terminal signal peptide followed by the insulin B-chain, the proinsulin C-peptide, and finally the insulin A-chain (Fig. 9.1). Pairs of basic residues are located between the B-chain and the C-peptide and between the C-peptide and the A-chain, Arg^{31}–Arg^{32} and Lys^{64}–Arg^{65} respectively. The intact preproinsulin, which presumably is found in a randomly coiled configuration, can be detected for a short period in metabolic labeling studies with suitable pulse–chase techniques (Patzelt *et al.*, 1978; Albert & Permutt, 1979). As described in more detail later in this chapter, the signal peptide is rapidly removed and the three disulfide bridges are correctly formed as the peptide folds into the proinsulin configuration in the cisternae of the rough endoplasmic reticulum. The processing of proinsulin to insulin has been studied both in cells and in test tubes for many years (for reviews, see Steiner *et al.*, 1972, 1974; Steiner, 1984). Basically, the conversion is performed by a trypsin-like cleavage on the C-terminal side of the dibasic sequences, followed by the removal of the basic residues by a carboxypeptidase B-like enzyme. Recent, meticulous experiments, using HPLC characterization of radiolabeled intermediate forms of proinsulin obtained from pulse–chase experiments, indicate that the processing site between the C-peptide and the A-chain of proinsulin is cleaved significantly earlier than the processing site between the B-chain and the C-peptide (Davidson *et al.*, 1988). The basic residues are in all cases removed rapidly and efficiently, as cleaved intermediate forms of proinsulin extended with basic residues are not observed in notable amounts (Given *et al.*, 1985; Davidson *et al.*, 1988).

(ii) **The peptide products** The result of the proinsulin processing is the insulin molecule and the proinsulin C-peptide. It is well established that insulin and the C-peptide are released together from the B-cell, and that determination of C-peptide secretion is a valuable tool in clinical and physiological studies (Faber & Binder, 1986). However, despite much effort, the C-peptide has not been assigned any convincing postsecretory function as a hormone (Steiner, 1978). The other parts of preproinsulin are involved in the cellular precursor processing: the signal peptide secures the translocation of the precursor, and the dibasic sequences serve as recognition sites for the processing enzymes. It has been envisioned that the C-peptide in the proinsulin molecule is also important in the biosynthetic process, e.g. in the assembly of its flanking peptides, the A- and B-chains (Steiner, 1978). Nevertheless, normal insulin can be produced, albeit in yeast, from a proinsulin where the whole 31 amino acid C-peptide is replaced by just two residues, Leu–Gln, still preserving the two pairs of basic residues on each side (Thim *et al.*, 1987). It was already known from the X-ray crystallographic structure of insulin that the C-peptide was much too long just to serve as bridge from the B- to the A-chain (Steiner,

1978). Recent experiments with expression of mutant proinsulins in higher eukaryotic cells show that a proinsulin in which the C-peptide is reduced to six amino acid residues is not secreted from these cells, but is presumably retained in the ER (K. Docherty, personal communication). Thus the C-peptide seems to be important in ensuring that the proinsulin is correctly transported within the eukaryotic cell (but apparently not in the yeast cell). Whether it serves as a sorting epitope or just covers up part of the insulin molecule which otherwise would react with retarding molecules in the ER (see §9.3.2.) remains to be determined.

Insulin is member of a family of homologous peptides, which includes insulin-like growth factor I (IGF-I), insulin-like growth factor II (IGF-II), and relaxin. All these peptides have B- and A-chain-like regions bound together by disulfide bridges in a similar arrangement to that of the insulin molecule (Blunder & Humbel, 1980). However, in IGF-I and -II no dibasic conversion sites are found to flank the 'C-peptide' region, which accordingly is an integral part of the secreted, biologically active peptides.

(iii) Mutant proinsulins ('natural site-directed mutagenesis') Since the description of the first insulin mutant, insulin Chicago (Tager *et al.*, 1979), Tager and coworkers have elegantly characterized a series of mutants in the human insulin gene, including mutants which interfere with the processing of proinsulin (for review see Tager, 1984; and Chapter 13). Based on material from blood samples only, and through protein chemical manipulations, they were able to point out that the mutation had occurred in the arginyl residue of the conversion site, Lys^{64}–Arg^{65}, in proinsulins Tokyo and Boston (Robbins *et al.*, 1981). This was confirmed by cloning and sequence determination of the gene (Shibasaki *et al.*, 1985). The mutation led to the secretion of a two-chained intermediate form of proinsulin in which the C-peptide was still attached to the N-terminus of the A-chain of the insulin molecule (Fig. 9.2). Thus, the dibasic-specific processing enzyme is truly dibasic-specific, since the mutant Lys^{64}–His^{65} sequence is not cleaved.

Another interesting, but not totally elucidated, mutant is proinsulin Providence, $[Asp^{B10}]$proinsulin (Gruppuso *et al.*, 1984; Chan *et al.*, 1987). In this case, the mutation is situated relatively remote from both of the processing sites; nevertheless, the mutation is associated with the occurrence of a single-chain proinsulin in the circulation (Fig. 9.2). It has been suggested that the inhibition of proinsulin conversion is related to altered folding or self-association of the mutated proinsulin molecule (Chan *et al.*, 1987). Studies in transgenic mice expressing high levels of $[Asp^{B10}]$proinsulin have shown that a substantial amount of the mutant proinsulin

(a) Human proinsulin Tokyo/Boston

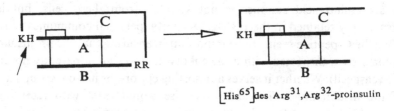

$\left[\mathrm{His}^{65}\right]$des Arg^{31},Arg^{32}-proinsulin

(b) Human proinsulin Providence

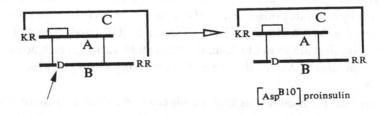

$\left[\mathrm{Asp}^{B10}\right]$proinsulin

(c) Ratfish insulin

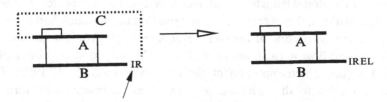

(d) Lamprey insulin

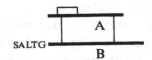

Fig. 9.2. Mutant human proinsulins and 'phylogenetic' insulins with
bearings on precursor processing. (a) Human proinsulin
Tokyo/Boston: exchange of Arg^{65} with histidine, H, prevents
processing at the dibasic site. (b) Human proinsulin Providence:
substitution of histidine B^{10} with aspartic acid, D, totally prevents

escapes the processing mechanism and is secreted constitutively from the islet cells through the non-regulated pathway (Carroll et al., 1988). Interestingly, [Asp^{B10}]insulin has been shown to be a superactive insulin analog (Schwartz et al., 1987a).

Phylogenetic investigation has revealed two insulin variants with bearings on precursor processing. The ratfish insulin is extended in the C-terminal end of the B-chain, presumably because Arg^{31}, normally occurring in proinsulins, has been substituted with an isoleucine residue (Fig. 9.2) (Conlon et al., 1986). Like the proinsulins Tokyo and Boston, the ratfish insulin underlines the fact that substitutions of basic residues in the conversion sites prevent cleavage of the precursors. In the ratfish insulin, an alternative processing occurs after a leucine residue within the C-peptide (Conlon et al., 1986). This processing is probably analogous to the chymotryptic cleavage described as part of the cellular processing of other C-peptides (Markussen & Sundby, 1972; Tager et al., 1973). The insulin of the lamprey is extended in its N-terminus with five residues which are homologous with the last five residues of the hagfish signal peptide (Plisetskaya et al., 1988; Chan et al., 1981). Apparently the signal peptidase chooses to cleave the lamprey precursor five residues upstream from the normal site. Whether this is due to the substitution of the alanine residue of the processing site with a glycine residue or to some more general change in the structure of the signal peptide remains to be shown. However, cleavage of the prolactin signal peptide was moved three residues downstream by substitution of the threonine of the signal peptide cleavage site by the metabolically active threonine analog, hydroxynorvaline (Hortin & Boime, 1981).

9.2.2. The pancreatic polypeptide precursor

The precursor for PP is, like the precursor for insulin, rather simple. Like proinsulin, proPP was already characterized in respect of structure and processing pattern (Schwartz et al., 1980; Schwartz & Tager, 1981) before the primary structure of the precursor was deduced from the cloned cDNA structure (Boel et al., 1984; Leiter et al., 1984; Takeuchi & Yamada, 1985; Yamamoto et al., 1986).

Caption to Fig. 9.2 (cont.)
proinsulin processing, through elusive means. (c) Ratfish insulin: substitution of Arg^{31} with isoleucine, I, prevents the processing at the dibasic site. The proinsulin is processed at an alternative site in the C-peptide, leading to the indicated C-terminally B-chain extended insulin. (d) Lamprey insulin: an N-terminally B-chain extended insulin, probably produced by alternative cleavage of the signal peptide.

(i) Precursor structure and processing pattern The long signal peptide, 29 residues, constitutes almost one third of the 95 residue, rather short primary translation product of the PP gene (Fig. 9.3). The PP sequence makes up the N-terminal half of proPP and is separated from the C-terminal peptide by a combined cleavage and amidation signal, Gly–Lys–Arg, of which the glycine residue serves as nitrogen donor for the formation of the amide group on the preceding PP sequence, as discussed later in this chapter. The C-terminal part of the precursor is in most mammals further processed at a monobasic cleavage site, Pro–Arg, giving rise to a stable, secreted product designated pancreatic icosapeptide (Schwartz & Tager, 1981; Schwartz et al., 1984). Thus proPP is subject to two different types of proteolytic attack, one monobasic-specific and one dibasic-specific. The kinetics in the processing of the PP precursor is known in great detail, owing to the fact that viable and metabolically active PP cells can be isolated in large numbers, especially from the duodenal part of the canine pancreas (Schwartz et al., 1980). In such primary cell cultures, pulse–chase experiments have demonstrated that the conversion takes place sequentially at the two sites. The dibasic processing follows a time course much like that of the proinsulin conversion, whereas the monobasic processing occurs later (Schwartz, 1987). In fact, there is no evidence indicating that proPP is cleaved at the monobasic site only. In the cell cultures the dibasic and monobasic processing can also be differentiated in another way, as the cells during the first days of culture gradually lose their ability to cleave at the monobasic conversion site whereas they retain full capacity to perform dibasic cleavage (Schwartz, 1987). A potential conversion site, Arg^{25}–Arg^{26}, in the middle of the PP structure, is not used for processing; in the same way a potential glycosylation site in the human PP precursor, Asn–Ala–Thr, is not glycosylated either (T. Schwartz, unpublished observations). Both of these apparent signal epitopes are situated in an α-helix in the PP molecule, which probably means that although they look like processing sites in the amino acid sequence, they are not in fact true signal epitopes in the folded structure of the precursor.

(ii) The peptide products PP is today a well-established hormone of the pancreas, which regulates the exocrine pancreatic secretion and the biliary motility, conceivably through interaction with receptors on nerve cells (Schwartz, 1983; Schwartz et al., 1987b). However, the function of the C-terminal part of the PP precursor is just as undetermined as that of the C-peptide of proinsulin. Like proinsulin C-peptide, the C-terminal part of

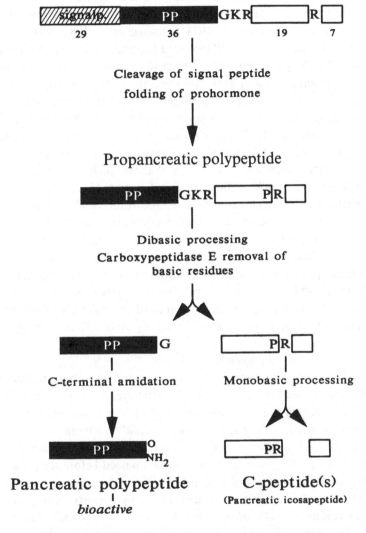

Fig. 9.3. Scheme for the organization and processing pattern for the pancreatic polypeptide (PP) precursor. In the prepropancreatic polypeptide the number of amino acid residues is indicated below the different peptide components. The single-letter code for amino acids is used for the conversion sites: Lys, K; Arg, R; Gly, G; Pro, P. The order of dibasic and monobasic processing indicated in the scheme is based on dynamic biosynthesis studies.

proPP does not seem to have any postsecretory physiological function (T. Schwartz, unpublished observations). The structures of both peptides are rather poorly conserved during evolution as compared with the PP and insulin structures (Steiner, 1978; Schwartz *et al.*, 1984; Yamamoto *et al.*, 1986). In the case of proPP the extra peptide is at least necessary in order to make the precursor long enough to function in the ribosomal–signal recognition particle system. However, the C-terminal peptide could very well, like the proinsulin-C peptide, serve a role in the targeting of the precursor, or rather in the prevention of e.g. the ER retention.

PP is member of a family of regulatory peptides, including neuropeptide Y (NPY) and peptide YY (PYY). These peptides are more homologous in their three-dimensional structure than in their amino acid sequence (Glover *et al.*, 1985). They are only between 45 and 70% homologous when all amino acids are considered; however, all the key amino acids of importance for the special folded configuration, the PP-fold, are conserved in all peptides. The precursors for the three peptides are also very similar (Boel *et al.*, 1984; Minth *et al.*, 1984; Leiter *et al.*, 1987) with the PP-like peptide occupying the N-terminal half of the propeptide separated from a C-terminal extension peptide by a combined cleavage and amidation signal. The C-terminal peptides of the NPY and PYY precursors have not yet been thoroughly tested for physiological functions.

9.2.3. The glucagon precursor

The first fragment of proglucagon was characterized in 1973 (Tager & Steiner, 1973) and its structure, with glucagon bearing a C-terminal extension starting with two basic residues, was one of the first confirmations of the hypothesis that peptide hormones are derived from precursors processed at dibasic sequences. Almost a decade of vigorous studies of glucagon-containing peptides passed before the publication of the next proglucagon fragment, the 69 amino acid glicentin molecule, which contains glucagon flanked by dibasic conversion sites and further extensions on both sides (Thim & Moody, 1982). However, two years before, metabolic pulse–chase labeling of pancreatic islets had demonstrated that the real proglucagon has a molecular mass of about 18 kDa and is processed to glucagon and a 10 kDa fragment with similar kinetics to the processing of proinsulin (Patzelt *et al.*, 1979). This was confirmed when the structure of the whole precursor was deduced from the cloned cDNA (Lund *et al.*, 1982; Bell *et al.*, 1983a, b; Lopez *et al.*, 1983; Heinrich *et al.*, 1984). Surprisingly, it was revealed that the large proglucagon is a multi-copy or perhaps rather a multi-function precursor, with a long

C-terminal domain containing one, or even two, extra glucagon-like peptides, depending on the species.

(i) Precursor structure and processing pattern The complex structure of the 180 amino acid primary translation product of the mammalian glucagon gene is depicted in Fig. 9.4. Besides the amino-terminal signal peptide, the precursor contains three glucagon-like sequences: glucagon, glucagon-like peptide I (GLP-I), and glucagon-like peptide II (GLP-II). These four peptides are separated by three intervening peptides, of which the nomenclature (for historical reasons) is: glucagon-related polypeptide (GRPP) (also called N-peptide), intervening peptide I (IVP-I) and intervening peptide II (IVP-II). These peptides are all joined by dibasic sequences. A dibasic sequence is also located within the glucagon molecule itself; however, this is apparently not used as a conversion site.

The complexity of the glucagon precursor is further underlined by the fact that it is synthesized in different tissues and here gives rise to different peptides. The structure of the glucagon gene opened possibilities for alternative RNA processing leading to the expression of different peptides in different cells (see Chapter 4). However, this possibility was ruled out since the mRNA turned out to be identical in the pancreas and intestine (Mojsov et al., 1986; Novak et al., 1987; Holst et al., 1987a). Before the precursor structure was known, similarities and differences among glucagon-like peptides in the pancreas and the gut had already been described through fastidious immunological studies performed by many different groups (for review see Holst, 1983). After it was realized that these more or less well-characterized peptides arose through tissue-specific processing of a single glucagon precursor, this phenomenon was further studied with new and more specific tools. Both metabolic labeling techniques and immunological methods based on antisera raised against synthetic peptides designed from different parts of the precursor structure have been used. A more of less clear picture has been revealed, which will be described below. There are many, although minor, discrepancies among the published results; these probably reflect differences in the specificity of antisera and thereby the accuracy with which the different assays quantitate the many possible cleavage products. Furthermore, species differences should also be taken into consideration; rat, pig and man are studied by the different groups.

In the pancreas Pulse–chase experiments performed by the group in Chicago (Patzelt et al., 1979) demonstrated a processing pattern which,

Preproglueagon

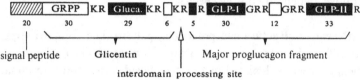

signal peptide   Glicentin

interdomain processing site

Major proglucagon fragment

(a)   pancreas

proglucagon

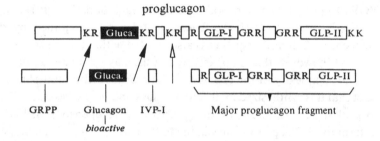

GRPP   Glucagon   IVP-I
bioactive

Major proglucagon fragment

(b)   intestine

proglucagon

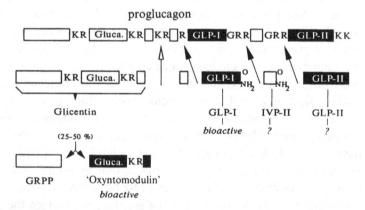

Fig. 9.4. Scheme for the organization and tissue-specific processing of the glucagon precursor. The number of amino acid residues is indicated below the different peptide components of preproglucagon. As discussed in the text, the glucagon precursor is considered to be a two-domain precursor composed of an N-terminal glicentin domain and a C-terminal 'major proglucagon fragment' separated by an 'interdomain processing site', Lys^{70}–Arg^{71} (KR), indicated by the open arrow; this processing site is apparently cleaved in all tissues. The

applied to the true proglucagon sequence, is indicated in Fig. 9.4. The glucagon precursor is apparently cleaved exclusively at the Lys–Arg sequences in the islets, thereby giving rise to glucagon and three possible inactive fragments: GRPP, IVP-I, and the whole, large C-terminal domain, called MPGF (major proglucagon fragment). Patzelt and coworkers have further substantiated this through characterization of the large C-terminal domain by radiosequencing and by determination of its amino acid composition (Patzelt & Schug, 1981; Patzelt & Schlitz, 1984). Small amounts of glicentin, the N-terminal domain of proglucagon, are present in the pancreas (Sheikh *et al.*, 1985; Mojsov *et al.*, 1986) but in this tissue probably only serves as a biosynthetic intermediate in the formation of glucagon and GRPP, which are released together from the isolated perfused pancreas (Moody *et al.*, 1981). Even the C-terminal hexapeptide of glicentin, IVP-I, has been isolated from the pancreas, and its co-release with glucagon has also been described (Yanaihara *et al.*, 1985a, b). The *O*-glycosylation of proglucagon in the pancreas is described later in the chapter.

In the intestine There are no metabolic labeling studies on proglucagon available from extrapancreatic tissues, so the description of the processing pattern in the intestine has to rely on static, mainly immunological data. However, these are consistent with a scheme in which the precursor, as in the pancreas, is cleaved initially in the middle (Fig. 9.4), but in the intestine it is mainly the C-terminal domain which is further processed. In contrast to the pancreas, conversion takes place at all the Arg–Arg sequences and even at a previously unnoticed single arginine residue within the GLP-I sequence. In tissue extracts, antisera specific for the different peptides have shown a more or less complete conversion of the C-terminal proglucagon domain to the following peptides: GLP-I, mainly in the form

Caption for Fig. 9.4 (*cont.*)

> peptide components shown in black are the potentially bioactive peptides. (*a*) Processing in the pancreas: the N-terminal glicentin domain is further processed to the bioactive glucagon, plus glucagon-related polypeptide (GRPP) and intervening peptide I (IVP-I). (*b*) Processing in the intestine: the major proglucagon fragment is further processed to the bioactive glucagon-like peptide I (GLP-I), intervening peptide II (IVP-II), and glucagon-like peptide II (GLP-II). In GLP-I the black symbol only includes the bioactive, real GLP-I, which corresponds to the 7–36 fragment of the originally suggested GLP-I sequence (see text). Some further processing of the glicentin domain to oxyntomodulin and GRPP is indicated.

of truncated GLP-I, i.e. GLP-I$^{7-36}$ amide; GLP-II; and finally, an amidated form of IVP-II (Ørskov et al., 1986; Mojsov et al., 1986; Holst et al., 1987b; Buhl et al., 1988). Approximately 25–50% of the glicentin molecules are further processed (or perhaps glicentin is further processed in 25–50% of the cells?) to GRPP and the C-terminally extended form of glucagon which was originally described by Tager & Steiner (1973). However, in the porcine intestine this molecule, which affects gastric acid secretion, is called 'oxyntomodulin' (Bataille et al., 1982).

Proglucagon is in this review regarded as a two-domain precursor composed of an N-terminal glicentin domain and a C-terminal MPGF domain. Both in the pancreatic islets and in the endocrine cells of the intestine the two domains are disconnected by cleavage at Lys^{70}–Arg^{71}, which could be termed the 'interdomain conversion site' (Fig. 9.4). It is suggested that this interdomain conversion site is particularly susceptible to endoproteolytic attack, owing to a special three-dimensional configuration in the intact proglucagon molecule; and/or that a general, tissue-unspecific protease cleaves this site, and this site only. The separated domains are then treated differently in different cells. In the pancreas, only the N-terminal glicentin domain, with its Lys–Arg conversion sites, is further processed; whereas in the intestine, it is mainly the C-terminal domain, with its Arg–Arg sites and even a single arginyl conversion site, which is further processed. The simplest explanation for the different processing patterns would be tissue-specific expression of different conversion enzymes, indicated by the difference in the type of dibasic pairs. This is an attractive but still very speculative explanation.

(ii) **The peptide products** Through the tissue-specific processing of proglucagon, described above, different peptide structures are activated in the pancreas and in the gut (Fig. 9.4). From the mammalian pancreas, glucagon appears to be the only bioactive peptide liberated from proglucagon. In the intestine, however, several peptides with biological potentials are activated and secreted, one of which acts on the islets.

> GLP-I: Based on the fact that, for example, fish glucagon genes do not code for the N-terminal six amino acids found in mammalian GLP-Is (Lund et al., 1982; Conlon et al., 1985, 1987) it has been argued that the real GLP-I is the truncated form, GLP-I$^{7-36}$ (Schwartz, 1986). It is also the GLP-I$^{7-36}$ which is the truly active peptide as a potent insulinotropic hormone (Schmidt et al., 1985b; Holst et al., 1987b; Mojsov et al., 1987). Recently, specific

receptors for this peptide have been described on insulinoma cells in culture (Ørskov & Nielsen, 1988; Göke & Conlon, 1988). It is therefore likely that the real GLP-I, i.e. the truncated form of GLP-I in mammals, acts as an 'incretin', i.e. an insulin-stimulating hormone from the intestine. In fish the picture is different: cleavage at the interdomain conversion site described above, directly activates the GLP-I peptide (because there is no secluding N-terminal extension and no IVP-II and GLP-II); accordingly, GLP-I is found in very high amounts in the fish pancreas (Plisetskaya et al., 1986). Interestingly, in fish, as opposed to mammals, GLP-I is a potent glyconeogenetic hormone (Mommsen et al., 1987).

GLP-II: Because of its homology with glucagon, and thereby other biologically active peptides, GLP-II is currently under vigorous investigation for possible physiological function.

IVP-II: The small IVP-II is also being tested for different activities, mainly owing to the fact that it appears to be C-terminally amidated (Drucker et al., 1986; Mojsov et al., 1986; Buhl et al., 1988), a modification which has been believed only to occur on biologically active peptides (Tatemoto & Mutt, 1978).

GRPP: This peptide was discovered at a pharmaceutical company several years ago (Thim & Moody, 1982) and has accordingly been tested thoroughly for biological actions, but without success. It is therefore likely that GRPP serves a purpose in the structure and function of the proglucagon molecule in analogy with the C-peptides of proinsulin and proPP.

Oxyntomodulin: This C-terminally extended form of glucagon has, like glucagon itself, an inhibitory effect on the secretion of gastric acid (Bataille et al., 1982).

9.2.4. The somatostatin precursor

The precursor for the 14 amino acid somatostatin, a peptide with widespread and interesting physiological actions, was studied eagerly in 1979–80 (Noe et al., 1979; Patzelt et al., 1979, 1980; Shields, 1980; Lauber et al., 1979). The isolation and chemical characterization of a mammalian, N-terminally extended form of somatostatin, somatostatin-28 (Pradayrol et al., 1980; Schally et al., 1980; Esch et al., 1980) was published at almost the same time as the structure of the whole fish precursor was deduced from the cloned somatostatin cDNA (Goodman et al., 1980a; Hobart et al., 1980). These last studies took advantage of the fact that the endocrine pancreas in certain fish is found concentrated in separate glands, so-called Brockmann bodies, which have proved to be an enriched and often very fruitful starting material both for (for example) cell-free translation studies (Shields, 1980; Goodman *et al.*, 1980*c*; Warren & Shields, 1984*b*) and for cloning studies. The first mammalian somatostatin precursors were cloned a few years later (Goodman *et al.*, 1982, 1983; Shen *et al.*, 1982; Funckes *et al.*, 1983; Shen & Rutter, 1984; Tavianini *et al.*, 1984).

(i) Precursor structure and processing pattern Somatostatin-14, and therefore also somatostatin-28, are located at the far C-terminal end of the 92 amino acid prosomatostatin (Fig. 9.5). In all systems studied the somatostatin precursor gives rise to either somatostatin-14 or somatostatin-28 through a presumably relatively rapid and direct cleavage of the precursor in front of either of these two peptides, both of which are biologically active. Besides the dibasic processing site in the middle of somatostatin-28, i.e. just preceding somatostatin-14, there are no other pairs of basic residues within the mammalian precursor. The conversion site between somatostatin-28 and the N-terminal part of the precursor is a monobasic one. Unfortunately, metabolic labeling studies in mammals have not been as detailed as those for insulin and PP, and there is no real information available on the kinetics and order of the different cleavages in the pancreas. The somatostatin gene is expressed in several tissues of the body (see Chapter 5), and the precursor is, like the glucagon precursor, processed in a tissue- or rather cell-specific manner.

In the mammalian islets The majority of biologically active somatostatin produced and released from the pancreas is somatostatin-14 (Arimura *et al.*, 1975; Patel & Reichlin, 1978; Newgard & Holst, 1981; Trent & Weir, 1981). By radioimmunoassays developed against the other parts of the somatostatin precursor, it has been shown that a variable degree of proteolysis takes place conceivably at the different basic residues (Benoit *et al.*, 1982; Holst *et al.*, 1988). The monobasic conversion site between somatostatin-28 and the N-terminal domain of the precursor seems to be partly cleaved. From the isolated perfused porcine pancreas somatostatin-14 is released, together with the 1–12 fragment of somatostatin-28 and a couple of large N-terminal precursor fragments with and without the 1–12 sequence of somatostatin-28 attached (Holst *et al.*, 1988; Patel & O'Neil, 1988).

In the intestine Somatostatin-28 is the major end-product of the precursor processing (Patel *et al.*, 1981; Trent & Weir, 1981; Baskin & Ensinck, 1984;

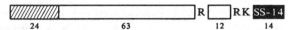

preprosomatostatin

(a) pancreas

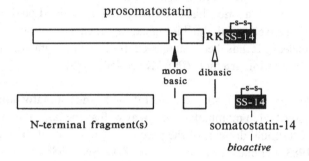

prosomatostatin

N-terminal fragment(s)

somatostatin-14

(b) intestine

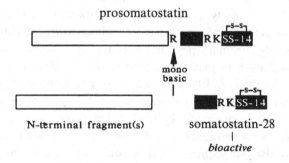

prosomatostatin

N-terminal fragment(s)

somatostatin-28

Fig. 9.5. Scheme for the organization and tissue-specific processing of the somatostatin precursor. The number of amino acid residues is indicated below the different components of preprosomatostatin. (a) Processing in the pancreas: somatostatin-14 is excised probably directly from the prohormone at a dibasic conversion site, and the large N-terminal precursor fragment is cleaved at a monobasic processing site (in front of the 1–12 fragment of somatostatin-28). (b) Processing in the intestine: somatostatin-28 is excised by cleavage at the monobasic conversion site. In the pancreas, and possibly also in the intestine, there is some degree of partial cleavage of the N-terminal precursor fragment at single arginyl residues (not indicated).

Baldissera *et al.*, 1985*a*; Patel & O'Neil, 1988); this observation indicates that in the gut there is no cleavage at the only dibasic conversion site of the precursor, but conceivably full cleavage at the monobasic site. In accordance with this, the majority of somatostatin released from isolated perfused preparations of porcine intestines is in the form of somatostatin-28 (Baldissera *et al.*, 1985*b*). Two N-terminal fragments of prosomatostatin, 1–32 and 1–10, have been found in intestinal extracts (Schmidt *et al.*, 1985*a*; Benoit *et al.*, 1987). These fragments would be excised from the precursor at non-basic cleavage sites, after Leu–Leu and Gln–Phe respectively, and the biological significance of these fragments is still unclear. It has been suggested that such fragments are extraction or degradation artefacts (Patel & O'Neil, 1988).

In the brain Both somatostatin-14 and somatostatin-28 have been described in mammalian brain, as well as the large major fragments of the N-terminal domain of the precursor (Benoit *et al.*, 1982, 1984; Wu *et al.*, 1983; Penman *et al.*, 1983; Patel & O'Neil, 1988).

In fish islets The anglerfish islets, i.e. Brockmann bodies, have been used by several groups for studies on the somatostatin biosynthesis (Noe *et al.*, 1979; Noe, 1981; Noe & Spiess, 1983; Goodman *et al.*, 1980*a*, *c*; Shields, 1980). Recently, Andrews and coworkers have elegantly demonstrated how detailed information can be obtained by the combined use of DNA cloning, ordinary peptide chemistry, and peptide analysis by fast atom bombardment mass spectrometry (FAB MS) (Andrews & Dixon, 1987; Andrews *et al.*, 1987). FAB MS can give an accurate mass of the isolated peptide, and thereby demonstrate subtle posttranslational modifications which are not obvious in the cDNA structure, and changes which often are lost in conventional peptide chemical methods. Furthermore, FAB MS can also give sequence information. With these techniques, the group from Perdue were able to show that in anglerfish the two different somatostatin precursors, derived from two different genes (see Chapter 5), are processed differently. Preprosomatostatin I is fully processed at the dibasic conversion site preceding somatostatin-14, and at least partly processed at the monobasic site in front of somatostatin-28 and at another monobasic site in the middle of the N-terminal domain of the precursor (Andrews & Dixon, 1987). In preprosomatostatin-II the dibasic conversion site in front of somatostatin-14 is *not* used, whereas Lys[120] (i.e. Lys[9] of somatostatin-14) is modified to a hydroxylysine residue. The precursor is processed at the monobasic conversion site in front of somatostatin-28 and, somewhat surprisingly, at a dibasic site found in the middle of this anglerfish precursor (Andrews *et al.*, 1987). In both precursors the signal peptide cleavage site could also be determined in agreement with previous labeling studies (Spiess & Noe, 1985; Noe *et al.*, 1986). Although there are methods for predicting the cleavage sites for signal peptides fairly precisely (Von Heijne, 1983) it is still necessary that this be actually determined by protein chemical means. In all the characterized peptides the C-terminal basic residues, remnants of the endoproteolytic processing sites, had been removed efficiently, and it was furthermore shown that an N-terminal glutamine residue had been converted to a pyroglutamic residue (Andrews & Dixon, 1987; Andrews *et al.*, 1987). Similar peptides were also described in metabolic labeling studies using intact anglerfish Brockmann bodies (Noe *et al.*, 1986). The reason for the detailed description of these studies in fish islets is that they represent the state of the art in qualitative characterization of peptide precursor processing on a static basis. In order to get a full picture of the conversion process, such exquisite studies should probably be combined with dynamic labeling studies or more quantitative studies.

(ii) The peptide products Only somatostatin-14 and somatostatin-28 have convincingly been assigned physiological effects. Although somatostatin can influence the secretion of the other islet hormones, it is likely that it plays a role as a true hormone released from the pancreas (Kawai *et al.*, 1982). Most of the somatostatin in the gut probably acts as a paracrine regulator of intestinal functions, as it is released from cellular processes of specialized neuroparacrine cells (Larsson *et al.*, 1979). Nevertheless, somatostatin-28 is a major molecular form in the circulation, and it is the one which increases after a meal (Baldissera *et al.*, 1983). The other fragments of the somatostatin precursor have not yet been shown to have any physiological effects.

9.3. Cell biology and biochemistry of peptide precursor processing

'Precursor processing' used to be almost synonymous with 'dibasic conversion'. However, as shown in Table 9.1, the four precursors of the pancreatic islets are subject to many different posttranslational modifications, etc. before the final, active peptide is ready to be secreted. In this section the hormone precursor is followed from its synthesis, through cellular compartments and biochemical modifications, until it reaches the mature secretory granule in the form of an activated peptide

Table 9.1. *Posttranslational modifications of precursors for islet cell hormones*

	Insulin	Pancreatic polypeptide	Glucagon	Somatostatin
Signal peptide cleavage	X	X	X	X
Disulfide formation	X			X
O-glycosylation			X	X[a]
Dibasic processing	X	X	X	X
Monobasic processing	X[b]	X	X[c]	X
Removal of basic residues	X	X	X	X
C-terminal amidation		X	X[c]	

[a] In the catfish.
[b] In the dog.
[c] In the intestine.

hormone. Some of these events (§9.3.1–3.4) are general cellular processes whereas others (§9.3.5–3.8) are more or less unique for neuroendocrine cells.

9.3.1. Translocation and removal of the signal peptide

The amino-terminal end of all peptide precursors constitutes the so-called 'signal peptide', which ensures that the peptide is targeted to the cisternae of the endoplasmic reticulum. This means that the precursor is brought from the cytosol across the ER membrane, which in this respect is equivalent to the plasma membrane. Because all further transport takes place within vesicles or cisternae, the peptide can be considered from now on to be on the 'outside' of the cell, albeit within a membrane-bound compartment. The translocation process, with binding of the signal peptide to the signal recognition particle etc. is a general and well-characterized cellular process (Blobel, 1980). It should be noted that the signal peptide is an interesting 'signal epitope' in the sense that it is in fact the special three-dimensional configuration, an overall hydrophobic α-helical structure in the middle portion of the peptide, which is important for its recognition (Von Heijne, 1985). Thus, different signal peptides are homologous and can substitute for each other, mainly in the way that they can form similar chemical configuration, although they are not particularly homologous in their simple amino acid sequence. After the signal peptide has served its purpose it is removed by an enzyme, the signal peptidase, which is a membrane-bound, stationary ER protease. The signal peptidase

cleaves the signal peptide from the rest of the precursor after an amino
acid with a small side chain which is situated in a certain distance from the
hydrophobic core of the signal peptide (Von Heijne, 1983).

9.3.2. Peptide folding and formation of disulfide bridges

As the nascent peptide precursor starts to fold up, the disulfide
bridges of (for example) preproinsulin and preprosomatostatin start to
form. The formation of disulfide bridges is an enzyme-catalyzed process
performed by a resident protein of the ER lumen, the protein disulfide
isomerase (PDI) (for review see Freedman, 1984). PDI is an enzyme which
catalyses the formation and breakage of disulfide bonds in proteins, i.e. it
catalyses thiol–disulfide interchange. This interchange helps thiol-
containing proteins out of configurations containing incorrect disulfide
bonds, configurations which are often open and energetically unfavorable. The interchange continues until the final, most stable conformation,
which is kinetically accessible, is reached. The concept of cellular protein
folding being a relatively passive function of the proteins has been
challenged by the discovery and characterization of the heat shock
proteins, especially the 'glucose regulated protein-78', which is the major
resident, luminal protein of the ER (Pelham, 1986). This protein, which is
identical to the heavy-chain binding protein (BiP) of lymphoid cells (Bole
et al., 1986; Munro & Pelham, 1986), binds incorrectly folded proteins and
protein subunits until they have folded up correctly or have found a
corresponding subunit. For example, unglycosylated and incorrectly disulfide-bonded prolactin binds to this protein in the ER (Kassenbrock et
al., 1988). The nascent unfolded prolactin precursor is apparently not
bound to BiP, at least not in the very high-affinity associated state. The
release of proteins from BiP is energy-dependent and the function of BiP
in catalyzing protein folding is thus even more interesting. It could be
suggested that the function of some of the C-peptides or intervening
peptides of the hormone precursors, in fact, is to cover up certain epitopes
of the hormone which otherwise would bind to BiP and thus, by masking
these epitopes, the C-peptides ensure that the precursor is allowed to leave
the ER.

9.3.3. O-glycosylation

Shortly after its synthesis and the removal of the signal peptide,
proglucagon increases in apparent molecular mass (Patzelt et al., 1979).
Such a change is usually due to glycosylation at asparagine residues found
in a consensus sequence, a signal epitope, for so-called N-linked glycosy-

lation, Asn–X–Thr/Ser. However, no such sequence is found in progluca-gon. Recently, Patzelt & Weber (1986) showed that proglucagon is in fact glycosylated, but that the sugars are attached through a serine residue in the large C-terminal domain of the precursor. This O-linked glycosy-lation, with only relatively few sugar residues, happens very rapidly and may even occur cotranslationally (Patzelt & Weber, 1986). The signal epitope which determines the O-linked glycosylation of that particular serine residue is not known, just as the function of the O-linked sugar is not clear. A similar glycosylation has previously only been demonstrated in a particular subset of catfish somatostatin (Andrews et al., 1984).

9.3.4. Transport from the ER to the trans-Golgi network

From the ER the prohormone is transported to the cis-Golgi cisternae and further through the Golgi stack to the trans-Golgi network, where the major sorting of proteins to (for example) secretory granules takes place (Griffiths & Simons, 1986). Most of this transport occurs in small vesicles that bud off from one compartment and fuse with the following compartment. Recent studies indicate that the driving force for protein transport from the ER to the trans-Golgi network is mainly a rapid bulk flow, and that no particular signal epitope is needed for forward transport (Rothman, 1987). Resident proteins of the lumen of the ER, like the protein disulfide isomerase and the BiP protein discussed above, are instead held back by the action of a special retention signal epitope. This epitope, a C-terminal tetrapeptide sequence Lys–Asp–Glu–Leu (KDEL), is found on all luminal ER proteins and can make non-ER proteins stay in the ER if it is attached to them by protein-engineering techniques (Munro & Pelham, 1987). The transport of the prohormone from the ER and through the Golgi is energy-dependent, probably owing to the energy requirement of the vesicular budding and fusing process.

9.3.5. Sorting to secretory vesicles

In the trans-Golgi network, which is the last, tubular cisternae of the Golgi compartment (Griffiths & Simons, 1986), the prohormone is sorted to the secretory granules and thus separated from lysosomal proteins and from constitutively secreted proteins. The dense secretory granules are the export vesicles of the so-called regulated secretory pathway of the cell (Kelly, 1985; Orci et al., 1987a). The sorting of the prohormone is, both in time and place within the cell, tightly connected to the other main event, the endoproteolytic processing. This has been demonstrated in an impressive series of already classic papers by Orci and coworkers (Orci *et al.*, 1984*a*, *b*, 1985*a*, *b*, 1987*a–c*). The connection is so close that it could be suggested that the mechanisms behind the sorting and the major processing are closely related or interconnected, if not one and the same.

(i) Evidence for active sorting of prohormone In the mature, dense-core secretory granule, the secretory proteins are 200-fold concentrated compared with the ER and the Golgi area. This fact has been taken as an indication that prohormones are actively sorted and thereby concentrated in these vesicles (Kelly, 1985; Rothman, 1987). There is a series of experimental evidence in favor of active sorting for secretory vesicles (Kelly, 1985; Moore, 1986). In neuroendocrine cells, which have both constitutive and regulated secretory pathways, transfection studies with vectors coding for hormone or neuropeptide precursors (see below), have shown that the prohormone, e.g. proinsulin, is sorted correctly to the secretory granules (Moore *et al.*, 1983*b*; Hellerman *et al.*, 1984; Moore & Kelly, 1985; Schweitzer & Kelly, 1985). Even trypsinogen, a secretory granule protein of the exocrine pancreatic cell, is sorted to secretory granules of the regulated pathway in transfected endocrine cells (Burgess *et al.*, 1985). However, not all secretory proteins introduced into endocrine cells by transfection are routed to the secretory granules. When endocrine cells are transfected with a vector that codes for a truncated transmembrane protein (a viral protein from which the transmembrane part has been deleted), this protein is sorted to the vesicles of the constitutive pathway (Moore & Kelly, 1985). However, when the truncated viral protein, through molecular biology manipulations, is joined to the structure of human growth hormone, the hybrid protein is sorted to the secretory granules of the regulated pathway (Moore & Kelly, 1986). Apparently, a signal epitope of the growth hormone 'drags' the truncated viral protein to the regulated pathway. All these studies indicate that prohormones display some kind of sorting epitope that governs their routing to secretory granules. However this hypothetical sorting epitope is still very elusive. There is no apparent sequence homology among all known prohormones when longer stretches of amino acids are considered. All prohormones do have dibasic sequences, although some of them are not cleaved, as for example in growth hormone. However, there is no dibasic sequence in (for example) trypsinogen (MacDonald *et al.*, 1982), which in endocrine cells is sorted in parallel with the prohormones (Burgess *et al.*, 1985). If sorting to the regulated pathway is indeed an active process, as the evidence today indicates, then the sorting epitope

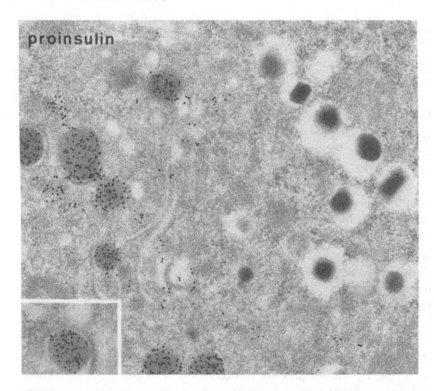

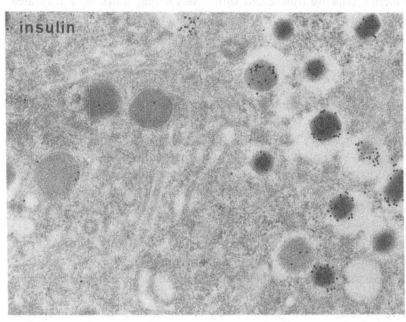

should probably be looked for in the three-dimensional structures of the prohormones, which unfortunately are still not known. However, the three-dimensional structure of many exocrine secretory proteins has been determined.

(ii) Clathrin-coated precursor vesicles The *trans*-Golgi network is the place where the packaging of prohormones takes place, as demonstrated for proinsulin (Orci *et al.*, 1987*a*). In the last cisternae of the Golgi complex, proinsulin accumulates in buddings which are coated with clathrin (Fig. 9.6) in analogy with the coated pits on the surface of the cell in which receptor-mediated endocytosis takes place (Orci *et al.*, 1984*a, b*, 1985*a*). This homology obviously suggests that some kind of receptor-mediated sorting also takes place in relation to the formation of secretory vesicles. By using monoclonal antisera that are totally specific for proinsulin and insulin, respectively, Orci and coworkers have demonstrated that proinsulin immunoreactivity is found only in the clathrin-coated buddings of the *trans*-Golgi network and in the newly formed, immature vesicles which still have a patchy clathrin coat (Orci *et al.*, 1985*b*, 1987*b*). In the clathrin-coated immature granules, proinsulin immunoreactivity disappears and insulin immunoreactivity appears, indicating that this is the cellular compartment where the conversion of proinsulin to insulin takes place (Fig. 9.6) (Orci *et al.*, 1985*b*, 1987*c*; Steiner *et al.*, 1987). This phenomenon has also been described, although not in quite as much detail, in the glucagon cells of the islets (Ravazzola *et al.*, 1983, 1984). The newly formed secretory vesicles lose their clathrin coat rapidly, just as the endocytotic vesicles do (Orci *et al.*, 1985*a*). When islet cells are treated with the carboxylic ionophore monensin, proinsulin conversion is prevented and proinsulin immunoreactivity accumulates in the clathrin-coated part of the Golgi area and in clathrin-coated vesicles (Orci *et al.*, 1984*a*).

Caption to Fig. 9.6

Fig. 9.6. Electron micrographic demonstration of the cellular compartment of proinsulin processing. Consecutive thin serial sections of a B-cell, immunostained with proinsulin- or insulin-specific monoclonal antibodies. The antibodies are detected by the protein A colloidal gold method. The proinsulin-specific antibody is directed against the conversion site of proinsulin, including the B–C junction, Arg^{31}–Arg^{32}. Note the reverse staining pattern of proinsulin-rich clathrin-coated granules (immature secretory granules) and insulin-rich non-coated granules (mature secretory granules). The inset details the clathrin coat on a proinsulin-rich granule (from Orci *et al.*, 1987*b*, by courtesy of Professor Lelio Orci).

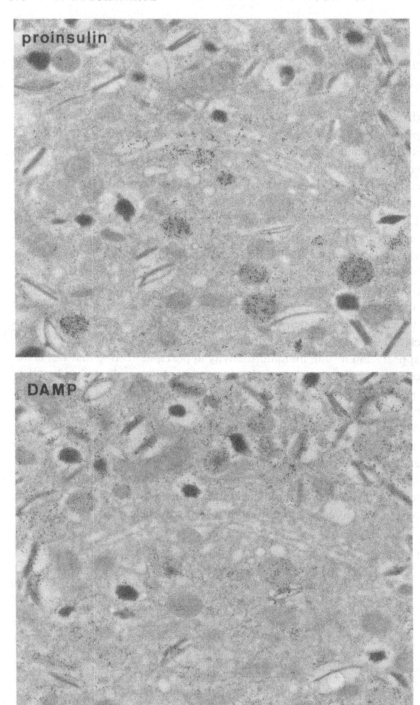

This further underlines the importance of the clathrin-coated immature secretory granule in the processing of prohormones.

(iii) Acidification and processing The interior of mature secretory granules is acidic in all endocrine cells studied, including islet cells (Mellman *et al.*, 1986). Recently it was demonstrated, by immunohistochemistry and the use of an antibody directed against a probe which accumulates in cellular organelles with low pH, that the clathrin-coated immature secretory granule also has a low pH (Orci *et al.*, 1986, 1987c). Insulin, i.e. the final conversion product, was only observed in vesicles that were clearly acidic (Fig. 9.7) and below a certain threshold of acidity; a direct relationship was observed between acidification and the amount of mature insulin in the granules (Orci *et al.*, 1987c). Diffusible bases like chloroquine, which prevents the intravesicular fall in pH, have been reported also to prevent sorting by making the cells secrete unprocessed prohormones through the non-regulated or constitutive export route (Moore *et al.*, 1983a). Acidification of immature granules in the *trans*-Golgi network seems to be characteristic for vesicles of the regulated pathway irrespectively of the pH of the mature granule. In exocrine cells the mature granules are near neutral; but, importantly, the immature or condensing granules are acidic (Orci *et al.*, 1987b). There are a number of possible explanations to the mechanism behind the intravesicular fall in pH, including the activation of a proton pump (Mellman *et al.*, 1986).

In conclusion, there is in the islet cells, as well as in other endocrine cells, a close association between: (i) formation of clathrin-coated, immature secretory vesicles; (ii) sorting of prohormones; (iii) acidification of the vesicles; and (iv) processing of the prohormone. However, the precise interrelationship between the different events is not totally clear.

Caption to Fig. 9.7

Fig. 9.7. Acidification of secretory granules demonstrated by immunohistochemistry. Consecutive thin serial sections immunostained with proinsulin- or dinitrophenol-monoclonal antibodies. The dinitrophenol antibodies bind to DAMP, a diffusible base, which accumulates in acid compartments. The proinsulin-rich, clathrin-coated granules have a low to moderate degree of DAMP staining, whereas the proinsulin-poor, non-coated granules (some with a distinct crystalloid core) show moderate to high labeling with DAMP antibodies, indicating a more acidic pH in these granules (from Orci *et al.*, 1986, by courtesy of Professor Lelio Orci).

9.3.6. Endoproteolytic cleavage of the prohormone

The prohormones are processed by limited proteolysis at basic residues. As discussed above, morphological evidence indicates that the endoproteolytic conversion occurs in the clathrin-coated immature secretory granules forming in the *trans*-Golgi network. Today, twenty years after Steiner's discovery of proinsulin, the proteolytic mechanism behind the prohormone processing is still not characterized in higher eukaryotes. It is not even known for certain whether one or more enzyme(s) are involved. The classical conversion site is a dibasic sequence (Tager *et al.*, 1980; Docherty & Steiner, 1982); however, within recent years it has become clear that processing at single basic residues is also important, e.g. in three of the islet cell hormones (Table 9.1) (Schwartz, 1986). Although the actual enzymatic mechanisms may be related, the disbasic- and monobasic-specific conversion processes are described separately, because the limited knowledge available today indicates that they are distinct mechanisms.

(i) **Dibasic-specific processing** All combinations of basic residues are used in dibasic conversion sites, although there is a preference for Lys–Arg sequences (Schwartz *et al.*, 1983). There is, however, no consensus sequence of amino acids surrounding the pairs of basic residues which are used as processing sites, and which distinguish them from pairs which are not used. It has been suggested that processing sites in the precursors are situated in turns in between segments of α-helical and β-sheet structure (Geisow, 1978; Rholam *et al.*, 1986). At least it is likely that dibasic pairs that are not used as conversion sites are restricted in their movement within the precursor structure, as illustrated by the non-cleaved pair of basic residues situated in an α-helix in the PP precursor (see above).

Many different, more or less well-characterized enzymes have over the years been suggested as being *the* prohormone-converting enzyme. However, only in yeast has a dibasic-specific conversion enzyme been firmly established. Through the aid of a naturally occurring mutant strain in which the α-mating factor and the killer toxin were not processed normally from their respective precursors, a structural gene for a dibasic-specific enzyme, the *kex*2 enzyme, was cloned (Julius *et al.*, 1984). This enzyme is a membrane-bound, Ca^{2+}-dependent endoprotease with a neutral pH optimum and a specificity for dibasic sequences, especially Lys–Arg (Julius *et al.*, 1984). However, it has not been possible (by nucleotide hybridization techniques, etc.) to find any homologous counterpart to this enzyme in higher eukaryotes.

In extracts from islets and from other endocrine and neuronal tissues, a series of enzymes have been described which have some specificity for basic residues and thus could be candidates for the elusive converting enzyme(s) (Yip, 1971; Ole-Moi Yoi et al., 1979; Fletcher et al., 1981; Docherty et al., 1982, 1983, 1984; Lindberg et al., 1984; Loh et al., 1985; Gomez et al., 1985; Parish et al., 1986; Cromlish et al., 1986a, b; Davidson et al., 1987, 1988; Mackin & Noe, 1987b). Furthermore, enzymes like tissue plasminogen activator and members of the large kallikrein family have also been suggested (Virji et al., 1980; Mason et al., 1983). The enzymes cover nearly the whole spectrum of known types of enzyme, and there is no pattern suggesting that the different groups are, or have been, on the same lead. Most of the studies have in fact led back to well-known (for example, lysosomal) enzymes, or to nowhere. Using conventional techniques, the task of isolating and purifying conversion enzymes has turned out to be almost insurmountably great. The true converting enzymes are probably found only in small amounts compared with all the other enzymes in the organism with similar specificity for basic residues. It should be noted that another secretory granule enzyme, the amidating enzyme, which even has a unique activity to be assayed for, was only recently fully characterized (see below). Nevertheless, the latest candidate(s) as a proinsulin conversion enzyme, described by Hutton and coworkers, fulfills many of the criteria of a true conversion enzyme (Davidson et al., 1987, 1988). From secretory granules of insulinoma cells they have isolated two enzymes which are both Ca^{2+}-dependent endoproteases with acidic pH optima. One cleaves exclusively the Arg–Arg sequence in proinsulin (type I); the other preferentially cleaves the Lys–Arg sequence (type II). The pH and Ca^{2+} requirement of the type-I protease suggests that this enzyme is only active in the secretory granules, in contrast to the type-II endoprotease which would probably be active both in the Golgi complex and in the secretory vesicles (Davidson et al., 1988). Kinetic biosynthesis studies from the same group support the conjecture of differential timing in the processing of the two dibasic sites in proinsulin (Davidson et al., 1988). The morphological evidence indicates that both conversion sites in proinsulin are cleaved in the clathrin-coated immature secretory vesicles (Orci et al., 1986; Steiner et al., 1987). Thus the type-1 Ca^{2+}-dependent protease of Hutton and coworkers, which is active at the low pH and high Ca^{2+} concentrations found in the secretory granule (Andersson et al., 1982), fits directly into the current concept of a proinsulin, and possibly a general precursor processing enzyme. Acceptance of the type-II protease as a precursor converting enzyme, active in the Golgi area, is dependent on morphological data supporting the concept that cleavage actually starts in this compartment. It should be noted that very little hard evidence is available on this subject and that details in the general concept could change if careful morphological studies like those of Orci and coworkers, and detailed kinetic biosynthesis studies like those of Hutton and coworkers, were performed on other cell systems. This furthermore emphasizes that we should not be too dogmatic in establishing and following criteria which prospective converting enzymes should meet, just as we should not designate any enzymatic activity which happens to degrade a particular peptide substrate as a conversion enzyme.

(ii) **Monobasic-specific processing** As the structures of many precursors were characterized, mainly by deduction from cDNA sequences, it became apparent that processing at single basic residues was a general and apparently distinct conversion mechanism (Schwartz, 1986). In the peptide precursors of the islets, monobasic processing is responsible for the generation of the insulinotropic hormone GLP-I$^{7-36}$, somatostatin-28, and the pancreatic icosapeptide. In the dog the proinsulin C-peptide is also very efficiently cleaved at a monobasic site (Kwok et al., 1983). Monobasic processing appears to be a distinct conversion mechanism as, for example, proinsulins, in which one of the basic residues is substituted with a non-basic residue, are not converted in the B-cell (Robbins et al., 1981; Conlon et al., 1986). Furthermore, the dibasic processing in isolated secretory granules from insulinomas is inhibited by dibasic- but not by monobasic-specific enzyme inhibitors (Hutton et al., 1987a). The monobasic processing is also different from the dibasic processing, at least in the PP cell of the islet, as the monobasic conversion takes place after the dibasic cleavage and it is thus likely to occur in the mature secretory granule (Schwartz, 1987).

Monobasic processing in most cases occurs after single arginyl residues but in a few cases after a lysyl residue. There is no consensus sequence of amino acids surrounding the monobasic processing site; however, the configuration of the peptide backbone seems to be important for the presentation of certain basic residues as processing sites which distinguish them from the many single basic residues which are not recognized by the conversion enzyme (Schwartz, 1986). This is underlined by the frequent occurrence, just before or after monobasic processing sites, of the imino acid proline, with its restricted peptide-bound angles (Schwartz, 1986).

In summary, endoproteolytic processing of the islet cell prohormones occurs at basic residues. The converting enzymes are still not characterized; however, there appear to be at least two kinds, dibasic- and monobasic-specific ones. It is likely that the distinction between dibasic and monobasic processing enzymes just indicates that there is more than one type of enzyme, and that families of processing enzymes, including both monobasic and dibasic ones, will appear when they are eventually isolated and structurally characterized.

9.3.7. Removal of C-terminal basic residues

When the basic residues have served their purpose as signals for the endoproteolytic enzyme(s) they are removed. According to Steiner's original scheme for precursor processing, the endoprotease would cleave after the two basic residues and a carboxypeptidase B-like enzyme would then be needed to remove the exposed basic residues (Oyer et al., 1981; Nolan et al., 1971; Steiner et al., 1974). It has been suggested that the endoproteases could cleave in between the two basic residues of a conversion site (which some of the uncharacterized enzyme preparations do, in vitro); an aminopeptidase with specificity for basic residues has, in this context, been described (Gainer et al., 1984). However, when secretory granules from insulin cells are treated with a specific inhibitor of the carboxypeptidase B-like enzyme (Fricker et al., 1983), a biosynthetic intermediate form accumulates, which by treatment with carboxypeptidase B can be turned into true insulin (Davidson et al., 1987). Thus, the endoproteolytic attack must occur on the C-terminal side of the basic pair, at least in insulin, and a carboxypeptidase B-like enzyme has to be part of the peptide-generating machinery of endocrine cells.

Under the names enkephalin convertase and carboxypeptidase H, this carboxypeptidase B-like enzyme was isolated and characterized from several tissues, including islets (Hook et al., 1982; Fricker & Snyder, 1982, 1983; Docherty & Hutton, 1983; Hook & Loh, 1984; Suppatone et al., 1984; Davidson & Hutton, 1987; Mackin & Noe, 1987a) and the enzyme was eventually cloned under the name carboxypeptidase E (CPE) (Fricker et al., 1986; Fricker, 1988b). Recent, partial characterization of the amino acid sequence of the carboxypeptidase B-like enzyme from an insulinoma has revealed that this enzyme, called carboxypeptidase H, is identical to the CPE (J.C. Hutton & L.D. Fricker, personal communication). CPE is found in peptide-producing tissues only; in subcellular fractionations from such tissues it is associated with the peptide-containing secretory granules. Importantly, it is active at the low pH found in these vesicles (Fricker, 1988a).

PROCESSING OF GRANULAR PROTEINS

(a) COOH-terminal amidation enzyme

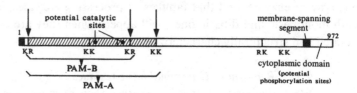

(b) carboxypeptidase E

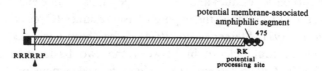

(c) chromogranin A or β-granin

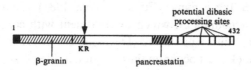

Fig. 9.8. Processing of secretory granule proteins. The structures of three proteins from islet cell secretory granules are shown in a schematic form. The known processing sites are indicated by large vertical arrows; signal peptides are indicated in black. (a) COOH-terminal amidating enzyme: the transmembrane segment is indicated in black, and the major secreted forms of the enzyme, PAM-A and PAM-B, by hatching. (b) Carboxypeptidase E: the secreted enzyme is indicated by hatching, and the C-terminal segment, which has been proposed to form a membrane-associated amphiphilic α-helix, is indicated after the proposed dibasic cleavage site. (c) Chromogranin A: β-granin, which forms the N-terminal, well-conserved part of chromogranin A, and pancreastatin, another chromogranin fragment, are shown by hatching. Dibasic sequences that are potential processing sites are indicated by vertical lines.

CPE, which is homologous to carboxypeptidase A and B, is synthesized as a 475 amino acid precursor which gives rise to the 434 amino acid, glycosylated enzyme (Fricker, 1988a, b). The processing of the zymogen involves the removal of both a signal peptide and a small N-terminal propeptide (Fig. 9.8). Interestingly, four adjacent arginyl residues immediately precede the N-terminus of the mature enzyme, which itself starts with Arg–Pro. Thus the processing site in the zymogen for CPE resembles that of (for example) the insulin receptor, in which the two subunits are separated by four basic residues (Ebina et al., 1985), and that of members of the complement family (Belt et al., 1985). The integrity of the tetrabasic processing site seems to be important for correct processing, as indicated by a naturally occurring mutant in the insulin receptor gene in which a substitution of Arg number four with Ser leads to the production of a single-chain inactive protein (Yoshimasa et al., 1988). The proline residue following the processing site in CPE could indicate that in this case too the presentation of the peptide backbone is important for the processing (Schwartz, 1986).

The CPE is found mainly as a soluble protein in the secretory vesicle, but also in a membrane-associated form with a somewhat larger apparent molecular mass (Fricker, 1988a, b). It has been suggested that the CPE is initially bound to the membrane through an amphiphilic α-helical segment in the C-terminus, and that this part of the molecule is cleaved from the main part by processing at a dibasic processing site (Fricker, 1988b). Another possibility would be that the signal peptide is not cleaved from the precursor and serves as a membrane-anchor until the processing at the tetrabasic site liberates the soluble enzyme.

9.3.8. Formation of C-terminal amide function

In the pancreatic islets, formation of a C-terminal amide is involved in the activation of the PP molecule. This modification is essential for the biological activity of the hormone (Schwartz, 1983) as it is for many other amidated peptides (Tatemoto & Mutt, 1978). The intervening peptide II (IVP-II) of the glucagon precursor is also amidated, but this peptide is only excised from the glucagon precursor and modified in the intestine. The substrate for the amidating enzyme is a glycine-extended intermediate form of the peptide hormone (Suchanek & Kreil, 1977) which is produced by the consecutive action of the dibasic-specific endoprotease and the CPE at a combined cleavage and amidation site, e.g. Gly–Lys–Arg as found in the PP precursor. Thus the amidating enzyme is the last in the production line in the endocrine cell and its place of action is the mature secretory granule (Eipper & Mainz, 1988). The amidating enzyme was isolated and characterized as an enzyme dependent on copper, oxygen, and ascorbate, which uses the nitrogen of the glycyl residue to form the amide function on the preceding amino acid while the rest of the glysyl residue is cleaved off as glyoxylate (Bradbury *et al.*, 1982; Eipper *et al.*, 1983). Recently, both the mammalian and the homologous *Xenopus laevis* enzyme have been cloned (Eipper *et al.*, 1987; Ohsuye *et al.*, 1988). The enzyme is made as a 972 amino acid precursor with a membrane-spanning segment close to its presumed cytosolic, C-terminal end (Fig. 9.8). The previously characterized forms of the enzyme correspond to two overlapping N-terminal fragments of approximately 400 and 350 amino acids, PAM (peptidylglycine α-amidating monooxygenase) A and PAM B respectively. After the signal peptide is removed these molecules can be cleaved from the zymogen, possibly at dibasic processing sites (Eipper *et al.*, 1987; Eipper & Mainz, 1988). A histidine-rich segment, possibly the site for copper binding, is located within the structure. On the short cytosolic part are found potential phosphorylation sites (Eipper *et al.*, 1987).

9.3.9. The mature secretory granule

Through the biochemical modifications and cell biological events described above, the peptide hormone is activated and routed to the mature secretory granule, ready to be released when the cell is appropriately stimulated.

(i) **Secretory granule proteins and their processing** The mature secretory granule of (for example) the B-cell contains more than 150 identifiable proteins (Hutton, 1984). The synthesis of some of these proteins is regulated in parallel with the synthesis of proinsulin and not with the general synthesis of proteins within the cell, as recently shown for a minor protein of the insulin secretory granule (Grimaldi *et al.*, 1987). In the parathyroid it has been known since the earliest biosynthetic studies that a so-called 'secretory protein-I' was synthesized along with PTH in almost equimolar amounts (Cohn & Elting, 1983). In the B-cell a similar protein, β-granin, was recently discovered (Sopwith *et al.*, 1984). When the structures of these proteins were characterized it was revealed that they both correspond to chromogranin A, a quantitatively major protein of the chromaffin granules of the adrenal medulla (Cohn *et al.*, 1982; Hutton *et al.*, 1985). It appears today that the secretory granules of all endocrine cells contain at least one of the chromogranins of which chromogranin A and B, also called secretogranin I, recently have been cloned (Benedum *et al.*, 1986, 1987; Iacangelo *et al.*, 1986). β-granin, found in the B-cell of the islets, corresponds to the N-terminal, conserved fragment of chromogranin A (Fig. 9.8) (Hutton *et al.*, 1985). In fact the processing of chromogranin A to β-granin occurs in complete parallel to the processing of proinsulin to insulin, involving dibasic cleavage and CPE removal of basic residues (Hutton *et al.*, 1987*a*, *b*). The role of β-granin and chromogranins in endocrine cells has not been delineated. However, it has been suggested that they may function as Ca^{2+} binding, amine binding, osmoregulator, or sorting proteins, or that they help in solubilizing the granule content during secretion (Sharp & Richards, 1977; Helle *et al.*, 1985; Filicitas & Gratzl, 1986; Benedum *et al.*, 1987). The chromogranins contain many dibasic sequences and could serve as precursor molecules, as in the generation of β-granin; in fact, a fragment of chromogranin, called pancreastatin, which is released by monobasic processing, has been described (Fig. 9.8) (Tatemoto *et al.*, 1986; Huttner & Benedum, 1987; Eiden, 1987). As noted by many groups, the chromogranins contain many acidic residues and these are often in clusters. Because sequences of acidic residues are prone to form α-helices at low pH (Piszkiewicz, 1974), it could be speculated that during vesicular acidification, a pH-induced change in chromogranin structure would be important for its function in the newly formed secretory granules.

The processing enzymes are a group of granular proteins for which the function is known. Of these, at least the CPE and the amidating enzyme are active in the mature secretory granule. Interestingly, both enzymes are, during their synthesis, either membrane-spanning or membrane-associated (Fig. 9.8); however, both enzymes are processed along with the prohormones, resulting in soluble enzymes which are released from the endocrine cells in parallel with the hormone (Mainz & Eipper, 1984). The chromogranins are apparently also processed and eventually secreted in parallel with the hormone.

A picture is emerging in which not only the peptide precursors but also zymogens for processing enzymes and other granular proteins are all subject to sorting and sequential, coordinated processing and thereby activation (of each other).

(ii) Morphological heterogeneity of secretory granules The peptide products are found in very concentrated form in the mature secretory granule. The amount of water is barely enough for the hydration of the peptides and proteins (Fulton, 1982). In fact, in the B-cell granules, for example, the insulin is localized in typical microcrystals composed of different packings of zinc-stabilized hexamers, as shown in Fig. 9.7 (Greider *et al.*, 1969). The proinsulin C-peptide and the other secretory granule proteins are then found in the surrounding electrolucent part of the granules (Michael *et al.*, 1987). A similar non-homogeneous organization of different precursor fragments is also found in the glucagon cell of the pancreatic islets (Ravazzola & Orci, 1980).

(iii) The fate of the secretory granule – secretion or degradation A single B-cell contains approximately 13 000 mature secretory granules, of which only around 10% may be released during a one hour period of active stimulation (Howell & Tyhurst, 1984). Thus, the lifespan of a dense granule is to be measured in hours and days, in contrast to the vesicles of the constitutive export route, which are released in minutes (Kelly, 1985). The secretory granules of the regulated export route are either released by exocytosis or degraded by fusion with lysosomes, forming crinophagic or multivesicular bodies (Orci *et al.*, 1984c). Because the crystalloid insulin is relatively resistant to proteolytic degradation (Halban *et al.*, 1987), such crinophagic bodies contain only insulin immunoreactivity and no C-peptide (Orci *et al.*, 1984c). In pulse–chase experiments, the ratio of intracellular insulin to C-peptide can rise to 9 : 1 owing to degradation of C-peptide; however, when such cells are stimulated, insulin and C-peptide are still released together in a ratio of approximately 1 : 1 (Rhodes & Halban, 1988).

The intracellular degradation of peptide precursors should probably be borne in mind when different, maybe unexpected, fragments of precursors are determined by immunological means or are isolated and sequenced. Without dynamic biosynthesis studies, it is difficult to distinguish between the limited proteolysis, which is part of the precursor processing, and degradation, which is part of the housekeeping activity of the cell.

9.4. Precursor processing in heterologous systems

Processing of hormone precursors has recently also been studied in cellular systems where the genes for the hormones are not normally expressed. By using the hormone cDNA or gene built into expression vectors, it has been possible to drive cell lines, cells in transgenic animals and yeast or prokaryotic cells to synthesize (pre)prohormones either for scientific or for production purposes.

9.4.1. Precursor processing in transfected cell lines

The first transfection experiments were performed in COS cells, a non-neuroendocrine cell line, using DNA constructions in which the cDNA for preproinsulin was placed under the control of the SV40 promoter. The cells were studied after 2–3 days, at which time proinsulin was produced by transient expression of the transfected plasmids (Lomedico, 1982; Laub & Rutter, 1983). The production of unprocessed prohormone in non-neuroendocrine cell lines has been confirmed in transient expression experiments as well as in stably transfected clones for pancreatic polypeptide in CHO cells (Boel et al., 1987), somatostatin in 3T3 cells (Sevarino et al., 1987), glucagon in BHK cells (Drucker et al., 1986), enkephalin in BSC-40 cells (Thomas et al., 1986), and vasopressin in BHK cells (Cwikel & Habener, 1987). However, in an early study, Warren & Shields (1984b) did observe an apparent processing of prosomatostatin in COS cells. Recently, in a series of transfected CHO cells, some degree of cleavage of proNPY was in fact found in all clones, although the apparent processing at the best was around 30%.[3] Conceivably, the processing sites in certain precursors, like the prosomatostatin and proNPY, are rather susceptible to proteolytic attack by proteases with specificity for basic residues; such enzymes are found in the secretory pathway of all cells (for the processing of, for example, the insulin receptor precursor).

True processing of a prohormone, proinsulin, was first observed in a stably transfected clone derived from the mouse corticotroph cell line, AtT-20 (Moore et al., 1983b). It appears today that cell lines that store and process large amounts of their endogenous product, like the RIN insulinoma cells and the AtT-20 cells, can process any of the precursors which have so far been introduced to them by transfection (Drucker et al., 1986; Thomas et al., 1986; Dickerson et al., 1987; Sevarino et al., 1987; Cwikel & Habener, 1987; unpublished[3]). This result indicates that the processing enzyme(s) are relatively non-substrate-specific. On the other hand, neuroendocrine cell lines which are not very efficient in their endogenous processing, like the GH_4 and PC-12 cells, do not process exogenous precursors very well either (Drucker et al., 1986; Thomas et al., 1986; Sevarino et al., 1987; Cwikel & Habener, 1987; unpublished[3]). Nevertheless, GH_4 cells do remove the small propeptide from proPTH; however, this is also cleaved by transfected non-endocrine (3T3) cells indicating that the terminally situated cleavage site of proPTH, which consists of three basic residues, is very susceptible to proteolytic cleavage (Hellerman et al., 1984). Cells such as GH_4 cells, which cannot process a transfected peptide precursor by themselves, can be driven to do so by cotransfection with a vector expressing the yeast kex2 gene (Thomas et al., 1988).

The transfection experiments in neuroendocrine cell lines have opened

[3] B.S. Wulff, M.M.T. O'Hare & T.W. Schwartz, unpublished observations.

up, for protein engineering, experiments in which the signal epitopes of the hormone precursors that determine the posttranslational modifications etc. can be characterized by site-directed mutagenesis.

9.4.2. Processing of peptide precursors in transgenic animals

The major problem in interpreting results from transfection experiments in cell lines is that such cells are transformed and thereby not normal. However, the main conclusion from these experiments, that the processing enzyme(s) are relatively unspecific in their substrate requirements, has been confirmed in studies with transgenic mice. By introduction of a vector holding the somatostatin cDNA into fertilized eggs, transgenic mice were produced that synthesized preprosomatostatin under the control of the metallothionein promoter in various tissues (Low et al., 1985). In these animals, a tissue-specific processing of the somatostatin precursor was observed. Prosomatostatin was only processed to a limited extent in the liver and kidney, whereas a normal, almost total processing to somatostatin-14 was found in the pituitary (Low et al., 1985). The artificial gene was expressed very efficiently in gonadotrophs in the pituitary (Low et al., 1986a); in primary cultures of such cells prepared from transgenic mice, it was shown that the somatostatin-14 entered the regulated secretory pathway, as it was released in parallel to LH and FSH in response to the specific stimulus LHRH (Low et al., 1986b).

9.4.3. Processing of peptide precursors in yeast and prokaryotes

As the first hormone, somatostatin was produced in E. coli by the expression of a synthetic oligonucleotide sequence placed in conjunction with the β-galactosidase gene (Itakura et al., 1977). Today, prokaryotes are used generally for the production of proinsulin as well as the separate production of insulin A- and B-chains (Chance et al., 1981; Frank et al., 1981; Frank & Chance, 1983). The formation of the mature insulin molecule is not possible in the prokaryotes and thus insulin is produced either by in vitro conversion of proinsulin as originally described by Oyer et al. (1971) or through in vitro combination of the two separately produced chains (Chance et al., 1981).

In contrast to prokaryotes, the eukaryotic yeast cells produce substances such as their endogenous pheromones and killer toxins through dibasic and carboxypeptidase B-like proteolytic processing of precursors. This capacity was utilized in the production of proinsulin in yeast cells (Thim et al., 1986). Proinsulin was synthesized as part of a large precursor molecule consisting of the signal peptide and the propeptide from the α-mating factor of the host cells, separated from the proinsulin sequence by a dibasic cleavage site. The proinsulin molecule was efficiently cleaved from the chimeric molecule, probably by the yeast *kex2* protease (Julius *et al.*, 1984; Thim *et al.*, 1986). Although proinsulin was almost totally processed by the yeast cells, very little true insulin was produced. Most of the products were found in the form of conversion intermediates with around half of the molecules still having the C-peptide attached to either the B- or the A-chain. The other half of the major products were of the size of insulin but with one or two arginyl residues attached to the B-chain (Thim *et al.*, 1986). The NOVO group has made a series of C-peptide deletions in order to optimize the production of insulin from yeast. If the C-peptide is replaced by a single pair of basic residues no intracellular conversion is observed and Thim and coworkers conclude that a C-peptide is needed to the presentation of the conversion sites (Thim *et al.*, 1986). However, when the C-peptide is reduced to six residues, Arg–Arg–Leu–Gln–Lys–Arg, both a high level of expression and a reasonably good *in vivo* conversion to two-chained insulins is found (Thim *et al.*, 1986, 1987). These informations cannot, however, be directly transferred to islet cells, both because the cells *per se* are different, but also because the B-cell initially deals with a very different prohormone, the relatively small, true proinsulin and not the much larger artificial, chimeric prohormone.

9.5. Summary

The biochemical and cell biological events in the processing of the peptide hormone precursors are closely interconnected as the different, consecutive steps in the precursor processing occur while the precursor is routed through specific compartments of the cell, each with a distinct biochemical composition. The most crucial step in the generation of the bioactive peptide from the larger precursor is the endoproteolytic cleavage at dibasic or monobasic conversion sites, a cleavage mechanism which seems to be unique to neuroendocrine cells. For one of the proinsulin processing sites, this conversion has been conclusively established to occur in the clathrin-coated, immature secretory granules which are formed in the *trans*-Golgi network, as shown in Fig. 9.6. At this point in the precursor processing there is an intimate, but still not structurally clarified, interrelationship between the presumably active sorting of the precursor to the immature secretory granule, the acidification of the newly formed granule, and the endoproteolytic cleavage of the precursor.

In yeast the dibasic specific conversion enzyme, *kex2*, has been cloned; however, in higher eukaryotes the corresponding enzyme is still elusive.

Two other enzymes, both involved in peptide biosynthesis, have been structurally characterized: the carboxypeptidase E and the amidating enzyme. Both enzymes are made as larger membrane-associated precursors, which are processed at basic residues to give the active, soluble enzyme which is secreted along with the peptide hormones. Other secretory granule proteins, such as the chromogranins, are also processed at basic residues in parallel with the peptide precursors. Apparently, in the endocrine cells there is a coordinated sequence of sorting and processing, not only of the peptide precursors but also of other secretory granule proteins including the zymogens for the processing enzymes.

The basis for the tissue-specific processing of the glucagon and somatostatin precursors, which leads to the production of different biologically active peptides from the same precursor, is unknown. The elucidation of this phenomenon probably awaits the cloning and subsequent characterization of the control of the expression of genes for the endoproteolytic processing enzymes. The signal epitopes, i.e. the three-dimensional structural elements of the precursors that determine the correct cellular handling and processing, are also not known. However, such epitopes are likely to be characterized through protein engineering with site-directed mutagenesis combined with transfection experiments in neuroendocrine cells. In this connection it will be important to regard peptide precursors not as a combination of solid and hatched bars, as in the figures of the present chapter, nor as strings of amino acids, but rather as dynamic, three-dimensional chemical substances.

References

Albert, S.G. & Permutt, M.M. (1979). Proinsulin precursors in catfish pancreatic islets. *Journal of Biological Chemistry* **254**, 3483–92.

Andersson, T., Berggren, P.O., Gylfe, E. & Hellman, B. (1982). Amounts and distribution of intracellular magnesium and calcium in pancreatic β-cells. *Acta Physiologica Scandinavica* **114**, 235–41.

Andrews, P.C. & Dixon, J.E. (1987). Isolation of products and intermediates of pancreatic prosomatostatin: use of fast atom bombardment mass spectrometry as an aid in analysis of prohormone processing. *Biochemistry* **26**, 4853–61.

Andrews, P.C., Nichols, R. & Dixon, J.E. (1987). Post-translational processing of preprosomatostatin-II examined using fast atom bombardment mass spectrometry. *Journal of Biological Chemistry* **262**, 12692–9.

Andrews, P.C., Pubols, M.H., Hermodson, M.A., Sheares, B.T. & Dixon, J.E. (1984). Structure of the 22-residue somatostatin from cat-fish. An *O*-glycosylated peptide having multiple forms. *Journal of Biological Chemistry* **259**, 13267–72.

Arimura, A., Sato, H., DuPont, A., Nishi, N. & Schally, A.V. (1975). Somatostatin: abundance of immunoreactive hormone in rat stomach and pancreas. *Science* **189**, 1007–9.

Baldissera, F.G.A., Holst, J.J., Jensen, S.L. & Krarup, T. (1985*a*). Distribution and molecular forms of peptides containing somatostatin immunodeterminants in extracts from the entire gastrointestinal tract of man and pig. *Biochimica et Biophysica Acta* **838**, 132–43.

Baldissera, F.G.A., Munoz-Perez, M.A. & Holst, J.J. (1983). Somatostatin 1-28 circulates in human plasma. *Regulatory Peptides* **6**, 63–9.

Baldissera, F.G.A., Nielsen, O.V. & Holst, J.J. (1985*b*). The intestinal mucosa preferentially releases somatostatin-28 in pigs. *Regulatory Peptides* **11**, 251–62.

Baskin, D.G. & Ensinck, J.W. (1984). Somatostatin in epithelial cells of intestinal mucosa is present primarily as somatostatin 28. *Peptides* **5**, 615–21.

Bataille, D., Tatemoto, K., Gespach, C., Jörnvall, H., Rosselin, G. & Mutt, V. (1982). Isolation of glucagon-37 (bioactive enteroglucagon/oxyntomodulin) from porcine jejuno-ileum. Characterization of the peptide. *FEBS Letters* **146**, 79–86.

Bell, G.I., Sanchez-Pescador, R., Laybourn, P.J. & Najarian, R.C. (1983*a*). Exon duplication and divergence in the human preproglucagon gene. *Nature* **304**, 368–71.

Bell, G.I., Santerre, R.F. & Mullenbach, G.T. (1983*b*). Hamster preproglucagon contains the sequence of glucagon and two related peptides. *Nature* **302**, 716–18.

Bell, G.I., Swain, W.F., Pictet, R., Cordell, B., Goodman, H.M. & Rutter, W.J. (1979). Nucleotide sequence of a cDNA clone encoding human preproinsulin. *Nature* **282**, 525–7.

Belt, K.T., Carroll, M.C. & Porter, R.R. (1985). The structural basis of the multiple forms of human complement component C4. *Cell* **36**, 907–16.

Benedum, U.M., Baeuerle, P.A., Konecki, D.S., Frank, R., Powell, J., Mallet, J. & Huttner, W.B. (1986). The primary structure of bovine chromogranin A: a representative of a class of acidic secretory proteins common to a variety of peptidergic cells. *EMBO Journal* **5**, 1495–502.

Benedum, U.M., Lamouroux, A., Konecki, D.S., Rosa, P., Hille, A., Baeuerle, P.A., Frank, R., Lottspeich, F., Mallet, J. & Huttner, W.B. (1987). The primary structure of human secretogranin I (chromogranin B): comparison with chromogranin A reveals homologous terminal domains and a large intervening variable region. *EMBO Journal* **6**, 1203–11.

Benoit, R., Böhlen, P., Esch, F. & Ling, N. (1984). Neuropeptides derived from prosomatostatin that do not contain the somatostatin-14 sequence. *Brain Research* **311**, 23–9.

Benoit, R., Ling, N., Alford, B. & Guillemin, R. (1982). Seven peptides derived from prosomatostatin in rat brain. *Biochemical and Biophysical Research Communications* **107**, 944–50.

Benoit, R., Ling, N. & Esch, F. (1987). A new prosomatostatin-derived peptide reveals a pattern for prohormone cleavage at monobasic sites. *Science* **238**, 1126–9.

Binder, C., Hartling, S.G. & Faber, O.K. (1986). Proinsulin. *Diabetes Annual* **2**, 240–7.

Blobel, G. (1980). Intracellular protein topogenesis. *Proceedings of the National Academy of Sciences of the U.S.A.* **77**, 1496–500.

Blundel, T.L. & Humbel, R.E. (1980). Hormone families: pancreatic hormones and homologous growth factors. *Nature* **287**, 781–7.

Boel, E., Berkner, K.L., Nexø, B.A. & Schwartz, T.W. (1987). Expression of human pancreatic polypeptide precursors from a dicistronic mRNA in mammalian cells. *FEBS Letters* **219**, 181–9.

Boel, E., Schwartz, T.W., Norris, K.E. & Fiil, N.P. (1984). A cDNA encoding a small common precursor for human pancreatic polypeptide and pancreatic icosapeptide. *EMBO Journal* **3**, 909–12.

Bole, D.G., Hendershot, L.M. & Kearney, J.F. (1986). Posttranslational association of immunoglobulin heavy chain binding protein with nascent heavy chains in nonsecreting and secreting hybridomas. *Journal of Cell Biology* **102**, 1558–66.

Bradbury, A.F., Finnie, M.D.A. & Smyth, D.G. (1982). Mechanism of C-terminal amide formation by pituitary enzymes. *Nature* **298**, 686–8.

Buhl, T., Thim, L., Kofod, H., Ørskov, C., Harling, H. & Holst, J.J. (1988). Naturally occurring products of proglucagon 111–160 in the porcine and human small intestine. *Journal of Biological Chemistry* **263**, 8621–4.

Burgess, T.L., Craik, C.S. & Kelly, R.B. (1985). The exocrine protein trypsinogen is targeted into the secretory granules of an endocrine cell line: studies by gene transfer. *Journal of Cell Biology* **101**, 639–45.

Carrol, R.J., Hammer, R.E., Chan, S.J., Swift, H.H., Rubenstein, A.H. & Steiner, D.F. (1988). A mutant human proinsulin is secreted from islets of Langerhans in increased amounts via an unregulated pathway. *Proceedings of the National Academy of Sciences of the U.S.A.* **85**, 8943–7.

Chan, S.J., Emdin, S.E., Kwok, S., Kramer, J.M., Falkmer, S. & Steiner, D.F. (1981). Messenger RNA sequence and primary structure of preproinsulin in a primitive vertebrate, the Atlantic hagfish. *Journal of Biological Chemistry* **256**, 7595–602.

Chan, S.J., Keim, P. & Steiner, D.F. (1976). Cell-free synthesis of rat preproinsulins: characterization and partial amino acid sequence determination. *Proceedings of the National Academy of Sciences of the U.S.A.* **73**, 1964–8.

Chan, S.J., Seino, S., Gruppuso, P.A., Schwartz, R. & Steiner, D.F. (1987). A mutation in the B chain coding region is associated with impaired proinsulin conversion in a family with hyperproinsulinemia. *Proceedings of the National Academy of Sciences of the U.S.A.* **84**, 2194–7.

Chance, R.E., Ellis, R.M. & Bromer, W.W. (1968). Porcine proinsulin: characterization and amino acid sequence. *Science* **161**, 165–7.

Chance, R.E., Hoffmann, J.A., Kroeff, E.P., Johnson, M.G., Schirmer, E.W., Bromer, W.W., Ross, M.J. & Wetzel, R. (1981). The production of human insulin using recombinant DNA technology and a new chain combination procedure. In *Peptides – synthesis, structure and function* (ed. D.H. Rich and E. Gross), pp. 721–8. Rockford, Illinois: Pierce Chemical Co. Publications.

Cohn, D.V. & Elting, J. (1983). Biosynthesis, processing, and secretion of parathormone and secretory protein-1. *Recent Progress in Hormone Research* **39**, 181–209.

Cohn, D.V., Zangerle, R., Fischer-Colbrie, R., Chu, L.L.H., Elting, J.J., Hamilton, J.W. & Winkler, H. (1982). Similarity of secretory protein I from parathyroid gland to chromogranin A from adrenal medulla. *Proceedings of the National Academy of Sciences of the U.S.A.* **79**, 6056–9.

Conlon, J.M., Dafgård, E., Falkmer, S. & Thim, L. (1986). The primary structure of ratfish insulin reveals an unusual mode of proinsulin processing. *FEBS Letters* **208**, 445–50.

Conlon, J.M., Falkmer, S. & Thim, L. (1987). Primary structure of three fragments of proglucagon from the pancreatic islets of the daddy sculpin (*Cottus scorpius*). *European Journal of Biochemistry* **164**, 117–22.

Conlon, J.M., Hansen, H.F. & Schwartz, T.W. (1985). A truncated glucagon-like peptide I from torpedo pancreas. *Regulatory Peptides* **11**, 94.

Cromlish, J.A., Seidah, N.G. & Chrétien, M. (1986*a*). A novel serine protease (IRCM-serine protease 1) from porcine neurointermediate and anterior pituitary lobes. *Journal of Biological Chemistry* **261**, 10850–6.

Cromlish, J.A., Seidah, N.G. & Chrétien, M. (1986*b*). Selective cleavage of human ACTH, β-lipotropin, and the N-terminal glycopeptide at pairs of basic residues by IRCM-serine protease 1. *Journal of Biological Chemistry* **261**, 10859–70.

Cwikel, B.J. & Habener, J.F. (1987). Provasopressin-neurophysin II processing is cell-specific in heterologous cell lines expressing a metallothionein–vasopressin fusion gene. *Journal of Biological Chemistry* **262**, 14235–40.

Davidson, H.W. & Hutton, J.C. (1987). The insulin-secretory-granule carboxypeptidase H. *Biochemical Journal* **245**, 575–82.

Davidson, H.W., Peshavaria, M. & Hutton, J.C. (1987). Proteolytic conversion of proinsulin into insulin. *Biochemical Journal* **246**, 279–86.

Davidson, H.W., Rhodes, C.J. & Hutton, J.C. (1988). Intraorganellar calcium and pH control proinsulin cleavage in the pancreatic β cell via two distinct site-specific endopeptidases. *Nature* **333**, 93–6.

Dickerson, I.M., Dixon, J.E. & Mains, R.E. (1987). Transfected human neuropeptide Y cDNA expression in mouse pituitary cells. *Journal of Biological Chemistry* **262**, 13646–53.

Docherty, K., Carroll, R.J. & Steiner, D.F. (1982). Conversion of proinsulin to insulin: involvement of a 31,500 molecular weight thiol protease. *Proceedings of the National Academy of Sciences of the U.S.A.* **79**, 4613–17.

Docherty, K., Carroll, R.J. & Steiner, D.F. (1983). Identification of a 31,500 molecular weight islet cell protease as cathepsin B. *Proceedings of the National Academy of Sciences of the U.S.A.* **80**, 3245–9.

Docherty, K. & Hutton, J.C. (1983). Carboxypeptidase activity in the insulin secretory granule. *FEBS Letters* **162**, 137–41.

Docherty, K., Hutton, J.C. & Steiner, D.F. (1984). Cathepsin B-related proteases in the insulin secretory granule. *Journal of Biological Chemistry* **259**, 6041–4.

Docherty, K. & Steiner, D.F. (1982). Post-translational proteolysis in polypeptide hormone biosynthesis. *Annual Review of Physiology* **44**, 625–38.

Drucker, D.J., Mosjov, S. & Habener, J.F. (1986). Cell-specific post-translational processing of preproglucagon expressed from a metallothionein–glucagon fusion gene. *Journal of Biological Chemistry* **261**, 9637–43.

Ebina, Y., Ellis, L., Jarnagin, K., Edery, M., Graf, L., Clauser, E., Ou, J., Masiarz, F., Kan, Y.W., Goldfine, I.D., Roth, R.A. & Rutter, W.J. (1985). The human insulin receptor cDNA: the structural basis for hormone-activated transmembrane signalling. *Cell* **40**, 747–58.

Eiden, L.E. (1987). Is chromogranin a prohormone? *Nature* **325**, 301.

Eipper, B.A. & Mains, R.E. (1988). Peptide alpha-amidation. *Annual Review of Physiology* **50**, 333–44.

Eipper, B.A., Mains, R.E. & Glembotski, C.C. (1983). Identification in pituitary tissue of a peptide alpha-amidation activity that acts on glycine-extended peptides and requires molecular oxygen, copper, and ascorbic acid. *Proceedings of the National Academy of Sciences of the U.S.A.* **80**, 5144–8.

Eipper, B.A., Park, L.P., Dickerson, I.M., Keutmann, H.T., Thiele, E.A., Rodriguez, H., Schofield, P.R. & Mains, R.E. (1987). Structure of the precursor to an enzyme mediating COOH-terminal amidation in peptide biosynthesis. *Molecular Endocrinology* **1**, 777–90.

Esch, F., Böhlen, P., Ling, N., Benoit, R., Brazeu, P. & Guillemin, R. (1980). Primary structure of ovine hypothalamic somatostatin 28 and somatostatin 25. *Proceedings of the National Academy of Sciences of the U.S.A.* **77**, 6827–31.

Faber, O.K. & Binder, C. (1986). C-peptide: an index of insulin secretion. *Diabetes/Metabolism Reviews* **2**, 169–83.

Filicitas, U.R. & Gratzl, M. (1986). Chromogranins, wide-spread in endocrine and nervous tissue, bind Ca^{2+}. *FEBS Letters* **195**, 327–30.

Fletcher, D.J., Quigley, J.P., Bauer, E. & Noe, B.D. (1981). Characterization of proinsulin- and proglucagon-converting activities in isolated islet secretory granules. *Journal of Cell Biology* **90**, 312–22.

Frank, B.H. & Chance, R.E. (1983). Two routes for producing human insulin utilizing recombinant DNA technology. *Münchener Medizinisches Wochenschrift* **125** (suppl. 1), S14–S20.

Frank, B.H., Pettee, J.M., Zimmerman, R.E. & Burck, P.J. (1981). The production of human proinsulin and its transformation to human insulin and C-peptide. In *Peptides –*

synthesis, structure and function (ed. D.H. Rich and E. Gross), pp. 721–8. Rockford, Illinois: Pierce Chemical Co. Publications.

Freedman, R.B. (1984). Native disulphide band formation in protein biosynthesis: evidence for the role of protein disulphide isomerase. *Trends in Biochemical Sciences* **9**, 438–41.

Fricker, L.D. (1988*a*). Carboxypeptidase E. *Annual Review of Physiology* **50**, 309–21.

Fricker, L.D. (1988*b*). Sequence analysis of the carboxypeptidase E precursor. In *Molecular biology of brain and endocrine peptidergic systems* in the series *Biochemical endocrinology* (ed. K.W. McKerns & M. Chrétien). New York: Plenum Press.

Fricker, L.D., Evans, C.J., Esch, F.S. & Herbert, E. (1986). Cloning and sequence analysis of cDNA for bovine carboxypeptidase E. *Nature* **323**, 461–4.

Fricker, L.D., Plummer, T.H. Jr, & Snyder, S.H. (1983). Enkephalin convertase: potent, selective, and irreversible inhibitors. *Biochemical and Biophysical Research Communications* **111**, 994–1000.

Fricker, L.D. & Snyder, S.H. (1982). Enkephalin convertase: purification and characterization of a specific enkephalin-synthesizing carboxypeptidase localized to adrenal chromaffin granules. *Proceedings of the National Academy of Sciences of the U.S.A.* **79**, 3886–90.

Fricker, L.D. & Snyder, S.H. (1983). Purification and characterization of enkephalin convertase, an enkephalin-synthesizing carboxypeptidase. *Journal of Biological Chemistry* **258**, 10950–5.

Fulton, A.B. (1982). How crowded is the cytoplasma. *Cell* **30**, 345–7.

Funckes, C.L., Minth, C.D., Deschenes, R., Magazin, M., Tavianini, M.A., Sheets, M., Collier, K., Weith, H.L., Aron, D.C., Roos, B.A. & Dixon, J.E. (1983). Cloning and characterization of a mRNA-encoding rat prepro-somatostatin. *Journal of Biological Chemistry* **258**, 8781–7.

Gainer, H., Russel, J.T. & Loh, Y.P. (1984). An aminopeptidase activity in bovine pituitary secretory vesicles that cleaves the N-terminal arginine from beta-lipotropin[60–65]. *FEBS Letters* **175**, 135–9.

Geisow, M.J. (1978). Polypeptide secondary structure may direct the specificity of prohormone conversion. *FEBS Letters* **87**, 111–14.

Given, B.D., Cohen, R.M., Shoelson, S.E., Frank, B.H., Rubenstein, A.H. & Tager, H.S. (1985). Biochemical and clinical implications of proinsulin conversion intermediates. *Journal of Clinical Investigation* **76**, 1398–405.

Glover, I.D., Barlow, D.J., Pitts, J.E., Wood, S.P., Tickle, I.J., Blundell, T.L., Tatemoto, K., Kimmel, J.R., Wollmer, A., Strassburger, W. & Zhang, Y.S. (1985). Conformational studies on the pancreatic polypeptide hormone family. *European Journal of Biochemistry* **142**, 379–85.

Gomez, S., Gluschankof, P., Morel, A. & Cohen, P. (1985). The somatostatin-28 convertase of rat brain cortex is associated with secretory granule membranes. *Journal of Biological Chemistry* **260**, 10541–5.

Goodman, R.H., Aron, D.C. and Roos, B.A. (1983). Rat pre-prosomatostatin. Structure and processing by microsomal membranes. *Journal of Biological Chemistry* **258**, 5570–3.

Goodman, R.H., Jacobs, J.W., Chin, W.W., Lund, P.K., Dee, P.C. & Habener, J.F. (1980*a*). Nucleotide sequence of a cloned structural gene coding for a precursor of pancreatic somatostatin. *Proceedings of the National Academy of Sciences of the U.S.A.* **77**, 5869–73.

Goodman, R.H., Jacobs, J.W., Dee, P.C. & Habener, J.F. (1982). Somatostatin-28 encoded in a cloned cDNA obtained from a rat medullary thyroid carcinoma. *Journal of Biological Chemistry* **257**, 1156–9.

Goodman, R.H., Lund, P.K., Barnett, F.H. & Habener, J.F. (1980*b*). Intestinal preprosomatostatin. *Journal of Biological Chemistry* **256**, 1499–501.

Goodman, R.H., Lund, P.K., Jacobs, J.W. & Habener, J.F. (1980*c*). Pre-prosomatostatins

– products of cell-free translations of messenger RNAs from angler fish islets. *Journal of Biological Chemistry* **255**, 6549–52.

Greider, M.H., Howell, S.L. & Lacy, P.E. (1969). Isolation and properties of secretory granules from rat islets of Langerhans. *Journal of Cell Biology* **41**, 162–6.

Griffiths, G. & Simons, K. (1986). The *trans*-Golgi network: sorting at the exit site of the Golgi complex. *Science* **234**, 438–43.

Grimaldi, K.A., Siddle, K. & Hutton, J.C. (1987). Biosynthesis of insulin secretory granule membrane proteins. *Biochemical Journal* **245**, 567–73.

Gruppuso, P.A., Gordon, P., Kahn, C.R., Cornblath, M., Zeller, W.P. & Schwartz, R. (1984). Familial hyperproinsulinemia due to a proposed defect in conversion of proinsulin to insulin. *New England Journal of Medicine* **311**, 629–34.

Göke, R. & Conlon, J.M. (1988). Receptors for glucagon-like peptide-I(7–36)amide on rat insulinoma-derived cells. *Journal of Endocrinology* **116**, 357–62.

Halban, P.A., Mutkoski, R., Dodson, G. & Orci, L. (1987). Resistance of the insulin crystal to lysosomal proteases: implications for pancreatic B-cell crinophagy. *Diabetologia* **30**, 348–53.

Heinrich, G., Gros, P., Lund, P.K., Bentley, R.C. & Habener, J.F. (1984). Pre-proglucagon messenger ribonucleic acid: nucleotide and encoded amino acid sequences of the rat pancreatic complementary deoxyribonucleic acid. *Endocrinology* **115**, 2176–81.

Helle, K.B., Reed, R.K., Pihl, K.E. & Serck-Hanssen, G. (1985). Osmotic properties of the chromogranins and relation to osmotic pressure in catecholamine storage granules. *Acta Physiologica Scandinavica* **123**, 21–8.

Hellerman, J.G., Cone, R.C., Potts, J.T., Jr, Rich, A., Mulligan, R.C. & Kronenberg, H.M. (1984). Secretion of human parathyroid hormone from rat pituitary cells infected with a recombinant retrovirus encoding preproparathyroid hormone. *Proceedings of the National Academy of Sciences of the U.S.A.* **81**, 5340–4.

Hobart, P., Crawford, R., Shen, L.P., Picket, R. & Rutter, W.J. (1980). Cloning and sequence analysis of cDNAs encoding two distinct somatostatin precursors found in the endocrine pancreas of anglerfish. *Nature* **288**, 137–41.

Holst, J.J. (1983). Gut glucagon, enteroglucagon, gut glucagonlike immunoreactivity, glicentin – current status. *Gastroenterology* **84**, 1602–13.

Holst, J.J., Baldissera, F.G.A., Skak-Nielsen, T., Seier-Poulson, S. & Nielsen, O.V. (1988). Processing and secretion of prosomatostatin by the pig pancreas. *Pancreas* **3**(6), 653–61.

Holst, J.J., Olsen, J. & Lund, T. (1987*a*). Are the enteroglucagons encoded by the glucagon gene? *Diabetologia* **30**, 532A.

Holst, J.J., Ørskov, C. & Schwartz, T.W. (1987*b*). Truncated glucagon-like peptide I, an insulin-releasing hormone from the distal gut. *FEBS Letters* **211**, 169–74.

Hook, V.Y.H., Eiden, L.E. & Brownstein, M.J. (1982). A carboxypeptidase processing enzyme for enkephalin precursors. *Nature* **295**, 341–2.

Hook, V.Y.H. & Loh, Y.P. (1984). Carboxypeptidase B-like converting enzyme activity in secretory granules of rat pituitary. *Proceedings of the National Academy of Sciences of the U.S.A.* **81**, 2776–80.

Hortin, G. & Boime, I. (1981). Miscleavage at the presequence of rat preprolactin synthesized in pituitary cells incubated with a threonine analog. *Cell* **24**, 453–61.

Howell, S.L. & Tyhurst, M. (1984). Insulin secretion: the effector system. *Experientia* **40**, 1098–105.

Huttner, W.B. & Benedum, U.M. (1987). Chromogranin A and pancreastatin. *Nature* **325**, 305.

Hutton, J.C. (1984). Secretory granules. *Experientia* **40**, 1091–8.

Hutton, J.C., Davidson, H.W. & Peshavaria, M. (1987*a*). Proteolytic processing of chromo-granin A in purified insulin granules. Formation of a 20 kDa N-terminal fragment (betagranin) by the concerted action of a Ca^{2+}-dependent endopeptidase and carboxypep-tidase H (EC 3.4.17.10). *Biochemical Journal* **244**, 457–64.

Hutton, J.C., Davidson, H.W., Grimaldi, K.A. & Peshavaria, M. (1987b). Biosynthesis of betagranin in pancreatic β-cells. *Biochemical Journal* **244**, 449–56.

Hutton, J.C., Hansen, F. & Peshavaria, M. (1985). β-Granins: 21 kDa co-secreted peptides of the insulin granule closely related to adrenal medullary chromogranin A. *FEBS Letters* **188**, 336–40.

Iacangelo, A., Affolter, H.-U., Eiden, L.E., Herbert, E. & Grimes, M. (1986). Bovine chromogranin A sequence and distribution of its messenger RNA in endocrine tissues. *Nature* **324**, 82–6.

Itakura, K., Hirose, T., Crea, R., Riggs, A.D., Heyneker, H.L., Bolivivar, F. & Boyer, H.W. (1977). Expression in *Escherichia coli* of a chemically synthesized gene for the hormone somatostatin. *Science* **198**, 1056–63.

Julius, D., Brake, A., Blair, L., Kunisawa, L.R. & Throner, J. (1984). Isolation of the putative structural gene for the lysine-arginine-cleaving endopeptidase required for processing of yeast prepro-ã-factor. *Cell* **37**, 1075–89.

Kassenbrock, C.K., Garcia, P.D., Walter, P. & Kelly, R.B. (1988). Heavy-chain binding protein recognizes aberrant polypeptides translocated in vitro. *Nature* **333**, 90–3.

Kawai, K., Ipp, E., Orci, L., Perrelet, A. & Unger, R.H. (1982). Circulating somatostatin acts on the islets of Langerhans by way of a somatostatin-poor compartment. *Science* **218**, 477–8.

Kelly, R.B. (1985). Pathways of protein secretion in eukaryotes. *Science* **230**, 25–32.

Kwok, S.C., Chan, S.J. & Steiner, D.F. (1983). Cloning and nucleotide sequence analysis of the dog insulin gene. *Journal of Biological Chemistry* **258**, 2357–62.

Larsson, L.-I., Golterman, N., de Magistris, L., Rehfeld, J.F. & Schwartz, T.W. (1979). Somatostatin cell processes as pathways for paracrine secretion. *Science* **205**, 1393–5.

Laub, O. & Rutter, W.J. (1983). Expression of the human insulin gene and cDNA in a heterologous mammalian system. *Journal of Biological Chemistry* **258**, 6043–50.

Lauber, M., Cannier, M. & Cohen, P. (1979). Higher molecular weight forms of immunoreactive somatostatin in mouse hypothalamic extracts: evidence of processing in vitro. *Proceedings of the National Academy of Sciences of the U.S.A.* **76**, 6004–8.

Leiter, A.B., Keutmann, H.T. & Goodman, R.H. (1984). Structure of a precursor to human pancreatic polypeptide. *Journal of Biological Chemistry* **259**, 14702–5.

Leiter, A.B., Montminy, M.R., Jamieson, E. & Goodman, R.H. (1985). Exons of the human pancreatic polypeptide gene define functional domains of the precursor. *Journal of Biological Chemistry* **260**, 13013–17.

Leiter, A.B., Toder, A., Wolfe, H.J., Taylor, I.L., Cooperman, S., Mandel, G. & Goodman, R.H. (1987). Peptide YY. Structure of the precursor and expression in exocrine pancreas. *Journal of Biological Chemistry* **262**, 12984–8.

Lindberg, I., Yang, H.Y.T. & Costa, E. (1984). Further characterization of an enkephalin-generating enzyme from adrenal medullary chromaffin granules. *Journal of Neurochemistry* **42**, 1411–18.

Loh, Y.P., Parish, D.C. & Tuteja, R. (1985). Purification and characterization of a paired basic residue-specific pro-opiomelanocortin converting enzyme from bovine pituitary intermediate lobe secretory vesicles. *Journal of Biological Chemistry* **260**, 7194–205.

Lomedico, P.T. (1982). Use of recombinant DNA technology to program eukaryotic cells to synthesize rat proinsulin: a rapid expression assay for cloned genes. *Proceedings of the National Academy of Sciences of the U.S.A.* **79**, 5798–802.

Lopez, L.C., Frazier, M.L., Su, C.-J., Kumar, A. & Saunders, G.F. (1983). Mammalian pancreatic preproglucagon contains three glucagon-related peptides. *Proceedings of the National Academy of Sciences of the U.S.A.* **80**, 5485–9.

Low, M.J., Hammer, R.E., Goodman, R.H., Habener, J.F., Palmiter, R.D. & Brinster, R.L. (1985). Tissue-specific posttranslational processing of pre-prosomatostatin encoded by a metallothionein-somatostatin fusion gene in transgenic mice. *Cell* **41**, 211–19.

Low, M.J., Lechan, R.M., Hammer, R.E., Brinster, R.L., Habener, J.F., Mandel, G. & Goodman, R.H. (1986a). Gonadotroph-specific expression of metallothionein fusion genes in pituitaries of transgenic mice. *Science* 231, 1002-4.

Low, M.J., Stork, P.J., Hammer, R.E., Brinster, R.L., Warhol, M.J., Mandel, G. & Goodman, R.H. (1986b). Somatostatin is targeted to the regulated secretory pathway of gonadotrophs in transgenic mice expressing a metallothionein-somatostatin gene. *Journal of Biological Chemistry* 261, 16260-3.

Lund, P.K., Goodman, R.H., Dee, P.C. & Habener, J.F. (1982). Pancreatic preproglucagon cDNA contains two glucagon-related coding sequences arranged in tandem. *Proceedings of the National Academy of Sciences of the U.S.A.* 79, 345-9.

MacDonald, R.J., Stary, S.J. & Swift, G.H. (1982). Two similar but nonallelic rat pancreatic trypsinogens. *Journal of Biological Chemistry* 257, 9724-32.

Mackin, R.B. & Noe, B.D. (1987a). Characterization of an islet carboxypeptidase B involved in prohormone processing. *Endocrinology* 120, 457-68.

Mackin, R.B. & Noe, B.D. (1987b). Direct evidence for two distinct prosomatostatin converting enzymes. *Journal of Biological Chemistry* 262, 6453-6.

Mainz, R.E. & Eipper, B.A. (1984). Secretion and regulation of two biosynthetic enzyme activities, peptidyl-glycine alpha-amidating monooxygenase and a carboxypeptidase, by mouse pituitary corticotrophic tumor cells. *Endocrinology* 115, 1683-90.

Markussen, J. & Sundby, F. (1972). Rat-proinsulin C-peptides. Amino-acid sequences. *European Journal of Biochemistry* 25, 153-62.

Mason, A.J., Evans, B.A., Cox, D.R., Shine, J. & Richards, R.I. (1983). Structure of mouse kallikrein gene family suggests a role in specific processing of biologically active peptides. *Nature* 303, 300-7.

Mellman, I., Fuchs, R. & Helenius, A. (1986). Acidification of the endocytic and exocytic pathways. *Annual Review of Biochemistry* 55, 663-700.

Michael, J., Carroll, R., Swift, H.H. & Steiner, D.F. (1987). Studies on the molecular organization of rat insulin secretory granules. *Journal of Biological Chemistry* 262, 16531-5.

Minth, C.D., Bloom, S.R., Polak, J.M. & Dixon, J.E. (1984). Cloning, characterization, and DNA sequence of a human cDNA encoding neuropeptide tyrosine. *Proceedings of the National Academy of Sciences of the U.S.A.* 81, 4577-81.

Mojsov, S., Heinrich, G., Wilson, I.B., Ravazzola, M., Orci, L. & Habener, J.F. (1986). Preproglucagon gene expression in pancreas and intestine diversifies at the level of post-translational processing. *Journal of Biological Chemistry* 261, 11880-9.

Mojsov, S., Weir, G.C. & Habener, J.F. (1987). Insulinotropin: glucagon-like peptide I(7-37) co-ended in the glucagon gene is a potent stimulator of insulin release in the perfused rat pancreas. *Journal of Clinical Investigation* 79, 616-19.

Mommsen, T.P., Andrews, P.C. & Plisetskaya, E.M. (1987). Glucagon-like peptides activate hepatic gluconeogenesis. *FEBS Letters* 219, 227-32.

Moody, A.J., Holst, J.J., Thim, L. & Jensen, S.L. (1981). Relationship of glicentin to proglucagon and glucagon in the porcine pancreas. *Nature* 289, 514-16.

Moore, H.-P.H. (1986). Factors controlling packaging of peptide hormones into secretory granules. *Annals of the New York Academy of Sciences* 493, 50-61.

Moore, H.-P.H., Gumbiner, B. & Kelly, R.B. (1983a). A subclass of proteins and sulfated macromolecules secreted by AtT-20 (mouse pituitary tumor) cells is sorted with adrenocorticotropin into dense secretory granules. *Journal of Cell Biology* 97, 810-17.

Moore, H.-P.H. & Kelly, R.B. (1985). Secretory protein targeting in a pituitary cell line: differential transport of foreign secretory proteins to distinct secretory pathways. *Journal of Cell Biology* 101, 1773-81.

Moore, H.-P.H. & Kelly, R.B. (1986). Re-routing of a secretory protein by fusion with human growth hormone sequences. *Nature* 321, 443-6.

Moore, H.-P.H., Walker, M.D., Lee, F. & Kelly, R.B. (1983b). Expressing a human proinsulin cDNA in a mouse ACTH-secreting cell: intracellular storage, proteolytic processing, and secretion on stimulation. *Cell* **35**, 531–8.

Munro, S. & Pelham, H.R.B. (1986). An Hsp70-like protein and the ER: identify with the 78 kd glucose-regulated protein and immunoglobulin heavy chain binding protein. *Cell* **46**, 291–300.

Munro, S. & Pelham, H.R.B. (1987). A C-terminal signal prevents secretion of luminal ER proteins. *Cell* **48**, 899–907.

Newgard, C.B. & Holst, J.J. (1981). Heterogeneity of somatostatin-like immunoreactivity (SLI) in extracts of porcine, canine, and human pancreas. *Acta Endocrinologica* **96**, 564–72.

Noe, B.D. (1981). Synthesis of one form of pancreatic islet somatostatin predominates. *Journal of Biological Chemistry* **256**, 9397–400.

Noe, B.D., Andrews, P.C., Dixon, J.E. & Spiess, J. (1986). Cotranslational and posttranslational proteolytic processing of preprosomatostatin-I in intact islet tissue. *Journal of Cell Biology* **103**, 1205–11.

Noe, B.D., Fletcher, D.J. & Spiess, J. (1979). Evidence for the existence of a biosynthetic precursor for somatostatin. *Diabetes* **28**, 724–30.

Noe, B.D. & Spiess, J. (1983). Evidence for biosynthesis and differential post-translational proteolytic processing of different (pre)prosomatostatins in pancreatic islets. *Journal of Biological Chemistry* **258**, 1121–8.

Nolan, C., Margoliash, E., Peterson, J.D. & Steiner, D.F. (1971). The structure of bovine proinsulin. *Journal of Biological Chemistry* **246**, 2780–95.

Novak, U., Wilks, A., Buell, G. & McEwen, S. (1987). Identical mRNA for preproglucagon in pancreas and gut. *European Journal of Biochemistry* **164**, 553–8.

Ohsuye, K., Kitano, K., Wada, Y., Fuchimura, K., Tanaka, S., Mizuno, K. & Matsuo, H. (1988). Cloning of cDNA encoding a new peptide C-terminal alpha-amidating enzyme having a putative membrane-spanning domain from *Xenopus laevis* skin. *Biochemical and Biophysical Research Communications* **150**, 1275–81.

Ole-Moi Yoi, O., Seldin, D.C., Spragg, J., Pinkus, G.S. & Austen, K.F. (1979). Sequential cleavage of proinsulin by human pancreatic kallikrein and a human pancreatic kinase. *Proceedings of the National Academy of Sciences of the U.S.A.* **76**, 3612–16.

Orci, L., Halban, P., Amherdt, M., Ravazzola, M., Vassalli, J.D. & Perrelet, A. (1984a). A clathrin-coated, Golgi-related compartment of the insulin secreting cell accumulates proinsulin in the presence of monensin. *Cell* **39**, 39–47.

Orci, L., Halban, P., Amherdt, M., Ravazzola, M., Vassalli, J.D. & Perrelet, A. (1984b). Nonconverted, amino acid analog-modified proinsulin stays in a Golgi-derived clathrin-coated membrane compartment. *Journal of Cell Biology* **99**, 2187–92.

Orci, L., Ravazzola, M., Amherdt, M., Louvard, D. & Perrelet, A. (1985a). Clathrin-immunoreactive sites in the Golgi apparatus are concentrated at the trans pole in polypeptide hormone-secreting cells. *Proceedings of the National Academy of Sciences of the U.S.A.* **82**, 5385–9.

Orci, L., Ravazzola, M., Amherdt, M., Madsen, O., Perrelet, A., Vassalli, J.-D. & Anderson, R.G.W. (1986). Conversion of proinsulin to insulin occurs coordinately with acidification of maturing secretory vesicles. *Journal of Cell Biology* **103**, 2273–81.

Orci, L., Ravazzola, M., Amherdt, M., Madsen, O., Vassalli, J.-D. & Perrelet, A. (1985b). Direct identification of prohormone conversion site in insulin-secreting cells. *Cell* **42**, 671–81.

Orci, L., Ravazzola, M., Amherdt, M., Perrelet, A., Powell, S.K., Quinn, D.L. & Moore, H.-P.H. (1987a). The trans-most cisternae of the Golgi complex: a compartment for sorting of secretory and plasma membrane proteins. *Cell* **51**, 1039–51.

Orci, L., Ravazzola, M., Amherdt, C., Yanaihara, C., Yanaihara, N., Halban, P., Renold,

A.E. & Perrelet, A. (1984c). Insulin, not C-peptide (proinsulin), is present in crinophagic bodies of the pancreatic B-cell. *Journal of Cell Biology* **98**, 222–8.

Orci, L., Ravazzola, M. & Anderson, R.G.W. (1987). The condensing vacuole of exocrine cells is more acidic than the mature secretory vesicle. *Nature* **326**, 77–9.

Orci, L., Ravazzola, M., Storch, M.-J., Anderson, R.G.W., Vassalli, J.-D. & Perrelet, A. (1987b). Proteolytic maturation of insulin is a post-Golgi event which occurs in acidifying clathrin-coated secretory vesicles. *Cell* **49**, 865–8.

Oyer, P.E., Cho, S., Peterson, J.D. & Steiner, D.F. (1971). Studies on human proinsulin. Isolation and amino acid sequence of the human pancreatic C-peptide. *Journal of Biological Chemistry* **246**, 1375–86.

Parish, D.C., Tuteja, R., Altstein, M., Gainer, H. & Loh, Y.P. (1986). Purification and characterization of a paired basic residue-specific prohormone-converting enzyme from bovine pituitary neural lobe secretory vesicles. *Journal of Biological Chemistry* **261**, 14392–7.

Patel, Y.C. & O'Neil, W. (1988). Peptides derived from the cleavage of prosomatostatin at carboxyl- and amino-terminal segments. Characterization of tissue and secreted forms in the rat. *Journal of Biological Chemistry* **263**, 745–51.

Patel, Y.C. & Reichlin, S. (1978). Somatostatin in hypothalamus, extrahypothalamic brain, and peripheral tissues of the rat. *Endocrinology* **102**, 523–30.

Patel, Y.C., Wheatley, T. & Ning, C. (1981). Multiple forms of immunoreactive somatostatin: comparison of distribution in neural and non-neural tissues and portal plasma in the rat. *Endocrinology* **109**, 1943–9.

Patzelt, C., Labrecque, A.D., Duguid, J.R., Carroll, R.J., Keim, P.S., Heinrikson, R.L. & Steiner, D.F. (1978). Detection and kinetic behavior of preproinsulin in pancreatic islets. *Proceedings of the National Academy of Sciences of the U.S.A.* **75**, 1260–4.

Patzelt, C. & Schlitz, E. (1984). Conversion of proglucagon in pancreatic alpha cells: The major endproducts are glucagon and a single peptide, the major proglucagon fragment, that contains two glucagon-like sequences. *Proceedings of the National Academy of Sciences of the U.S.A.* **81**, 5007–11.

Patzelt, C. & Schug, G. (1981). The major proglucagon fragment: an abundant islet protein and secretory product. *FEBS Letters* **129**, 127–30.

Patzelt, C., Tager, H.S., Carroll, R.J. & Steiner, D.F. (1979). Identification and processing of proglucagon in pancreatic islets. *Nature* **282**, 260–6.

Patzelt, C., Tager, H.S., Carroll, R.J. & Steiner, D.F. (1980). Identification of prosomatostatin in pancreatic islets. *Proceedings of the National Academy of Sciences of the U.S.A.* **77**, 2410–14.

Patzelt, C. & Weber, B. (1986). Early O-glycosidic glycosylation of proglucagon in pancreatic islets: an unusual type of prohormonal modification. *EMBO Journal* **5**, 2103–8.

Pelham, H.R.B. (1986). Speculations on the function of the major heat shock and glucose-regulated proteins. *Cell* **46**, 959–61.

Penman, E., Wass, J.A.H., Butler, M.G., Penny, E.S., Price, J., Wu, P. & Rees, L.H. (1983). Distribution and characterization of immunoreactive somatostatin in human gastrointestinal tract. *Regulatory Peptides* **7**, 53–65.

Piszkiewicz, D. (1974). The pH-dependent conformational change of gastrin. *Nature* **248**, 341–2.

Plisetskaya, E.M., Pollock, H.G., Elliott, W.M., Youson, J.H. & Andrews, P.C. (1988). Isolation and structure of Lamprey (*Petromyzon marinus*) insulin. *General and Comparative Endocrinology* **69**, 46–55.

Plisetskaya, E.M., Pollock, H.G., Rouse, J.B., Hamilton, J.W., Kimmel, J.R. & Gorbman, A. (1986). Isolation and structures of coho salmon (*Oncorhynchus kisutch*) glucagon and glucagon-like peptide. *Regulatory Peptides* **14**, 57–67.

Pradayrol, L., Jörnvall, H., Mutt, V. & Ribet, A. (1980). N-terminally extended somatostatin: the primary structure of somatostatin-28. *FEBS Letters* **109**, 55–8.

Ravazzola, M., Benoit, R., Ling, N., Guillemin, R. & Orci, L. (1983). Immunocytochemical localization of prosomatostatin fragments in maturing and mature secretory granules of pancreatic and gastrointestinal D cells. *Proceedings of the National Academy of Sciences of the U.S.A.* **80**, 215–18.

Ravazzola, M. & Orci, L. (1980). Glucagon and glicentin immunoreactivity are topologically segregated in the alpha-granule of the human pancreatic A-cell. *Nature* **284**, 66–8.

Ravazzola, M., Perrelet, A., Unger, R.H. & Orci, L. (1984). Immunocytochemical characterization of secretory granule maturation in pancreatic A-cells. *Endocrinology* **114**, 481–5.

Rhodes, C.J. & Halban, P.A. (1988). The intracellular handling of insulin-related peptides in isolated pancreatic islets. *Biochemical Journal* (in press).

Rholam, M.M., Nicolas, P. & Cohen, P. (1986). Precursors for peptide hormones share common secondary structures forming features at the proteolytic processing sites. *FEBS Letters* **207**, 1–6.

Robbins, D.C., Blix, P.M., Rubenstein, A.H., Kanazawa, Y., Kosaka, K. & Tager, H.S. (1981). A human proinsulin variant at arginine 65. *Nature* **291**, 679–81.

Rothman, J.E. (1987). Protein sorting by selective retention in the endoplasmic reticulum and Golgi stack. *Cell* **50**, 521–2.

Schally, A.V., Huang, W.Y., Chang, R.C.C., Arimura, A., Redding, T.W., Millar, R.P. & Hood, L.E. (1980). Isolation and structure of pro-somatostatin: putative somatostatin precursor from pig hypothalamus. *Proceedings of the National Academy of Sciences of the U.S.A.* **77**, 4489–93.

Schmidt, W.E., Mutt, V., Kratzin, H., Carlquist, M., Conlon, J.M. & Creutzfeldt, W. (1985a). Isolation and characterization of proSS$_{1-32}$, a peptide derived from the N-terminal region of porcine preprosomatostatin. *FEBS Letters* **192**, 141–6.

Schmidt, W.E., Siegel, E.G. & Creutzfeldt, W. (1985b). Glucagon-like peptide-1 but not glucagon-like peptide-2 stimulates insulin release from isolated rat pancreatic islets. *Diabetologia* **28**, 704–7.

Schwartz, T.W. (1983). Pancreatic polypeptide: a hormone under vagal control. *Gastroenterology* **85**, 1411–25.

Schwartz, T.W. (1986). The processing of peptide precursors: 'proline directed arginyl cleavage' and other monobasic processing mechanisms. *FEBS Letters* **200**, 1–10.

Schwartz, T.W. (1987). Cellular peptide processing after a single arginyl residue – studies on the common precursor for pancreatic polypeptide and pancreatic icosapeptide. *Journal of Biological Chemistry* **262**, 5093–9.

Schwartz, G.P., Burke, G.T. & Katsoyannis, P.G. (1987a). A superactive insulin: B10-aspartic acid insulin (human). *Proceedings of the National Academy of Sciences of the U.S.A.* **84**, 6408–11.

Schwartz, T.W., Gingerich, R.L. & Tager, H.S. (1980). Biosynthesis of pancreatic polypeptide. Identification of a precursor and a co-synthesized product. *Journal of Biological Chemistry* **255**, 11494–8.

Schwartz, T.W., Hansen, H.F., Håkanson, R., Sundler, F. & Tager, H.S. (1984). Human pancreatic icosapeptide: isolation, sequence, and immunocytochemical localization of the COOH-terminal fragment of the pancreatic polypeptide precursor. *Proceedings of the National Academy of Sciences of the U.S.A.* **81**, 708–12.

Schwartz, T.W., Sheikh, S.P. & O'Hare, M.M.T. (1987b). Receptors on phaeochromocytoma cells for two members of the PP-fold family – NPY and PP. *FEBS Letters* **225**, 209–14.

Schwartz, T.W. & Tager, H.S. (1981). Isolation and biogenesis of a new peptide from pancreatic islets. *Nature* **294**, 589–91.

Schwartz, T.W., Wittels, B. & Tager, H.S. (1983). Hormone precursor processing in the pancreatic islet. In *Peptides – structure and function* (ed. V.J. Hruby & D.H. Rich), pp. 229–38. Rockford, Illinois: Pierce Chemical Co. Publications.

Schweitzer, E.S. & Kelly, R.B. (1985). Selective packaging of human growth hormone into synaptic vesicles in a rat neuronal (PC12) cell line. *Journal of Cell Biology* **101**, 667–76.

Sevarino, K.S., Felix, R., Banks, C.M., Low, M.J., Montminy, M.R., Mandel, G. & Goodman, R.H. (1987). Cell-specific processing of preprosomatostatin in cultured neuroendocrine cells. *Journal of Biological Chemistry* **262**, 4987–93.

Sharp, R.R. & Richards, E.P. (1977). Molecular mobilities of soluble components in the aqueous phase of chromaffin granules. *Biochimica et Biophysica Acta* **497**, 260–71.

Sheikh, S.P., Baldissera, F.G.A., Karlsen, F.Ø. & Holst, J.J. (1985). Glicentin is present in the pig pancreas. *FEBS Letters* **179**, 1–6.

Shen, L.-P., Pictet, R.L. & Rutter, W.J. (1982). Human somatostatin I. Sequence of the cDNA. *Proceedings of the National Academy of Sciences of the U.S.A.* **79**, 4575–9.

Shen, L.-P. & Rutter, W.J. (1984). Sequence of the human somatostatin I gene. *Science* **224**, 168–70.

Shibasaki, Y., Kawakami, T., Kanazawa, Y., Akanuma, Y. & Takaku, F. (1985). Posttranslational cleavage of proinsulin is blocked by a point mutation in familial hyperproinsulinemia. *Journal of Clinical Investigation* **76**, 378–80.

Shields, D. (1980). In vitro biosynthesis of fish islet preprosomatostatin: evidence of processing and segregation of a high molecular weight precursor. *Proceedings of the National Academy of Sciences of the U.S.A.* **77**, 4074–8.

Sopwith, A.M., Hales, C.N. & Hutton, J.C. (1984). Pancreatic B-cells secrete a range of novel peptides besides insulin. *Biochimica et Biophysica Acta* **803**, 342–5.

Spiess, J. & Noe, B.D. (1985). Processing of an anglerfish somatostatin precursor to a hydroxylysine-containing somatostatin 28. *Proceedings of the National Academy of Sciences of the U.S.A.* **82**, 277–81.

Steiner, D.F. (1978). On the role of the proinsulin C-peptide. *Diabetes* **27**, 145–8.

Steiner, D.F. (1984). The Biosynthesis of Insulin: Genetic, Evolutionary & Pathophysiological aspects. *Harvey Lectures*, series 78, 191–228.

Steiner, D.F., Cunningham, D.D., Spigelman, L. & Aten, B. (1967). Insulin biosynthesis: evidence for a precursor. *Science* **157**, 697–700.

Steiner, D.F., Kemmler, W., Clark, J.L., Oyer, P.E. & Rubenstein, A.H. (1972). The biosynthesis of insulin. In *Handbook of physiology*, section 7 (*Endocrinology*), vol. 1 (ed. N. Freinkel & D.F. Steiner), pp. 175–98. Washington, D.C.: American Physiological Society.

Steiner, D.F., Kemmler, W., Tager, H.S. & Peterson, J.D. (1974). Proteolytic processing in the biosynthesis of insulin and other proteins. *Federation Proceedings* **33**, 2105–15.

Steiner, D.F., Michael, J., Houghten, R., Mathieu, M., Gardner, P.R., Ravazzola, M. & Orci, L. (1987). Use of a synthetic peptide antigen to generate antisera reactive with a proteolytic processing site in native human proinsulin: demonstration of cleavage within clathrin-coated (pro)secretory vesicles. *Proceedings of the National Academy of Sciences of the U.S.A.* **84**, 6184–8.

Steiner, D.F. & Oyer, P.E. (1967). The biosynthesis of insulin and a probable precursor of insulin by a human islet cell adenoma. *Proceedings of the National Academy of Sciences of the U.S.A.* **57**, 473–80.

Suchanek, G. & Kreil, G. (1977). Translation of mellitin messenger RNA in vitro yields a product terminating with glutaminylglycine rather than with glutaminamide. *Proceedings of the National Academy of Sciences of the U.S.A.* **74**, 975–8.

Suppatone, S., Fricker, L.D. & Snyder, S.H. (1984). Purification and characterization of a membrane-bound enkephalin-forming carboxypeptidase, 'enkephalin convertase'. *Journal of Neurochemistry* **42**, 1017–23.

Tager, H.S. (1984). Abnormal products of the human insulin gene. *Diabetes* **33**, 693–9.

Tager, H.S., Emdin, S.O., Clark, J.L. & Steiner, D.F. (1973). Studies on the conversion of proinsulin to insulin II: evidence for a chymotryptic-like cleavage of proinsulin C-peptide in the rat. *Journal of Biological Chemistry* **248**, 3476–82.

Tager, H.S., Given, B., Baldwin, D., Mako, M., Rubenstein, A., Olefsky, J., Kobayaski, M., Koltermann, O. & Poucher, P. (1979). A structurally abnormal insulin causing human diabetes. *Nature* **281**, 122–5.

Tager, H.S., Patzelt, C., Assoian, R.K., Chan, S.J.,Duguid, J.R. & Steiner, D.F. (1980). Biosynthesis of islet cell hormones. *Annals of the New York Academy of Sciences* **343**, 133–47.

Tager, H.S. & Steiner, D.F. (1973). Isolation of glucagon-containing peptide: primary structure of a possible fragment of proglucagon. *Proceedings of the National Academy of Sciences of the U.S.A.* **70**, 2321–5.

Takeuchi, T. & Yamada, T. (1985). Isolation of a cDNA clone encoding pancreatic polypeptide. *Proceedings of the National Academy of Sciences of the U.S.A.* **82**, 1536–9.

Tatemoto, K., Efendic, S., Mutt, V., Makk, G., Feinster, G.F. & Barachas, J.D. (1986). Pancreastatin, a novel pancreatic peptide that inhibits insulin secretion. *Nature* **324**, 476–8.

Tatemoto, K. & Mutt, V. (1978). Chemical determination of polypeptide hormones. *Proceedings of the National Academy of Sciences of the U.S.A.* **75**, 4115–19.

Tavianini, M.A., Hayes, T.E., Magazin, M.D., Minth, C.D. & Dixon, J.E. (1984). Isolation, characterization, and DNA sequence of the rat somatostatin gene. *Journal of Biological Chemistry* **259**, 11789–803.

Thim, L., Hansen, M.T., Norris, K., Hoegh, I., Boel, E., Forstrom, J., Ammerer, G. & Fiil, N.P. (1986). Secretion and processing of insulin precursors in yeast. *Proceedings of the National Academy of Sciences of the U.S.A.* **83**, 6766–70.

Thim, L., Hansen, M.T. & Sørensen, A.R. (1987). Secretion of human insulin by transformed yeast cell. *FEBS Letters* **212**, 307–12.

Thim, L. & Moody, A.J. (1982). Purification and chemical characterization of a glicentin-related pancreatic peptide (proglucagon fragment) from porcine pancreas. *Biochimica et Biophysica Acta* **703**, 134–41.

Thomas, G., Herbert, E. & Hruby, D.E. (1986). Expression and cell type-specific processing of human preproenkephalin with a vaccinia recombinant. *Science* **232**, 1641–3.

Thomas, G., Thorne, B.A., Thomas, L., Allen, R.G., Hruby, D.E., Fuller, R. & Thorner, J. (1988). Yeast KEX2 endoprotease correctly cleaves a neuroendocrine prohormone in mammalian cells. *Science* **241**, 226–30.

Trent, D.F. & Weir, G.C. (1981). Heterogeneity of somatostatin-like peptides in rat brain, pancreas, and gastrointestinal tract. *Endocrinology* **108**, 2033–7.

Virji, M.A.G., Vassalli, J.-D., Estensen, R.D. & Reich, E. (1980). Plasminogen activator of islets of Langerhans: modulation by glucose and correlation with insulin production. *Proceedings of the National Academy of Sciences of the U.S.A.* **77**, 875–9.

Von Heijne, G. (1983). Patterns of amino acids near signal-sequence cleavage sites. *European Journal of Biochemistry* **133**, 17–21.

Von Heijne, G. (1985). Ribosome-SRP-signal sequence interactions. *FEBS Letters* **190**, 1–5.

Warren, T.G. & Shields, D. (1984a). Cell-free biosynthesis of multiple preprosomatostatin: characterization by hybrid selection and amino-terminal sequencing. *Biochemistry* **23**, 2684–90.

Warren, T.G. & Shields, D. (1984b). Expression of preprosomatostatin in heterologous cells: biosynthesis, posttranslational processing and secretion of mature somatostatin. *Cell* **39**, 547–55.

Wu, P., Penman, E., Coy, D.H. & Rees, L.H. (1983). Evidence for direct production of somatostatin-28 in a phaeochromocytoma. *Regulatory Peptides* **5**, 219–33.

Yamamoto, H., Nata, K. & Okamoto, H. (1986). Mosaic evolution of prepropancreatic polypeptide. *Journal of Biological Chemistry* **261**, 6156–9.

Yanaihara, C., Matsumoto, T., Hong, Y.-M. & Yanaihara, N. (1985a). Isolation and chemical characterization of glicentin C-terminal hexapeptide in porcine pancreas. *FEBS Letters* **189**, 50–5.

Yanaihara, C., Matsumoto, T., Kadowaki, M., Iguchi, K. & Yanaihara, N. (1985b). Rat pancreas contains the proglucagon(64–69) fragment and arginine stimulates its release. *FEBS Letters* **187**, 307–10.

Yip, C.C. (1971). Bovine pancreatic enzyme catalyzing conversion of proinsulin to insulin. *Proceedings of the National Academy of Sciences of the U.S.A.* **68**, 1312–15.

Yip, C.C., Hew, C.-L. & Hsu, H. (1975). Translation of messenger ribonucleic acid from isolated pancreatic islets and human insulinomas. *Proceedings of the National Academy of Sciences of the U.S.A.* **72**, 4777–9.

Yoshimasa, Y., Seino, S., Whittaker, J., Kakehi, T., Kosaki, A., Kuzuya, H., Imura, H., Bell, G.I. & Steiner, D.F. (1988). Insulin-resistant diabetes due to a point mutation that prevents insulin proreceptor processing. *Science* **240**, 784–7.

Ørskov, C., Holst, J.J., Knuhtsen, S., Baldissera, F.G.A., Poulsen, S.S. & Nielsen, O.V. (1986). Glucagon-like peptides GLP-1 and GLP-2, predicted products of the glucagon gene, are secreted separately from pig small intestine but not pancreas. *Endocrinology* **119**, 1467–75.

Ørskov, C. & Nielsen, J.H. (1988). Truncated glucagon-like peptide-1 (proglucagon 78–107 amide), an intestinal insulin-releasing peptide, has specific receptors on rat insulinoma cells (RIN 5AH). *FEBS Letters* **229**, 175–8.

II

Molecular aspects of diabetes mellitus

10

The molecular basis of experimental diabetes

HIROSHI OKAMOTO

10.1. Introduction

Diabetes can be usefully studied at an experimental level by methods that make use of either surgical techniques or chemical agents.

Many of the acute metabolic derangements of severe human insulinopenic diabetes can be reproduced by the removal of insulin-producing pancreatic B-cells of the islets of Langerhans. This artificial form of diabetes was first produced by von Mering & Minkowski (1890) when they removed the pancreases of dogs. The same model of diabetes was adopted by Banting & Best (1922) for their historic observations on the hypoglycemic properties of a crude pancreatic extract of insulin.

Of chemical agents, alloxan (2,4,5,6-tetraoxohexahydropyrimidine) (Dunn et al., 1943) and streptozotocin (2-deoxy-2-(3-methyl-3-nitrosoureido)-D-glucopyranose) (Rakieten et al., 1963) have been found to be particularly instructive: they exert selective cytotoxic effects on pancreatic B-cells in animals and are extremely potent diabetogenic substances. Since the original discoveries were made by Dunn and Rakieten, alloxan and streptozotocin have been widely used to produce diabetes in experimental animals because of their particular advantages of specificity and convenience. How the B-cytotoxins produce impaired function and degenerative changes in the B-cell has been the subject of much debate in the literature (Dulin & Soret, 1977; Cooperstein & Watkins, 1981; Chang & Diani, 1985).

Recently, a hypothesis has been offered by Okamoto and coworkers, who present a unifying concept as to the diabetogenic effect of both alloxan and streptozotocin (Okamoto, 1981, 1985a; Okamoto & Yamamoto, 1983; Okamoto et al., 1988). These workers suggest that the principal action of B-cytotoxins on pancreatic B-cells is the induction of lesions in the DNA strand through the formation of free radicals. This injury induces DNA repair involving the action of poly(ADP-ribose) synthetase, which uses cellular NAD as a substrate. The depletion of the cellular pool of NAD results in the depression of B-cell functions, and the B-cell ultimately dies. Accordingly, this sequence of events can be interrupted either by the provision of free radical scavenging activity or by the inhibition of poly(ADP-ribose) synthetase; radical scavengers and inhibitors of the enzyme, such as nicotinamide, can prevent alloxan and streptozotocin diabetes. The hypothesis also attempts to explain the occurrence of B-cell tumors, insulinomas, in animals after administration of streptozotocin (or alloxan) and nicotinamide together (Rakieten *et al.*, 1971; Kazumi *et al.*, 1980; Yamagami *et al.*, 1985). More recently, it has been demonstrated that administration of poly(ADP-ribose) synthetase inhibitors to subtotally depancreatized rats induces regeneration of pancreatic B-cells, thereby ameliorating this kind of diabetes (Yonemura *et al.*, 1984; Okamoto, 1985*a, b*; Okamoto *et al.*, 1988). This chapter will deal primarily with recent studies on the molecular basis of experimental diabetes. A unifying model for B-cell damage and its prevention in toxin- or virus-induced and immune diabetes will also be introduced.

10.2. Historical background

Because the two highly B-cytotoxic compounds alloxan and streptozotocin are structurally different, the two substances were commonly thought not to act in an identical way. Certain observations, however, suggested great similarities between particular pathogenic reactions caused by these two chemicals (Rerup, 1970; Dulin & Soret, 1977). Such similarities are the B-cell chromatin clumping (Lazarus & Shapiro, 1972; Dulin & Soret, 1977) and the ability of nicotinamide to prevent diabetes induced by either substance. In 1950, Lazarow *et al.* reported that nicotinamide, in doses of 7.5 mmol per kilogram, injected intravenously immediately preceding a diabetogenic dose of alloxan, protected 23 out of 34 rats from alloxan diabetes (Lazarow *et al.*, 1950). Schein & Loftus (1968) found that injections of nicotinamide into mice either 15 min before or up to 2 h after streptozotocin treatment prevented the diabetogenic action.

Subsequently, a diabetogenic dose of streptozotocin was discovered to decrease the levels of NAD in mouse liver (Schein & Loftus, 1968) and pancreatic islets (Schein *et al.*, 1973; Hinz *et al.*, 1973; Hellerström *et al.*, 1974). These effects being prevented by nicotinamide (Dulin & Wyse, 1969; Anderson *et al.*, 1974; Gunnarsson, 1975), streptozotocin was considered to cause diabetes by impairing NAD synthesis in islets (Schein, 1969; Schein et al., 1973). On the other hand, Hinz et al. (1973) suggested that the streptozotocin-induced depression of mouse islet NAD content might be ascribed to increased NAD degradation. Doi (1975) and Kazumi et al. (1978) reported that picolinamide, an isomer of nicotinamide, also prevented streptozotocin diabetes in rats. Their observation suggested that the mechanism of streptozotocin diabetes and its prevention should be sought in the degradation of NAD, because picolinamide is structurally incapable of acting as a precursor in NAD synthesis via any biosynthetic routes (Preiss & Handler, 1957; Nishizuka & Hayaishi, 1963; Dietrich et al., 1966). Additionally, it has been found that nitrosourea compounds like streptozotocin increase the activity of poly(ADP-ribose) synthetase, an NAD-degrading enzyme, in some eukaryotic cells (Whish et al., 1975; Smulson et al., 1977). Poly(ADP-ribose) synthetase catalyzes chromatin-bound polymerization of an ADP-ribose moiety of NAD (Hayaishi & Ueda, 1977); evidence has accumulated that this enzyme plays a major role in NAD degradation in mammalian cells (Bock et al., 1968; Rechsteiner et al., 1976; Hayaishi & Ueda, 1977). In 1980, Yamamoto & Okamoto clearly demonstrated that poly(ADP-ribose) synthetase was present in rat pancreatic islet nuclei and that both picolinamide and nicotinamide were potent inhibitors of the islet enzyme. Further experiments with isolated islets showed that both picolinamide and nicotinamide could protect proinsulin synthesis, as well as the intracellular NAD level, against streptozotocin-induced depression (Yamamoto & Okamoto, 1980). A temporal correlation between an increase in poly(ADP-ribose) synthetase activity and a decrease in intracellular NAD content had been found in mouse leukemia cells treated with N-methyl-N-nitrosourea (Skidmore et al., 1979). Furthermore, Juarez-Salinas et al. (1979) had shown that an equimolar increase in the intracellular level of poly(ADP-ribose) coupled with an NAD reduction actually occurs in vivo. Therefore, in 1980, Yamamoto & Okamoto postulated that streptozotocin causes an increased flux from NAD through poly(ADP-ribose) by increasing poly(ADP-ribose) synthetase activity to depress islet NAD content and inhibit islet cell functions, including proinsulin synthesis, and that inhibitors of poly(ADP-ribose) synthetase, such as picolinamide and nicotinamide, prevent the diabetogenic action of streptozotocin by maintaining the intracellular NAD level (Yamamoto & Okamoto, 1980; Okamoto, 1981). This concept also explains the inability of nicotinic acid to prevent streptozotocin diabetes (Dulin & Wyse, 1969): nicotinic acid acts as a precursor for NAD synthesis but is proven not to inhibit poly(ADP-ribose) synthetase (Preiss *et al.*, 1971; Hayaishi & Ueda, 1977).

10.3. Streptozotocin and alloxan induce DNA strand breakage and poly(ADP-ribose) synthetase activity in pancreatic islets

Poly(ADP-ribose) synthetase activity is known to increase under the conditions that cause DNA damage, such as exposure to chemical carcinogens (Smulson *et al.*, 1977; Goodwin *et al.*, 1978; Berger *et al.*, 1980), nuclease treatment (Miller, 1975; Berger *et al.*, 1978) and UV- or γ-irradiation (Skidmore *et al.*, 1979; Berger *et al.*, 1980). Ohgushi *et al.* (1980) provided direct evidence that poly(ADP-ribose) synthetase purified from bovine thymus is activated only when the enzyme is bound to nicked or fragmented DNA. Furthermore, damage to the nuclei of B-cells, as evidenced by chromatin clumping, has been observed after treatment with alloxan and streptozotocin (Lazarus & Shapiro, 1972; Dulin & Soret, 1977). These observations led to the proposal that the streptozotocin-associated increase in islet nuclear poly(ADP-ribose) synthetase activity could be attributed to the prior induction of DNA lesions by the chemical (Okamoto, 1981). It was supposed that this was also true of alloxan (Okamoto, 1981).

In 1981, Yamamoto *et al.* demonstrated that streptozotocin and alloxan cause DNA strand breaks in pancreatic islets of rats (Yamamoto *et al.*, 1981*a*). These workers isolated islets from rat pancreas and incubated them with streptozotocin or alloxan for 5–20 min in Krebs–Ringer's bicarbonate medium. Velocity sedimentation of DNA from the islet cells was carried out in an alkaline sucrose gradient. DNA from islets incubated for 10 min without the diabetogenic agents was recovered as a single peak near the bottom of the gradient, the position at which undamaged DNA sediments (Fig. 10.1(*a*), (*e*)). However, after only 5 min incubation with 2 mM streptozotocin or 1 mM alloxan, a considerable amount of DNA sedimented as a broad peak in the middle of the gradient with a concomitant decrease in undamaged DNA (Fig. 10.1(*b*), (*f*)); after 10–20 min incubation, the DNA was almost completely fragmented (Fig. 10.1(*c*), (*d*), (*g*), (*h*)). The effect of the two agents on islet DNA fragmentation was dose-dependent. These results clearly indicate that streptozotocin and alloxan produce strand breaks in islet DNA.

Next, Yamamoto *et al.* prepared a nuclear fraction from islets incubated in conditions causing breaks in islet DNA, and assayed poly(ADP-ribose) synthetase activity (Yamamoto *et al.*, 1981*a*). As shown in Fig. 10.2(*a*, *b*), both streptozotocin (2 mM) and alloxan (1 mM) induced a 2- to

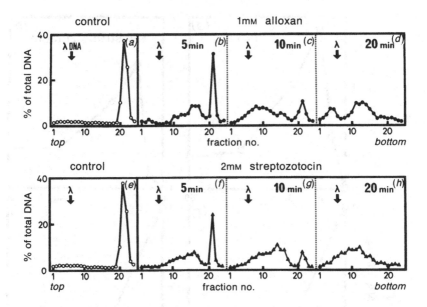

Fig. 10.1. Islet DNA strand breaks due to alloxan and streptozotocin (adapted from Yamamoto *et al.*, 1981*a*).

3-fold increase in islet poly(ADP-ribose) synthetase activity, with a peak at 10 min. The low (2- to 3-fold) increase in islet nuclear poly(ADP-ribose) synthetase activity was probably due to the high background of DNA breaks resulting from the isolation procedure of nuclei, which itself caused some DNA breaks. Cellular NAD content was reduced by either 2 mM streptozotocin or 1 mM alloxan within 20 min of incubation (Fig. 10.2(*c*, *d*)) and remained almost unaltered for 60 min. There was a striking temporal correlation between the decrease in the level of islet NAD and the increase in islet poly(ADP-ribose) synthetase activity. These results indicate that both alloxan and streptozotocin cause DNA strand breaks that result in an increase in poly(ADP-ribose) synthetase activity, thereby depleting islet NAD level.

The question now arises as to whether or not the biochemical events initiated by islet DNA breaks are actually induced *in vivo* by alloxan and streptozotocin administration. Yamamoto *et al.* (1981*b*) injected diabetogenic doses of alloxan and streptozotocin into rats via the tail vein and then isolated islets, analyzed islet DNA and determined islet NAD content. Alloxan (40 mg/kg) or streptozotocin (50 mg/kg) was injected intravenously into male Wistar rats. Pancreatic islets were isolated from

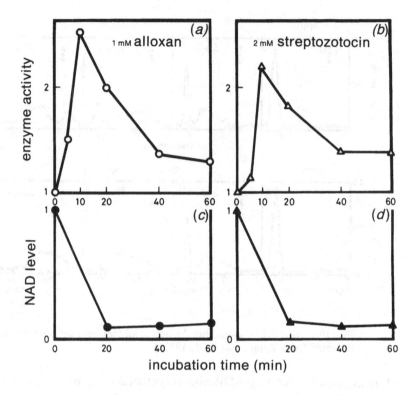

Fig. 10.2. Effect of alloxan and streptozotocin on islet poly(ADP-ribose) synthetase activity and NAD level (adapted from Yamamoto *et al.*, 1981*a*).

rats at 5–20 min after the injection, and submitted to alkaline sucrose gradient centrifugation. DNA of islets from untreated rats was recovered as a single peak near the bottom of the gradient, the position at which undamaged DNA sediments. Islet DNA from alloxan- or streptozotocin-treated rats was found to sediment as a broad peak in the middle of the gradient, with a concomitant decrease in undamaged DNA at 5–10 min after treatment. At 20 min after treatment, islet DNA was almost completely fragmented. DNA of pancreatic exocrine cells was essentially unaffected by either alloxan or streptozotocin treatment. Liver DNA was not affected by alloxan; however, it was fragmented by streptozotocin. Alloxan (40 mg/kg) and streptozotocin (50 mg/kg) led to a marked depletion in islet NAD content. Streptozotocin administration also decreased liver NAD to about 70% of the control; alloxan caused no significant change in liver NAD. The control NAD content of liver in untreated rats was about twice as great as that of islets when the value was calculated per microgram of DNA.

10.4. Protection by poly(ADP-ribose) synthetase inhibitors against the alloxan- and streptozotocin-induced decrease in islet NAD level and against the inhibition of islet functions

The above findings suggested a serial mechanism of action of alloxan and streptozotocin: alloxan and streptozotocin cause the DNA strand breaks which activate poly(ADP-ribose) synthetase, resulting in a decreased amount of NAD. Accordingly, it was expected that the inhibition of poly(ADP-ribose) synthetase would prevent a decrease in the NAD level. Nicotinamide and picolinamide have been shown to be potent inhibitors of islet nuclear poly(ADP-ribose) synthetase (Yamamoto & Okamoto, 1980). In addition, benzamide, 3-aminobenzamide, 3-nitrobenzamide, 3-methoxybenzamide, theophylline, and 3-isobutyl-1-methylxanthine (IBMX) are inhibitors of poly(ADP-ribose) synthetase (Uchigata et al., 1982). The concentrations at which these compounds cause 50% inhibition of the enzyme activity are as follows: benzamide, 4.5 μM; 3-aminobenzamide, 6.5 μM; 3-nitrobenzamide, 18 μM; 3-methoxybenzamide, 35 μM; theophylline, 65 μM; IBMX, 120 μM; picolinamide, 95 μM; nicotinamide, 120 μM. Benzamides were found to be the most potent inhibitors of islet poly(ADP-ribose) synthetase. The inhibitory ability of methylxanthines was similar to that of nicotinamide and picolinamide. These data are consistent with those obtained with non-islet cell nuclei (Preiss et al., 1971; Levi et al., 1978; Purnell & Whish, 1980; Oikawa et al., 1980). Poly(ADP-ribose) synthetase inhibitors such as nicotinamide and picolinamide can protect against the alloxan- as well as streptozotocin-induced decrease in islet NAD level (Yamamoto & Okamoto, 1980; Yamamoto et al., 1981a; Uchigata et al., 1982). Since NAD is the most abundant of cellular coenzymes and participates in many biological reactions in mammalian cells, the reduction in intracellular NAD to such a non-physiological level by alloxan and streptozotocin may severely affect islet cell functions. According to Gunnarsson et al. (1974), a marked depletion of islet NAD may be regarded as the primary molecular mechanism through which streptozotocin destroys B-cells. The ability of islets to synthesize proinsulin is considered to be a marker for the evaluation of the diabetogenicity of alloxan and streptozotocin. Uchigata et al. (1982) incubated rat pancreatic islets in the presence of alloxan or streptozotocin with or without the addition of the islet poly(ADP-ribose) synthetase inhibitors. Poly(ADP-ribose) synthetase inhibitors reversed the inhibition of proinsulin synthesis induced by alloxan and streptozotocin in a dose-dependent manner. The stronger inhibitors were found to protect against the inhibition of proinsulin synthesis at lower concentrations. The pretreatment of rats with poly(ADP-ribose) synthetase inhibitors was also found to protect against the alloxan- or streptozotocin-induced decrease in islet proinsulin synthesis (Uchigata *et al.*, 1983). 3-Aminobenzamide and nicotinamide were effective in preventing the inhibition of insulin release by streptozotocin and alloxan *in vitro* (Wilson *et al.*, 1984) as well as in protecting the rats from the diabetogenic effect of streptozotocin administered *in vivo* (Masiello *et al.*, 1985). The insulin content in poly(ADP-ribose) synthetase inhibitor-treated rat pancreas was restored; the glucagon content did not differ from that in normal rats (Shima *et al.*, 1987).

10.5 Mechanism of islet DNA strand breaks

It is important to establish how alloxan and streptozotocin induce the DNA strand breaks. It has been suggested that alloxan may work through the formation of the hydroxyl radical (OH·) (Heikkila *et al.*, 1976; Grankvist *et al.*, 1979; Fischer & Hamburger, 1980) which is produced by the interaction between superoxide (O_2^-·) and peroxide (H_2O_2) (Haber & Weiss, 1934; McCord & Day, 1978):

$$O_2^- \cdot + H_2O_2 \rightarrow OH \cdot + OH^- + O_2.$$

Superoxide dismutase and catalase catalyze the removal of O_2^-· and H_2O_2, respectively (Chance, 1947; McCord & Fridovich, 1969), and hence may inhibit the formation of OH·. Uchigata *et al.* (1982) showed that a combined administration of superoxide dismutase and catalase more effectively protects against alloxan-induced islet DNA breaks, as well as against proinsulin synthesis inhibition, than an administration of either of these scavenging enzymes alone. This result strongly suggests that it is OH·, rather than O_2^-· or H_2O_2, that attacks islet DNA. Brawn & Fridovich (1981) also showed that the hydroxyl radical breaks plasmid DNA. Grankvist *et al.* (1981) reported that *in vivo* injections of superoxide dismutase to mice act prophylactically against alloxan-induced diabetes, although whether the enzyme interacts with the oxygen radicals extracellularly or within islet cells remains to be elucidated. On the other hand, streptozotocin-induced islet DNA strand breaks were unaffected by the radical scavengers (Uchigata *et al.*, 1982). This is consistent with the result

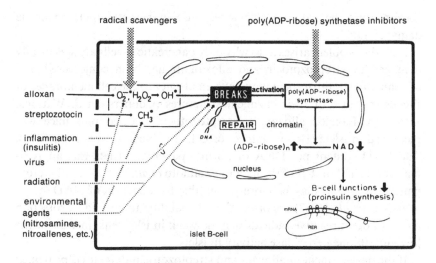

Fig. 10.3. The proposed mechanism of action of B-cytotoxic agents on pancreatic B-cells (adapted from Okamoto, 1981, 1985a, b; Okamoto et al., 1988).

reported by Gold et al. (1981) that superoxide dismutase failed to protect against the *in vivo* diabetogenic action of streptozotocin. Therefore, in contrast to alloxan, it is unlikely that streptozotocin acts on islet DNA through the formation of oxygen radicals. The streptozotocin-induced breakage of DNA is probably associated with the alkylating activity of the agent, as suggested with nitrosoureas (Cox & Irving, 1976; Erickson et al., 1977; Wilson et al., 1988).

10.6. The Okamoto model for B-cell damage and insulin-dependent diabetes in a new light

From the experimental results described above, Okamoto proposed a basic model for the action of alloxan and streptozotocin in the induction of experimental diabetes (Okamoto, 1981; Uchigata et al., 1982; Okamoto, 1985a). As shown in Fig. 10.3, the first step is the generation of free radicals by alloxan and streptozotocin, which attack DNA to produce strand breaks. Poly(ADP-ribose) synthetase then acts to repair the DNA breaks, consuming B-cell NAD. Since NAD is the most abundant of cellular coenzymes and participates in many biological reactions in mammalian cells, a severe reduction in intracellular NAD to non-physiological levels may adversely affect B-cell functions, including proinsulin synthesis and insulin secretion. Therefore, in diabetes induction, the

B-cells seem to be making a 'suicide response' in their attempt to repair the damaged DNA.

Here the problem arises as to why only pancreatic B-cells are specifically damaged by streptozotocin and alloxan. It has been conjectured that alloxan and streptozotocin may have an affinity to the cell membranes of B-cells, because of their chemical structures (Cooperstein & Watkins, 1981). ^{14}C-Labeled alloxan or streptozotocin injected into mice was recovered in islets (Hammarström & Ullberg, 1966; Tjälve et al., 1976). The NAD content per DNA of normal pancreatic islets was approximately one half of that of the liver (Yamamoto et al., 1981b) and therefore pancreatic B-cells may be more susceptible to reduction of NAD levels. Malaisse et al. (1982) reported that the ability to provide protection against potent reactive radicals may be weak in islet cells, in view of the low glutathione peroxidase activity in islets.

If the development of alloxan- and streptozotocin-diabetes is provoked by the mechanism shown in Fig. 10.3, then it is theoretically preventable through inhibition of the serial reactions originating from DNA strand breaks. One method is scavenging the radicals, which break DNA, by superoxide dismutase and catalase (as discussed previously) or by other radical scavengers. Another is inhibiting the poly(ADP-ribose) synthetase by specific inhibitors to prevent the sharp decrease in NAD levels and the subsequent reduction of B-cell functions such as proinsulin synthesis. It should be added that in certain conditions alloxan may also attack mitochondria and plasma membranes (Boquist, 1980; Malaisse, 1982). However, this effect of alloxan may be independent of the poly(ADP-ribose) synthetase-dependent reaction and thus may not be preventable by poly(ADP-ribose) synthetase inhibitors such as nicotinamide.

The two highly B-cytotoxic compounds, alloxan and streptozotocin, are structurally different, but their actions have proved to converge into a common pathway to induce DNA strand breaks, to activate poly(ADP-ribose) synthetase, to depress NAD levels and to inhibit B-cell functions. This seems to be of special importance in understanding the pathogenesis of insulin-dependent (Type I) diabetes, which has been thought to be caused by many different factors (Leslie, 1983).

As shown in Fig. 10.3, it is possible that inflammation (insulitis), viral infection, radiation, and chemical insult may independently or interactively produce DNA strand breaks, which can lead to a diabetic state. The DNA strands may be broken by viruses with an affinity for pancreatic B-cells or by the hydroxyl radical (OH·) produced in insulitis due to viral infection. During inflammation, large amounts of O_2^-· and hydroxyl radicals are produced (Badwey & Karnovsky, 1980). It is possible that islet cell antibodies (ICA) and islet cell surface antibodies (ICSA) (Nerup & Lernmark, 1981) specific against islet cells are produced as a result of the destruction of B-cells to provoke or exacerbate insulitis. In this case insulitis may be a cause, as well as a result, of insulin-dependent (Type I) diabetes. Irradiation may also cause DNA strand breaks. Repeated administrations of subdiabetogenic doses of streptozotocin to rats induced insulitis and a diabetic state similar to Type I diabetes (Rossini *et al.*, 1978; Leiter, 1982). The rodenticide *N*-3-pyridylmethyl-*N'*-nitrophenylurea (Vacor) is known to be diabetogenic in man (Karam *et al.*, 1980) and is chemically related to streptozotocin. The diabetogenic factor in Icelandic smoked mutton is most likely to be one or more of the *N*-nitroso compounds in the meat (Helgason *et al.*, 1982). Tsubouchi *et al.* (1981) have reported that a large dose of X-ray irradiation provoked necrosis of pancreatic islets in hamsters, and that the hamsters exhibited a diabetic state. Encephalomyocarditis virus (EMC) induces diabetes in certain inbred strains of mice by infecting and destroying pancreatic B-cells (Craighead, 1975). Virus-induced diabetes was enhanced by low doses of streptozotocin (Toniolo *et al.*, 1980).

Some types of diabetes in humans are also thought to be autoimmune diseases (Bottazzo *et al.*, 1987). Recent studies by Japanese investigators have established a new model of insulin-dependent diabetes with immunologic abnormalities, the NOD mouse (Makino *et al.*, 1980). This animal becomes spontaneously diabetic with clinical and pathological manifestations similar to those seen in human insulin-dependent diabetes. Recently it has been reported that diabetes in this animal can be prevented by treatment with immunosuppressive agents (Mori *et al.*, 1986), an immunomodulator (OK-432) (Okamoto, Sr, 1976; Toyota *et al.*, 1986), radical scavengers (Nomikos *et al.*, 1986), and poly(ADP-ribose) synthetase inhibitors (Yamada *et al.*, 1982; Nomikos *et al.*, 1986). Furthermore, Vague *et al.* (1987) reported that nicotinamide can extend the remission phase in human insulin-dependent diabetes. Thus, as shown in Fig. 10.4, Okamoto proposes that although insulin-dependent (Type I) diabetes can be caused by many different agents such as immunologic abnormalities, inflammatory tissue damage, alloxan and streptozotocin, the final pathway for the toxic agents is common (Okamoto *et al.*, 1988). This pathway involves the generation of free radicals, DNA damage, nuclear poly(ADP-ribose) synthetase activation and NAD depletion. The fall in cellular NAD inhibits the cellular activities and the B-cell ultimately dies. It should be noted here that the T-lymphocyte-mediated cytolytic process has been

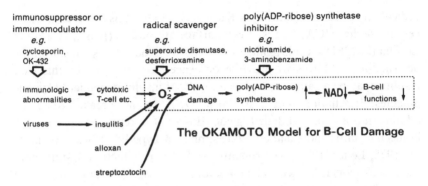

Fig. 10.4. A unifying model for B-cell damage and its prevention in toxin- or virus-induced and immune diabetes (the Okamoto model) (adapted from Okamoto *et al.*, 1988).

thought to be initiated in the target nucleus, which ultimately results in quantitative conversion of the target genome into small DNA fragments (Russell, 1983; Gromkowski *et al.*, 1986; Schmid *et al.*, 1986; Howell & Martz, 1987; Ucker, 1987). Murine cytotoxic T-lymphocytes contain a novel cytotoxin, leukalexin, which causes DNA fragmentation in several targets (Liu *et al.*, 1987).

10.7. Oncogenesis of B-cells and a unifying concept of the diabetogenic and oncogenic effects of B-cytotoxic agents

The induction of pancreatic islet cell tumors in rats by a combined administration of streptozotocin and nicotinamide was first reported by Rakieten *et al.* (1971). Immunochemical and biochemical studies indicated that the tumors were mainly composed of insulin-producing B-cells (Volk *et al.*, 1973; Kazumi *et al.*, 1978; Johnson *et al.*, 1982). Itoh & Okamoto revealed that the streptozotocin-nicotinamide-induced tumors contained as much proinsulin mRNA as did normal pancreatic islets, and succeeded in isolating proinsulin mRNA from the tumors (Itoh & Okamoto, 1977, 1980; Itoh *et al.*, 1979; Okamoto, 1981). Islet cell tumors were also found in rats given streptozotocin and picolinamide, an isomer of nicotinamide, and in rats given alloxan and nicotinamide (Kazumi *et al.*, 1978, 1980). However, how the combined administrations cause islet cell tumors was not understood.

As described above, streptozotocin and alloxan cause DNA strand breaks in islet B-cells to activate poly(ADP-ribose) synthetase (Yamamoto *et al.*, 1981*a*) while nicotinamide and picolinamide are potent inhibitors of islet poly(ADP-ribose) synthetase (Yamamoto & Okamoto,

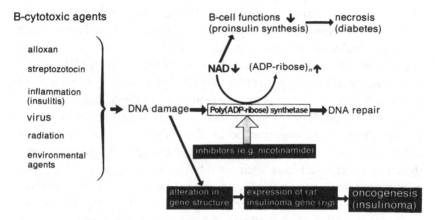

Fig. 10.5. A unifying concept of the diabetogenic and oncogenic effects of B-cytotoxic agents (adapted from Okamoto *et al.*, 1988).

1980; Uchigata *et al.*, 1982). As shown in Fig. 10.3, poly(ADP-ribose) synthetase inhibitors prevent NAD degradation, so that B-cell functions, including proinsulin synthesis, are therefore conserved. There is, however, a possibility that with blockage at this step, DNA breaks may remain. Using a hamster insulinoma cell line, Yamamoto & Okamoto (1982) demonstrated the inhibition by nicotinamide of the rejoining of DNA strands broken by streptozotocin. The inhibition or retardation of the DNA repair may increase the frequency of alteration in gene structure, which may subsequently result in the abnormal expression of certain genes that are involved in tumorigenesis. In 1985, Yamagami *et al.* investigated whether the combined administration of streptozotocin or alloxan with various poly(ADP-ribose) synthetase inhibitors to rats induces islet B-cell tumors and determined the proinsulin mRNA content of the tumors induced. As a result, they found that, in any combination examined, a high incidence of islet B-cell tumors containing significant amounts of proinsulin mRNA as well as B-granules was induced (Yamagami *et al.*, 1985). Furthermore, Takasawa *et al.* have constructed a B-cell tumor cDNA library, and through a comparative analysis with normal islet and B-cell tumor poly(A)$^+$ RNAs, they have recently found a novel gene that is activated in the B-cell tumors and seems to be involved in tumorigenesis (Takasawa *et al.*, 1986; Inoue *et al.*, 1987). The gene is designated *rig* (rat insulinoma gene); its nature is described in detail in Chapter 14.

A unifying concept is thus proposed by Okamoto and coworkers (Okamoto, 1985*a*; Okamoto *et al.*, 1988) for the diabetogenic and oncogenic effects of B-cytotoxic agents, as shown in Fig. 10.5. In the case of the induction of diabetes, damaged DNA activates poly(ADP-ribose) synthetase, resulting in the depletion of cellular NAD; hence B-cell functions, including proinsulin synthesis, may be impaired, leading to necrosis, which eventually induces diabetes. When the poly(ADP-ribose) synthetase inhibitor is present during this process, the activation of poly(ADP-ribose) synthetase and the depression of the cellular NAD level are prevented, and B-cell functions therefore proceed normally. However, prompt DNA repair does not occur. The inhibition or retardation of DNA repair may be the cause of increased frequency of alteration in gene structure. B-cells that have such alterations in their gene structure may exhibit abnormal expression of certain genes such as *rig* and be converted to tumor cells. The concept presented in Fig. 10.5 may enable the diabetogenic and oncogenic effects of B-cytotoxins to be understood in a comprehensive and unified manner.

10.8. B-cell regeneration in partly depancreatized rats

Current techniques for experimental diabetes in animals are mostly based on the use of B-cytotoxins such as alloxan and streptozotocin. An alternative approach, following von Mering & Minkowski (1890), is to perform a partial pancreatectomy. This model can also be utilized in a study of islet cell regeneration, and in 1984 Yonemura *et al.* demonstrated that administration of poly(ADP-ribose) synthetase inhibitors such as nicotinamide to 90% depancreatized rats induces regeneration of B-cells, thereby preventing the development of diabetes (Yonemura *et al.*, 1984; Okamoto *et al.*, 1985; Okamoto, 1985*a, b*). Male Wistar rats, weighing 180–200 g, were 90% depancreatized, and, beginning 7 days before the partial pancreatectomy and continuing postoperatively, nicotinamide (0.5 g/kg) and 3-aminobenzamide (0.05 g/kg) were intraperitoneally injected daily (Yonemura *et al.*, 1984). The control 90% depancreatized rats exhibited glucosuria 1–3 months after the 90% pancreatectomy, but in rats receiving nicotinamide or 3-aminobenzamide daily, the urinary glucose excretion decreased markedly. Plasma glucose levels before and after an intravenous glucose load, in the rats receiving the poly(ADP-ribose) synthetase inhibitors, also decreased significantly in comparison with those in the control rats. The islets of control rats without poly(ADP-ribose) synthetase inhibitors were less numerous, small in size, and had irregular contours; fibrotic degeneration and degranulation were frequently encountered, as reported previously (Martin & Lacy, 1963; Volk & Lazarus, 1964). The development of diabetes in the control 90% depancreatized animals is thought to be due to a reduced ability to regenerate B-cell mass and to synthesize insulin. In Yonemura's experiment, the islets in the remaining pancreases of rats which received the nicotinamide and 3-aminobenzamide injections were larger in size and more numerous (Yonemura et al., 1984; Okamoto, 1985b). Immunohistochemical findings indicated that the B-cells increased within the enlarged islets of the remaining pancreases of poly(ADP-ribose) synthetase inhibitor-treated rats, and that the number of cells staining for glucagon and somatostatin was unchanged (Yonemura et al., 1984; Okamoto, 1985a). It is reasonable to assume, then, that poly(ADP-ribose) synthetase inhibitors induced pancreatic B-cell regeneration in the 90% depancreatized rats. Results from an autoradiographic approach indicate that the proportion of labeled B-cell nuclei in the poly(ADP-ribose) synthetase inhibitor-administered rats 2–4 days after the pancreatectomy was about three times larger than in the control 90% depancreatized rats (Yonemura et al., 1988).

Recently, Terazono et al. (1988) have isolated regenerating islets from the remaining pancreases of 90% depancreatized and nicotinamide-treated rats and constructed a cDNA library from the islet poly(A)$^+$ RNA. In screening the regenerating islet-derived cDNA library, Terazono et al. (1988) came across a novel gene encoding a 165 amino acid protein. The gene, named reg (regenerating gene), was expressed in regenerating islets of 90% depancreatized and nicotinamide-administered rats but not in normal islets, liver, kidney, brain, insulinomas, or regenerating liver (Terazono et al., 1988). The increase in reg expression was temporally correlated with the increase in size of regenerating islets and the decrease in urinary glucose level of 90% depancreatized and nicotinamide-administered rats. Terazono et al. (1988) also found that a human pancreas-derived cDNA library contained a reg homolog, which coded for a protein quite similar to that encoded by rat reg. The hydrophobic signal sequences present in both human and rat reg proteins indicate the secretory nature of the protein. The structures of reg, reg proteins and genomic reg are described in Chapter 15. Further studies on the expression and function of the protein encoded by human reg may lead to new developments for the treatment of human diabetes.

10.9. Concluding remarks

In this chapter, the mechanisms of alloxan- and streptozotocin-induced diabetes and surgical diabetes in partly depancreatized animals have been discussed. A unifying model for B-cell damage and its prevention in toxin- or virus-induced and immune diabetes (the Okamoto model)

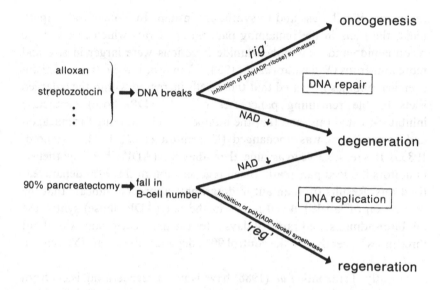

Fig. 10.6. The possible mechanisms for B-cell damage and its prevention in alloxan- and streptozotocin-diabetes and surgical diabetes.

has also been introduced. As shown in Fig. 10.3, B-cytotoxic agents such as alloxan and streptozotocin induce DNA strand breaks in islet B-cells, after which poly(ADP-ribose) synthetase acts to repair the DNA breaks, consuming B-cell NAD. The fall in cellular NAD inhibits the cellular activities. This sequence of events is regarded as the primary molecular mechanism behind B-cell necrosis. Fig. 10.4 shows the proposed Okamoto model. This model raises the possibility that insulin-dependent (Type I) diabetes can be prevented by suppressing immune reactions, scavenging free radicals, and inhibiting poly(ADP-ribose) synthetase activity. Such treatment with poly(ADP-ribose) synthetase inhibitors, however, should be carried out with great care and attention to possible B-cell oncogenesis, which could result from impairment of DNA repair and expression of the insulinoma gene, *rig*. In partly depancreatized animals, there has been shown to be an active regenerative process in the islet B-cells to meet an increased peripheral demand for insulin (Martin & Lacy, 1963; Hellerström & Swenne, 1985). The regenerative process, however, cannot continue and the B-cells degenerate. Evidence adduced in the last part of this chapter indicates that poly(ADP-ribose) synthetase inhibitors prevent experimental diabetes that is usually caused by partly removing the pancreases of rats. In this case, poly(ADP-ribose) synthetase inhibitors may also block NAD depletion, since the presence of gaps in the DNA during replication should increase the enzymic activity (Lönn & Lönn, 1985). Therefore, when a poly(ADP-ribose) synthetase inhibitor is present during the regenerative process, the cellular NAD level is maintained and B-cells continue to regenerate. Furthermore, replicative DNA synthesis increases in nicotinamide-treated islet B-cells (Sandler & Andersson, 1986); the poly(ADP-ribose) synthetase inhibitor may relieve restriction of DNA replication and so permit B-cell regeneration (Okamoto, 1985a; Okamoto et al., 1985, 1988). The activation of the regenerating gene, reg, may participate in B-cell regeneration. The possible mechanisms for B-cell damage and its prevention in alloxan- and streptozotocin-diabetes and surgical diabetes (caused by partly removing the pancreases of animals) are represented in Fig. 10.6.

References

Anderson, T., Schein, P.S., McMenamin, M.G. & Cooney, D.A. (1974). Streptozotocin diabetes. Correlation with extent of depression of pancreatic islet nicotinamide adenine dinucleotide. *Journal of Clinical Investigation* **54**, 672–7.

Badwey, J.A. & Karnovsky, M.L. (1980). Active oxygen species and the functions of phagocytic leukocytes. *Annual Review of Biochemistry* **49**, 695–726.

Banting, F.G. & Best, C.H. (1922). Pancreatic extracts. *Journal of Laboratory and Clinical Medicine* **7**, 464–72.

Berger, N.A., Sikorski, G.W., Petzold, S.J. & Kurohara, K.K. (1980). Defective poly(adenosine diphosphoribose) synthesis in xeroderma pigmentosum. *Biochemistry* **19**, 289–93.

Berger, N.A., Weber, G. & Kaichi, A.S. (1978). Characterization and comparison of poly(adenosine diphosphoribose) synthesis and DNA synthesis in nucleotide-permeable cells. *Biochimica et Biophysica Acta* **519**, 87–104.

Bock, K.W., Gäng, V., Beer, H.P., Kronau, R. & Grunicke, H. (1968). Localization and regulation of two NAD nucleosidases in Ehrlich ascites cells. *European Journal of Biochemistry* **4**, 357–63.

Boquist, L. (1980). A new hypothesis for alloxan diabetes. *Acta Pathologica, Microbiologica et Immunologica Scandinavica* **A88**, 201–9.

Bottazzo, G.F., Pujol-Borrell, R. & Gale, E.A.M. (1987). Autoimmunity and type I diabetes: bringing the story up to date. In *The diabetes annual*, vol. 3 (ed. K.G.M.M. Alberti & L.P. Krall), pp. 15–38. Amsterdam: Elsevier.

Brawn, K. & Fridovich, I. (1981). DNA strand scission by enzymatically generated oxygen radicals. *Archives of Biochemistry and Biophysics* **206**, 414–19.

Chance, B. (1947). An intermediate compound in the catalase–hydrogen peroxide reaction. *Acta Chemica Scandinavica* **1**, 236–67.

Chang, A.Y. & Diani, A.R. (1985). Chemically and hormonally induced diabetes mellitus. In *The diabetic pancreas* (2nd edn) (ed. B.W. Volk & E.R. Arquilla), pp. 415–38. New York: Plenum Medical Book Company.

Cooperstein, S.J. & Watkins, D. (1981). Action of toxic drugs on islet cells. In *The islets of Langerhans* (ed. S.J. Cooperstein & D. Watkins), pp. 387–425. New York: Academic Press.

Cox, R. & Irving, C.C. (1976). Effect of N-methyl-N-nitrosourea on the DNA of rat bladder epithelium. *Cancer Research* **36**, 4114–18.

Craighead, J.E. (1975). The role of viruses in the pathogenesis of pancreatic disease and diabetes mellitus. *Progress in Medical Virology* 19, 161–214.

Dietrich, L.S., Fuller, L., Yero, I.L. & Martinez, L. (1966). Nicotinamide mononucleotide pyrophosphorylase activity in animal tissues. *Journal of Biological Chemistry* 241, 188–91.

Doi, K. (1975). Studies on the mechanism of the diabetogenic activity of streptozotocin and on the ability of compounds to block the diabetogenic activity of streptozotocin. *Folia Endocrinologica Japonica* 51, 129–47.

Dulin, W.E. & Soret, M.G. (1977). Chemically and hormonally induced diabetes. In *The diabetic pancreas* (ed. B.W. Volk & K.F. Wellmann), pp. 425–65. New York: Plenum Press.

Dulin, W.E. & Wyse, B.M. (1969). Reversal of streptozotocin diabetes with nicotinamide. *Proceedings of the Society for Experimental Biology and Medicine* 130, 922–4.

Dunn, J.S., Sheehan, H.L. & McLetchie, N.G.B. (1943). Necrosis of islets of Langerhans *Lancet* i, 484–7.

Erickson, L.C., Bradley, M.O. & Kohn, K.W. (1977). Strand breaks in DNA from normal and transformed human cells treated with 1,3-bis(2-chloroethyl)-1-nitrosourea. *Cancer Research* 37, 3744–50.

Fischer, L.J. & Hamburger, S.A. (1980). Inhibition of alloxan action in isolated pancreatic islets by superoxide dismutase, catalase, and a metal chelator. *Diabetes* 29, 213–16.

Gold, G., Manning, M., Heldt, A., Nowlain, R., Pettit, J.R. & Grodsky, G.M. (1981). Diabetes induced with multiple subdiabetogenic doses of streptozotocin. Lack of protection by exogenous superoxide dismutase. *Diabetes* 30, 634–8.

Goodwin, P.M., Lewis, P.J., Davies, M.I., Skidmore, C.J. & Shall, S. (1978). The effect of gamma radiation and neocarzinostatin on NAD and ATP levels in mouse leukaemia cells. *Biochimica et Biophysica Acta* 543, 576–82.

Grankvist, K., Marklund, S., Sehlin, J. & Täljedal, I.-B. (1979). Superoxide dismutase, catalase and scavengers of hydroxyl radical protect against the toxic action of alloxan on pancreatic islet cells *in vitro*. *Biochemical Journal* 182, 17–25.

Grankvist, K., Marklund, S. & Täljedal, I.-B. (1981). Superoxide dismutase is a prophylactic against alloxan diabetes. *Nature* 294, 158–60.

Gromkowski, S.H., Brown, T.C., Cerutti, P.A. & Cerottini, J. (1986). DNA of human Raji target cells is damaged upon lymphocyte-mediated lysis. *Journal of Immunology* 136, 752–6.

Gunnarsson, R. (1975). Inhibition of insulin biosynthesis by alloxan, streptozotocin, and N-nitrosomethylurea. *Molecular Pharmacology* 11, 759–65.

Gunnarsson, R., Berne, C. & Hellerström, C. (1974). Cytotoxic effects of streptozotocin and N-nitrosomethylurea on the pancreatic B cells with special regard to the role of nicotinamide-adenine dinucleotide. *Biochemical Journal* 140, 487–94.

Haber, F. & Weiss, J. (1934). The catalytic decomposition of hydrogen peroxide by iron salts. *Proceedings of the Royal Society of London* A147, 332–51.

Hammarström, L. & Ullberg, S. (1966). Specific uptake of labelled alloxan in the pancreatic islets. *Nature* 212, 708–9.

Hayaishi, O. & Ueda, K. (1977). Poly(ADP-ribose) and ADP-ribosylation of proteins. *Annual Review of Biochemistry* 46, 95–116.

Heikkila, R.E., Winston, B., Cohen, G. & Barden, H. (1976). Alloxan-induced diabetes – evidence for hydroxyl radical as a cytotoxic intermediate. *Biochemical Pharmacology* 25, 1085–92.

Helgason, T., Ewen, S.W.B., Ross, I.S. & Stowers, J.M. (1982). Diabetes produced in mice by smoked/cured mutton. *Lancet* ii, 1017–22.

Hellerström, C., Andersson, A., Gunnarsson, R., Berne, C. & Asplund, K. (1974). Cytophysiology and metabolism of pancreatic B-cell. *Endocrinologia Experimentalis* 8, 115–26.

Hellerström, C. & Swenne, I. (1985). Growth pattern of pancreatic islets in animals. In *The diabetic pancreas* (2nd edn) (ed. B.W. Volk & E.R. Arquilla), pp. 53–79. New York: Plenum Medical Book Company.

Hinz, M., Katsilambros, N., Maier, V., Schatz, H. & Pfeiffer, E.F. (1973). Significance of streptozotocin induced nicotinamide-adenine-dinucleotide (NAD) degradation in mouse pancreatic islets. *FEBS Letters* **30**, 225–8.

Howell, D.M. & Martz, E. (1987). The degree of CTL-induced DNA solubilization is not determined by the human vs mouse origin of the target cell. *Journal of Immunology* **138**, 3695–8.

Inoue, C., Shiga, K., Takasawa, S., Kitagawa, M., Yamamoto, H. & Okamoto, H. (1987). Evolutionary conservation of the insulinoma gene *rig* and its possible function. *Proceedings of the National Academy of Sciences of the U.S.A.* **84**, 6659–62.

Itoh, N., Nose, K. & Okamoto, H. (1979). Purification and characterization of proinsulin mRNA from rat B-cell tumor. *European Journal of Biochemistry* **97**, 1–9.

Itoh, N. & Okamoto, H. (1977). Synthesis of preproinsulin with RNA preparation of streptozotocin-nicotinamide induced B-cell tumor. *FEBS Letters* **80**, 111–14.

Itoh, N. & Okamoto, H. (1980). Translational control of proinsulin synthesis by glucose. *Nature* **283**, 100–2.

Johnson, D.E., Dixit, P.K., Michels, J.E. & Bauer, G.E. (1982). Immunochemical identification of endocrine cell types in the streptozotocin nicotinamide-induced rat islet adenoma. *Experimental and Molecular Pathology* **37**, 193–207.

Juarez-Salinas, H., Sims, J.L. & Jacobson, M.K. (1979). Poly(ADP-ribose) levels in carcinogen-treated cells. *Nature* **282**, 740–1.

Karam, J.H., Lewitt, P.A., Young, C.W., Nowlain, R.E., Frankel, B.J., Fujiya, H., Freedman, Z.R. & Grodsky, G.M. (1980). Insulinopenic diabetes after rodenticide (Vacor) ingestion. A unique model of acquired diabetes in man. *Diabetes* **29**, 971–8.

Kazumi, T., Yoshino, G. & Baba, S. (1980). Pancreatic islet cell tumors found in rats given alloxan and nicotinamide. *Endocrinologia Japonica* **27**, 387–93.

Kazumi, T., Yoshino, G., Yoshida, Y., Doi, K., Yoshida, M., Kaneko, S. & Baba, S. (1978). Biochemical studies on rats with insulin-secreting islet cell tumors induced by streptozotocin: with special reference to physiological response to oral glucose load in the course of and after tumor induction. *Endocrinology* **103**, 1541–5.

Lazarow, A., Liambies, J. & Tausch, A.J. (1950). Protection against diabetes with nicotinamide. *Journal of Laboratory and Clinical Medicine* **36**, 249–58.

Lazarus, S.S. & Shapiro, S.H. (1972). Serial morphologic changes in rabbit pancreatic islet cells after streptozotocin. *Laboratory Investigation* **27**, 174–83.

Leiter, E.H. (1982). Multiple low-dose streptozotocin-induced hyperglycemia and insulitis in C57BL mice: influence of inbred background, sex, and thymus. *Proceedings of the National Academy of Sciences of the U.S.A.* **79**, 630–4.

Leslie, R.D.G. (1983). Causes of insulin dependent diabetes. *British Medical Journal* **287**, 5–6.

Levi, V., Jacobson, E.L. & Jacobson, M.K. (1978). Inhibition of poly(ADP-ribose) polymerase by methylated xanthines and cytokinins. *FEBS Letters* **88**, 144–6.

Liu, C., Steffen, M., King, F. & Young, J.D. (1987). Identification, isolation, and characterization of a novel cytotoxin in murine cytolytic lymphocytes. *Cell* **51**, 393–403.

Lönn, U. & Lönn, S. (1985). Accumulation of 10-kilobase DNA replication intermediates in cells treated with 3-aminobenzamide. *Proceedings of the National Academy of Sciences of the U.S.A.* **82**, 104–8.

Makino, S., Kunimoto, K., Muraoka, Y., Mizushima, Y., Katagiri, K. & Tochino, Y. (1980). Breeding of a non-obese diabetic strain of mice. *Experimental Animals* **29**, 1–13.

Malaisse, W.J. (1982). Alloxan toxicity to the pancreatic B-cell – a new hypothesis. *Biochemical Pharmacology* **31**, 3527–34.

Malaisse, W.J., Malaisse-Lagae, F., Sener, A. & Pipeleers, D.G. (1982). Determinants of the selective toxicity of alloxan to the pancreatic B cell. *Proceedings of the National Academy of Sciences of the U.S.A.* **79**, 927–30.

Martin, J.M. & Lacy, P.E. (1963). The prediabetic period in partially pancreatectomized rats. *Diabetes* **12**, 238–42.

Masiello, P., Cubeddu, T.L., Frosina, G. & Bergamini, E. (1985). Protective effect of 3-aminobenzamide, an inhibitor of poly(ADP-ribose) synthetase, against streptozotocin-induced diabetes. *Diabetologia* **28**, 683–6.

McCord, J.M. & Day, E.D., Jr (1978). Superoxide-dependent production of hydroxyl radical catalyzed by iron–EDTA complex. *FEBS Letters* **86**, 139–42.

McCord, J.M. & Fridovich, I. (1969). Superoxide dismutase. An enzymic function for erythrocuprein (Hemocuprein). *Journal of Biological Chemistry* **244**, 6049–55.

Miller, E.G. (1975). Stimulation of nuclear poly(adenosine diphosphate-ribose) polymerase activity from HeLa cells by endonucleases. *Biochimica et Biophysica Acta* **395**, 191–200.

Mori, Y., Suko, M., Okudaira, H., Matsuba, I., Tsuruoka, A., Sasaki, A., Yokoyama, H., Tanase, T., Shida, T., Nishimura, M., Terada, E. & Ikeda, Y. (1986). Preventive effects of cyclosporin on diabetes in NOD mice. *Diabetologia* **29**, 244–7.

Nerup, J. & Lernmark, A. (1981). Autoimmunity in Insulin-dependent Diabetes Mellitus. In *Diabetes mellitus* (ed. J.S. Skyler & G.F. Cahill), pp. 31–7. U.S.A.: Yorke Medical Books.

Nishizuka, Y. & Hayaishi, O. (1963). Studies on the biosynthesis of nicotinamide adenine dinucleotide. I. Enzymic synthesis of niacin ribonucleotides from 3-hydroxyanthranilic acid in mammalian tissues. *Journal of Biological Chemistry* **238**, 3369–77.

Nomikos, I.N., Prowse, S.J., Carotenuto, P. & Lafferty, K.J. (1986). Combined treatment with nicotinamide and desferrioxamine prevents islet allograft destruction in NOD mice. *Diabetes* **35**, 1302–4.

Ohgushi, H., Yoshihara, K. & Kamiya, T. (1980). Bovine thymus poly(adenosine diphosphate ribose) polymerase. Physical properties and binding to DNA. *Journal of Biological Chemistry* **255**, 6205–11.

Oikawa, A., Tohda, H., Kanai, M., Miwa, M. & Sugimura, T. (1980). Inhibitors of poly(adenosine diphosphate ribose) polymerase induce sister chromatid exchanges. *Biochemical and Biophysical Research Communications* **97**, 1311–16.

Okamoto, H. (1981). Regulation of proinsulin synthesis in pancreatic islets and a new aspect to insulin-dependent diabetes. *Molecular and Cellular Biochemistry* **37**, 43–61.

Okamoto, H. (1985a). Molecular basis of experimental diabetes: degeneration, oncogenesis, and regeneration of pancreatic B-cells of islets of Langerhans. *BioEssays* **2**, 15–21.

Okamoto, H. (1985b). The role of poly(ADP-ribose) synthetase in the development of insulin-dependent diabetes and islet B-cell regeneration. *Biomedica Biochimica Acta* **44**, 15–20.

Okamoto, H. & Yamamoto, H. (1983). DNA strand breaks and poly(ADP-ribose) synthetase activation in pancreatic islets – a new aspect to development of insulin-dependent diabetes and pancreatic B-cell tumors. In *ADP-ribosylation, DNA repair and cancer* (ed. M. Miwa, O. Hayaishi, S. Shall, M. Smulson & T. Sugimura), pp. 297–308. Tokyo: Japan Scientific Societies Press.

Okamoto, H., Yamamoto, H., Takasawa, S., Inoue, C., Terazono, K., Shiga, K. & Kitagawa, M. (1988). Molecular mechanism of degeneration, oncogenesis and regeneration of pancreatic B-cells of islets of Langerhans. In *Frontiers in diabetes research: lessons from animal diabetes II* (ed. E. Shafrir & A.E. Renold), pp. 149–57. London: John Libbey & Company.

Okamoto, H., Yamamoto, H. & Yonemura, Y. (1985). Poly(ADP-ribose) synthetase inhibitors induce islet B-cell regeneration in partially depancreatized rats. In *ADP-ribosylation of proteins* (ed. F.R. Althaus, H. Hilz & S. Shall), pp. 410–16. Berlin, Heidelberg: Springer-Verlag.

Okamoto, H., Sr (1976). Antitumor activity of streptolysin S-forming streptococci. In *Mechanisms in bacterial toxinology* (ed. A.W. Bernheimer), pp. 237–57. New York: John Wiley & Sons.

Preiss, J. & Handler, P. (1957). Enzymatic synthesis of nicotinamide mononucleotide. *Journal of Biological Chemistry* **225**, 759–70.

Preiss, J., Schlaeger, R. & Hilz, H. (1971). Specific inhibition of poly ADP ribose polymerase by thymidine and nicotinamide in HeLa cells. *FEBS Letters* **19**, 244–6.

Purnell, M.R. & Whish, W.J.D. (1980). Novel inhibitors of poly(ADP-ribose) synthetase. *Biochemical Journal* **185**, 775–7.

Rakieten, N., Gorden, B.S., Beaty, A., Cooney, D.A., Davis, R.D. & Schein, P.S. (1971). Pancreatic islet cell tumors produced by the combined action of streptozotocin and nicotinamide (35561). *Proceedings of the Society for Experimental Biology and Medicine* **137**, 280–3.

Rakieten, N., Rakieten, M.L. & Nadkarni, M.V. (1963). Studies on the diabetogenic action of streptozotocin. *Cancer Chemotherapy Reports* **29**, 91–8.

Rechsteiner, M., Hillyard, D. & Olivera, B.M. (1976). Magnitude and significance of NAD turnover in human cell line D98/AH2. *Nature* **259**, 695–6.

Rerup, C.C. (1970). Drugs producing diabetes through damage of the insulin secreting cells. *Pharmacological Reviews* **22**, 485–518.

Rossini, A.A., Williams, R.M., Appel, M.C. & Like, A.A. (1978). Sex differences in the multiple-dose streptozotocin model of diabetes. *Endocrinology* **103**, 1518–20.

Russell, J.H. (1983). Internal disintegration model of cytotoxic lymphocyte-induced target damage. *Immunological Reviews* **72**, 97–118.

Sandler, S. & Andersson, A. (1986). Long-term effects of exposure of pancreatic islets to nicotinamide in vitro on DNA synthesis, metabolism and B-cell function. *Diabetologia* **29**, 199–202.

Schein, P.S. (1969). 1-Methyl-1-nitrosourea and dialkylnitrosamine depression of nicotinamide adenine dinucleotide. *Cancer Research* **29**, 1226–32.

Schein, P.S., Cooney, D.A., McMenamin, M.G. & Anderson, T. (1973). Streptozotocin diabetes – further studies on the mechanism of depression of nicotinamide adenine dinucleotide concentrations in mouse pancreatic islets and liver. *Biochemical Pharmacology* **22**, 2625–31.

Schein, P.S. & Loftus, S. (1968). Streptozotocin: depression of mouse liver pyridine nucleotides. *Cancer Research* **28**, 1501–6.

Schmid, D.S., Tite, J.P. & Ruddle, N.H. (1986). DNA fragmentation: manifestation of target cell destruction mediated by cytotoxic T-cell lines, lymphotoxin-secreting helper T-cell clones, and cell-free lymphotoxin-containing supernatant. *Proceedings of the National Academy of Sciences of the U.S.A.* **83**, 1881–5.

Shima, K., Hirota, M., Sato, M., Numoto, S. & Oshima, I. (1987). Effect of poly(ADP-ribose) synthetase inhibitor administration to streptozotocin-induced diabetic rats on insulin and glucagon contents in their pancreas. *Diabetes Research and Clinical Practice* **3**, 135–42.

Skidmore, C.J., Davies, M.I., Goodwin, P.M., Halldorsson, H., Lewis, P.J., Shall, S. & Zia'ee, A.A. (1979). The involvement of poly(ADP-ribose) polymerase in the degradation of NAD caused by γ-radiation and N-methyl-N-nitrosourea. *European Journal of Biochemistry* **101**, 135–42.

Smulson, M.E., Schein, P., Mullins, D.W., Jr & Sudhakar, S. (1977). A putative role for nicotinamide adenine dinucleotide-promoted nuclear protein modification in the antitumor activity of N-methyl-N-nitrosourea. *Cancer Research* **37**, 3006–12.

Takasawa, S., Yamamoto, H., Terazono, K. & Okamoto, H. (1986). Novel gene activated in rat insolinomas. *Diabetes* **35**, 1178–80.

Terazono, K., Yamamoto, H., Takasawa, S., Shiga, K., Yonemura, Y., Tochino, Y. &

Okamoto, H. (1988). A novel gene activated in regenerating islets. *Journal of Biological Chemistry* **263**, 2111–14.

Tjälve, H., Wilander, E. & Johansson, E. (1976). Distribution of labelled streptozotocin in mice: uptake and retention in pancreatic islets. *Journal of Endocrinology* **69**, 455–6.

Toniolo, A., Onodera, T., Yoon, J.-W. & Notkins, A.L. (1980). Induction of diabetes by cumulative environmental insults from viruses and chemicals. *Nature* **288**, 383–5.

Toyota, T., Satoh, J., Oya, K., Shintani, S. & Okano, T. (1986). Streptococcal preparation (OK-432) inhibits development of type I diabetes in NOD mice. *Diabetes* **35**, 496–9.

Tsubouchi, S., Suzuki, H., Ariyoshi, H. & Matsuzawa, T. (1981). Radiation-induced acute necrosis of the pancreatic islet and the diabetic syndrome in the golden hamster. *International Journal of Radiation Biology and Related Studies in Physics, Chemistry and Medicine* **40**, 95–106.

Uchigata, Y., Yamamoto, H., Kawamura, A. & Okamoto, H. (1982). Protection by superoxide dismutase, catalase, and poly(ADP-ribose) synthetase inhibitors against alloxan- and streptozotocin-induced islet DNA strand breaks and against the inhibition of proinsulin synthesis. *Journal of Biological Chemistry* **257**, 6084–8.

Uchigata, Y., Yamamoto, H., Nagai, H. & Okamoto, H. (1983). Effect of poly(ADP-ribose) synthetase inhibitor administration to rats before and after injection of alloxan and streptozotocin on islet proinsulin synthesis. *Diabetes* **32**, 316–18.

Ucker, D.S. (1987). Cytotoxic T lymphocytes and glucocorticoids activate an endogenous suicide process in target cells. *Nature* **327**, 62–4.

Vague, Ph., Vialettes, B., Lassmann-Vague, V. & Vallo, J.J. (1987). Nicotinamide may extend remission phase in insulin-dependent diabetes. *Lancet* **i**, 619–20.

Volk, B.W. & Lazarus, S.S. (1964). Ultrastructure of pancreatic B cells in severely diabetic dogs. *Diabetes* **13**, 60–70.

Volk, B.W., Wellmann, K.F. & Brancato, P. (1973). Fine structure of rat islet cell tumors induced by streptozotocin and nicotinamide. *Diabetologia* **10**, 37–44.

von Mering, J. & Minkowski, O. (1890). Diabetes mellitus nach Pankreasexstirpation. *Archiv für Experimentelle Pathologie und Pharmakologie* **26**, 371–87.

Whish, W.J.D., Davies, M.I. & Shall, S. (1975). Stimulation of poly(ADP-ribose) polymerase activity by the antitumour antibiotic, streptozotocin. *Biochemical and Biophysical Research Communications* **65**, 722–30.

Wilson, G.L., Hartig, P.C., Patton, N.J. & LeDoux, S.P. (1988). Mechanisms of nitrosourea-induced β-cell damage. Activation of poly(ADP-ribose) synthetase and cellular distri-bution. *Diabetes* **37**, 213–16.

Wilson, G.L., Patton, N.J., McCord, J.M., Mullins, D.W. & Mossman, B.T. (1984). Mechanisms of streptozotocin- and alloxan-induced damage in rat B cells. *Diabetologia* **27**, 587–91.

Yamada, K., Nonaka, K., Hanafusa, T., Miyazaki, A., Toyoshima, H. & Tarui, S. (1982). Preventive and therapeutic effects of large-dose nicotinamide injections on diabetes associated with insulitis. An observation in nonobese diabetic (NOD) mice. *Diabetes* **31**, 749–53.

Yamagami, T., Miwa, A., Takasawa, S., Yamamoto, H. & Okamoto, H. (1985). Induction of rat pancreatic B-cell tumors by the combined administration of streptozotocin or alloxan and poly(adenosine diphosphate ribose) synthetase inhibitors. *Cancer Research* **45**, 1845–9.

Yamamoto, H. & Okamoto, H. (1980). Protection by picolinamide, a novel inhibitor of poly(ADP-ribose) synthetase, against both streptozotocin-induced depression of proinsulin synthesis and reduction of NAD content in pancreatic islets. *Biochemical and Biophysical Research Communications* **95**, 474–81.

Yamamoto, H. & Okamoto, H. (1982). Poly(ADP-ribose) synthetase inhibitors enhance streptozotocin-induced killing of insulinoma cells by inhibiting the repair of DNA strand breaks. *FEBS Letters* **145**, 298–302.

Yamamoto, H., Uchigata, Y. & Okamoto, H. (1981a). Streptozotocin and alloxan induce DNA strand breaks and poly(ADP-ribose) synthetase in pancreatic islets. *Nature* **294**, 284–6.

Yamamoto, H., Uchigata, Y. & Okamoto, H. (1981b). DNA strand breaks in pancreatic islets by *in vivo* administration of alloxan or streptozotocin. *Biochemical and Biophysical Research Communications* **103**, 1014–20.

Yonemura, Y., Takashima, T., Miwa, K., Miyazaki, I., Yamamoto, H. & Okamoto, H. (1984). Amelioration of diabetes mellitus in partially depancreatized rats by poly(ADP-ribose) synthetase inhibitors – evidence of islet B-cell regeneration. *Diabetes* **33**, 401–4.

Yonemura, Y., Takashima, T., Matsuda, Y., Miwa, K., Sugiyama, K., Miyazaki, I., Yamamoto, H. & Okamoto, H. (1988). Induction of islet B-cell regeneration in partially pancreatectomized rats by poly (ADP-ribose) synthetase inhibitors. *International Journal of Pancreatology* **3**, 73–82.

11

Class II histocompatibility genes and diabetes

DAVID OWERBACH

11.1. Introduction

Diabetes mellitus, a syndrome characterized by insufficient insulin secretion, hyperglycemia, and a propensity to develop universal microangiopathy, neuropathy and atherosclerosis, is a common condition affecting 2–4% of all Caucasians (Panzram & Zabel-Langenning, 1981). Insulin-dependent diabetes (IDDM) makes up 10–15% of all diabetes and is characterized by a selective B-cell destruction, very low if any insulin secretion, absolute requirement for exogenous insulin and often a young age of onset (although cases do occur at all ages). Microvascular disease becomes clinically evident 10–20 years after diagnosis of IDDM and constitutes a major cause of death, due to resulting renal insufficiency (Rossini & Chick, 1980). While improvements in treatment, such as intensive insulin and pump therapies, continue to be developed, and experimental treatment such as pancreatic transplantation and immunotherapy are proposed (Lacy et al., 1981; Rossini, 1983), a major goal must be prevention of the group of disorders we currently label as IDDM. At this time of clinical presentation, the majority of the IDDM diabetic pancreatic B-cells have undergone irreversible destruction (Cahill & McDevitt, 1981). Thus it is likely that potential interventions will have their greatest success if applied before the onset of clinically recognized disease, as has been demonstrated by immunologic interventions in appropriate animal models (Like et al., 1983). To accomplish this major goal of prevention, we must better understand the natural history, etiopathogenesis, and genetic susceptibility to IDDM, and develop accurate methods to detect genetically susceptible individuals in the population at large before the onset of clinically recognized disease.

The mechanisms which determine the loss of the pancreatic B-cells are

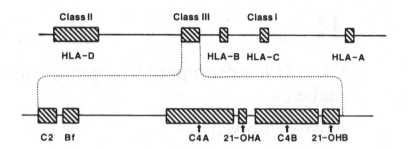

Fig. 11.1. Gene map of the MHC region on human chromosome 6. Abbreviations: Bf, factor B of complement; C2, second component of complement; C4A and C4B, fourth components of complement; 21-OHA and 21-OHB, 21-hydroxylase genes.

not known; however, it has been shown that the clinical onset of IDDM is associated with a number of immunologic abnormalities (Eisenbarth, 1986). In addition, it has long been recognized that genetic factors are of major importance in the etiology of IDDM. Insulin-dependent diabetes demonstrates familial aggregation (Wagener *et al.*, 1982) and twin studies have indicated that this familial aggregation is largely due to genetic predisposition to the disease (Barnett *et al.*, 1981). Studies in the past decade have demonstrated that a significant portion of the genetic susceptibility is provided by major histocompatibility complex (MHC) genes at or near the HLA-D subregion on chromosome 6: population studies have revealed that 90–95% or more of the IDDM patients have HLA DR3, or DR4, or both (Platz *et al.*, 1981; Wolf *et al.*, 1983) and family studies have shown that, in sibling pairs where both are affected with IDDM, such pairs tend to share both HLA haplotypes (Walker & Cudworth, 1980; Platz *et al.*, 1981).

Essential to an understanding of the etiology of IDDM is a more complete definition of the specific genetic susceptibilities. In this article, I shall review recent progress in characterizing the specific MHC genes involved in the susceptibility to IDDM in humans.

11.2. Role of the MHC in diabetes

The major histocompatibility complex (MHC), which comprises three groups of genes called class I, II and III respectively (Fig. 11.1), plays a critical role in regulating the immune response (Benacerraf, 1981). The class I genes encode antigens that serve as restricting elements in the T-cell-mediated killing of virally infected cells. The class II genes encode cell-surface molecules which function in antigen presentation to T-helper cells and control the extent of an immune response to individual foreign antigens. The class III genes encode several complement components, including properdin factor B (Bf), C2, C4A and C4B (Fig. 11.1). An outstanding feature of the MHC is the high degree of polymorphism found in class I, II and III loci (Bodmer & Bodmer, 1984).

In man, the MHC region is referred to as HLA; HLA classes I, II and III alleles have been associated with a number of autoimmune diseases, including IDDM (Stastny et al., 1983). In IDDM, the class I and III allelic associations are secondary (linkage disequilibrium) to a stronger association with class II (DR3 and DR4) encoded molecules (Nerup et al., 1987).

The class II molecules are heterodimeric (α and β chains) glycoproteins that are normally expressed on antigen-presenting cells (APCs) such as macrophages, B-lymphocytes and activated T-lymphocytes; they are not normally expressed on pancreatic B-cells. However, expression of class II molecules on islet cells of diabetic patients has been reported (Bottazzo et al., 1985; Foulis et al., 1986). Furthermore, human islet cells can be induced to express class II antigens in vitro by a combination of γ-interferon and tumor necrosis factor (TNF) (Pujol-Borrell et al., 1987). These results support a model for IDDM development that was originally proposed by Bottazzo and coworkers (Bottazzo et al., 1983) and is modified as follows. First, local production of γ-interferon and/or TNF α or β are produced upon viral infection. Second, class II antigens are induced on pancreatic B-cells; these then function as APCs in association with B-cell-specific autoantigens, and thus stimulate T-lymphocytes. This localized T-cell proliferation initiates an autoimmune reaction leading to the activation of antigen-specific cytotoxic T-lymphocytes, which destroy the pancreatic B-cells. Alternatively, the T-cell proliferations could lead to the production of cytokines with cytotoxic activities toward B-cells.

A second model for the role of class II molecules in IDDM has recently been proposed by Nerup et al. (1987). B-cell antigen, released due to B-cell injury (possibly due to viruses), is processed and presented by macrophages to T-helper lymphocytes. This initiates production of cytokines, of which interleukin-1 (IL-1) is B-cell cytotoxic. In this model, IL-1 production is controlled in a quantitative way by the specific class II histocompatibility genes.

Although the specific mechanisms for B-cell destruction in IDDM are currently unknown, it is clear that the class II histocompatibility genes regulate the specificity and magnitude of the immune response which leads to the B-cell destruction. Furthermore, the class II genes are necessary but

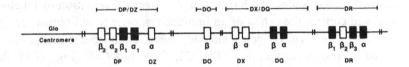

Fig. 11.2. The HLA-D region on human chromosome 6. Open boxes, genes for which no product has been detected; closed boxes, genes which are expressed.

not sufficient in themselves to target, precipitate and/or perpetuate the disease.

11.3. Class II gene organization and expression

The HLA-D region contains 14 or 15 different genes (Korman *et al.*, 1985; Hardy *et al.*, 1986; Andersson *et al.*, 1987) (Fig. 11.2). The DRβ1 and DRα gene products associate to form a cell-surface heterodimer. This molecule reacts with the alloantisera that define the major serological allotypes HLA-DR1 to DRw14. The relatively non-polymorphic DRβ3 and DRα gene products encode the DRw52 and DRw53 serological determinants, which are each associated with several DR allotypes. The DRβ2 gene is a pseudogene and is not transcribed into functional DRβ messenger RNA. A second DRβ pseudogene is present in some individuals (Andersson *et al.*, 1987). The DQα and DQβ genes encode the DQ serological specificities DQw1, DQw2 and DQw3. In addition, products of the DPβ1 and DPα1 locus genes associate to express DP antigens. The DP antigens are polymorphic and are the most centromeric of the class II loci, but they are not strongly associated with IDDM susceptibility (DeJongh *et al.*, 1984). Furthermore, DXα, DXβ, DPβ2 and DPα2 genes are not expressed, whereas the non-polymorphic DOβ and DZα genes are expressed but only in extremely small amounts. Therefore, the most likely MHC genes to be involved in IDDM susceptibility are the DRβ1, DQβ and/or DQα loci, which are highly polymorphic and are expressed on activated T-lymphocytes.

11.4. RFLP analysis and IDDM

The first studies using DNA markers for analysis of IDDM (Owerbach *et al.*, 1983*a*) used a cDNA probe for HLA-DQβ to analyze genomic DNA from IDDM and HLA-DR-matched healthy control individuals for restriction endonuclease fragment length polymorphisms (RFLPs). This procedure involves digesting genomic DNA with a restric-

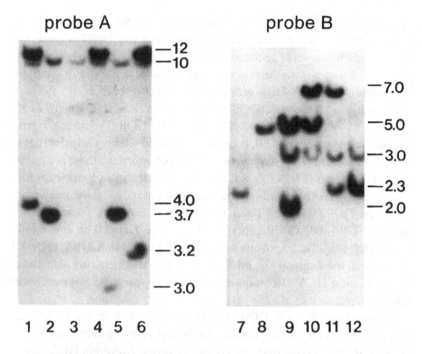

Fig. 11.3. Southern blot analysis using DQβ specific genomic probes. Lanes 1–6, BamHI digest, probe A; lanes 7–12, TaqI digest, probe B. The derivation of probes A and B has been previously described (Owerbach, 1987; Owerbach et al., 1987b).

tion endonuclease, separating the resulting DNA by molecular mass on agarose gels and transferring the DNA from the gel to a nitrocellulose or nylon membrane by the method of Southern (1975). The specific class II DNA sequences are detected by hybridization with ^{32}P-labeled DNA probes, stripping away the non-specifically bound probe and visualizing the sequences by autoradiography. A DQβ 3.7 kilobase (kb) BamHI sequence was rarely detected in IDDM patients (2%) but was frequently (30–40%) detected among healthy controls (even if they were HLA-DR4 positive). This was the first evidence that the DQβ genes may be more strongly associated with IDDM than serologically defined HLA-DR genes. Subsequent reports built upon this finding demonstrated two types of DQβ gene, one associated with the 3.7 kb fragment (DQw3.1 specificity) and the other with a BamHI 12.0 kb fragment (DQw3.2) (Owerbach et al., 1983b, 1984; Kim et al., 1985). The DQw3.1 specificity is frequently present in HLA-DR4 and DR5 non-diabetic individuals but is rare in HLA-DR4 IDDM patients (Owerbach et al., 1983a; Arnheim et al., 1985;

Cohen-Haguenauer *et al.*, 1985; Nepom *et al.*, 1986; Festenstein *et al.*, 1986; Michelsen & Lernmark, 1987; Henson *et al.*, 1987). Figure 11.3 shows a Southern blot of genomic DNA from Caucasoid blood doncr lymphocytes, digested with the restriction endonuclease *Bam*HI or *Taq*I and hybridized with HLA-DQβ genomic probes (Owerbach *et al.*, 1987*a, b*; Owerbach, 1987). Probe A (DQβ exon 2) reveals *Bam*HI fragments of 3.0, 3.2, 3.4, 3.7, 4.0, 10.0 and 12.0 kb; probe B (DQβ 3'-flanking region) detects *Taq*I sequences of 2.0, 2.3, 2.6, 3.0, 5.0 and 7.0 kb. Only the 10.0 kb *Bam*HI and 3.0 kb *Taq*I sequences are non-polymorphic; these are related to the DXβ locus (Fig. 11.2). All the other restriction endonuclease fragments are polymorphic and segregate as alleles of the DQβ locus. Furthermore, the *Taq*I 2.0 kb and *Bam*HI 12.0 kb fragments are associated with the HLA-DR4, DQw3.2 haplotype, whereas the *Taq*I 5.0 kb and *Bam*HI 3.7 kb endonuclease fragments are associated with HLA-DR4, DQw3.1. In addition, the *Taq*I 2.0 kb and *Bam*HI 12.0 kb fragments are also associated with some HLA-DR8 haplotypes which appear to be associated with IDDM (Owerbach *et al.*, 1987*a*).

Examination of HLA-DQβ RFLPs associated with HLA-DR2 has provided further evidence for the DQβ region being more specifically associated with IDDM than the HLA-DR genes. Although HLA-DR2 is rare in IDDM, the majority of HLA-DR2 IDDM patients (77%) have a DQβ type (AZH) which is rare (6%) in HLA-DR2 healthy individuals (Bach *et al.*, 1986; Böhme *et al.*, 1986; Segall *et al.*, 1986; Cohen *et al.*, 1986). The nucleotide sequence of the AZH β-chain gene is similar to that of DQw1 associated with HLA-DR1 (Todd *et al.*, 1987), which is positive or neutral with respect to IDDM susceptibility (Svejgaard *et al.*, 1986).

RFLP studies in IDDM have also been reported for DRα (Stetler *et al.*, 1985; Hitman *et al.*, 1987), DRβ (Arnheim *et al.*, 1985; Cohen-Haguenauer *et al.*, 1985; Owerbach *et al.*, 1986) and DXα polymorphisms (Festenstein *et al.*, 1986; Hitman *et al.*, 1986, 1987). The most interesting of these studies shows an increased frequency of a 2.1 kb *Taq*I DXα polymorphism in IDDM patients (Hitman *et al.*, 1986). Because the DXα locus does not appear to be expressed, this association appears to be due to linkage disequilibrium with the DQβ genes.

Additional evidence for the primary association of DQβ genes and secondary association of DRβ and DXα genes comes from a recent study of 27 Caucasoid families (D. Owerbach *et al.*, unpublished). Twenty-three families had a single diabetic proband; four had 2 diabetic patients each. The patient population under study is a homogeneous, young group of IDDM patients with onset of diabetes before 15 years of age. Further-

Table 11.1. *DNA haplotypes defined by RFLP analysis*

Haplotype	RFLP				
	DR	DQα	DXα	DQβ	DRβ
1	1	2.6	1.9	2.3	6.7, 4.4
2A	2	6.5	2.1	2.3	14, 1.9, 1.6
2B	2	6.5	1.9	2.3	1.9
2C	2	2.6	1.9	2.3	4.4
3A	3	4.6	2.1	5.0	12, 7.0, 4.0
3B	3	4.6	1.9	5.0	14, 7.0, 4.0
3C	3	6.5	2.1	2.3	12, 7.0, 4.0
4A	4	5.5	2.1	2.0	16, 6.5, 5.5, 3.0
4B	4	5.5	1.9	5.0	16, 6.5, 5.5, 3.0
4C	4	6.5	2.1	2.3	16, 6.5, 5.5, 3.0
4D	4	5.5	2.1	5.0	16, 6.5, 5.5, 3.0
5A	5	4.6	1.9	5.0	14, 6.7, 4.0
5B	5	4.6	1.9	7.0	14, 6.7, 4.0
6A	w6	6.5	1.9	2.3	12, 7.0, 4.0
6B	w6	2.6	1.9	2.3	14, 7.0, 4.0
7A	7	5.5	1.9	7.0	16, 7.2, 2.9
7B	7	5.5	1.9	7.0	16, 2.9
7C	7	5.5	1.9	2.6	16, 2.9
8A	w8	6.5	1.9	2.0	10
8B	w8	6.5	1.9	2.3	10
10	w10	2.6	1.9	2.3	4.4

*Taq*I polymorphisms in kilobases. The DQα cDNA probe detects both DQα and DXα RFLPs. The DQβ probe is derived from a genomic 2.0 kb sequence flanking the 3'-end of a DQβ gene. The DRβ probe is derived from exon 4 and flanking regions of a DRβ gene. DNA haplotypes 2C, 3C, 4C, 4D, 5B, 6B, 8B and 10 were present in only one family each; other DNA haplotypes were present in a number of families.

more, the non-diabetic siblings are all over 20 years of age. Ten micrograms of genomic DNA from the parents ($n = 54$), diabetic patients ($n = 31$) and non-diabetic siblings ($n = 68$) were digested with the restriction endonuclease *Taq*I and the digestion products were analyzed by electrophoresis on 1% agarose gels followed by Southern transfer to Zetabind nylon membranes (Owerbach *et al.*, 1987b). ^{32}P-labeled genomic probes DRβ (Owerbach *et al.*, 1986, 1987a, b) and DQβ (Owerbach *et al.*, 1987a, h; Owerbach, 1987) and cDNA probe DQα (Auffray *et al.*, 1982) were used to detect homologous class II sequences. Twenty-one different DNA haplotypes were deduced from the patterns of RFLPs detected in the families (Table 11.1). For example, four DNA haplotypes were associated with HLA-DR4. All contained identical DRβ RFLPs but varied with respect to DQβ, DXα or DQα polymorphisms (Table 11.1).

The 2A, 3A, 3B, 4A and 4B DNA haplotypes (Table 11.1) show a strong correlation with extended haplotypes defined by serologically defined extended MHC haplotypes (D. Owerbach et al., unpublished). The 3A, 3B and 4A DNA haplotypes are strongly associated with [HLA-B8, DR3, SC01], [HLA-B18, DR3, F1C30], and [HLA-B15, DR4, SC33], respectively, which are all strongly associated with susceptibility for IDDM. (The class III loci are designated as complotypes, i.e. 'SC01' represents Bf (factor B)*S, C2*C, C4A*QO, C4B*1 (Raum et al., 1984).) The 4B and 2A DNA haplotypes are strongly associated with the [HLA-Bw44, DR4, SC30] and [HLA-B7, DR2, SC31] extended MHC haplotypes respectively, which were shown to be protective for IDDM (Raum et al., 1984).

Differences between the IDDM-associated 4A and IDDM-protective 4B DNA haplotypes would indicate that genes defined by either the DXα 2.1 kb and/or DQβ 2.0 kb polymorphisms are important for susceptibility to IDDM. However, the DXα 2.1 kb RFLP is also present on the 2A DNA haplotype, which is negatively associated with IDDM (Table 11.1). Thus, the DXα polymorphism cannot be directly associated with IDDM susceptibility. In contrast, the 2.0 kb DQβ RFLP was present on DNA haplotypes 4A and 8A and appears to be associated with IDDM susceptibility (Owerbach et al., 1987a). These results implicate the DQβ locus, and specifically alleles of the DQβ locus on the 3A, 3B, 4A and 8A DNA haplotypes, as primarily associated with IDDM susceptibility.

11.5. Oligonucleotide probe analysis

Most of the polymorphism detected in RFLP studies resides in non-amino acid coding regions of the genes, including flanking regions, introns and untranslated regions. RFLP analysis is thus a relatively weak method of analysis when the sequences of importance are likely in the coding region and most often not detected by commonly used restriction endonucleases. Methods utilizing 19-mer oligonucleotide hybridization probes were originally used to differentiate single base pair mutations in the globin genes (Conner et al., 1983). To use this methodology for analyzing class II histocompatibility genes, nucleotide sequence information was first compared from diabetic (Owerbach et al., 1986, 1987b; Todd et al., 1987; Freeman et al., 1987) and non-diabetic individuals (Todd et al., 1987) having various DQβ, DQα and DRβ loci. In these genes, most of the coding region polymorphism is in exon 2, which codes

Table 11.2. *DQβ polymorphic residues in exon 2*

DQβ allele	DR antigen	amino acid residue																										
		13	14	26	28	30	37	38	45	46	47	52	53	55	57	66	67	70	71	74	75	77	84	85	86	87	89	90
w3.2	DR4	G	M	L	T	Y	Y	A	G	V	Y	P	L	P	A	E	V	R	T	E	L	T	Q	L	E	L	T	T
w3.1	DR4, DR5	A	—	Y	—	—	—	E	—	—	—	—	D	—	—	—	—	—	—	—	—	—	—	—	—	—	—	—
w2	DR3, DR7	—	—	S	S	I	V	—	E	F	L	—	D	I	—	K	A	V	R	—	—	R	—	—	—	—	—	—
w1	DR1	—	L	G	—	H	—	V	—	—	—	Q	R	V	—	—	—	G	A	S	V	R	E	V	A	Y	G	I

Nucleotide and amino acid sequence data are derived from the sources stated in the text.

Table 11.3. *DNA haplotypes detected with oligonucleotide probes*

DNA haplotype	probe								IDDM susceptibility
	α1	α3	α4	α7	β1	β3	β4A	β4B	
3A, 3B, 5B		+				+			positive
4A			+				+		positive
4D			+			+			positive
8A		+							positive
1, 2C, 8B, 10	+				+				neutral
4B			+				+	+	negative
5A		+					+	+	negative
2A-2, 2B, 3C, 4C, 6A, 6B	+								negative
2A-1	+							+	negative
7A, 7B				+		+			negative
7C				+		+			negative

for the amino-terminal domain. Table 11.2 shows the amino acid polymorphic sites derived from DQβ exon 2 DNA sequence analysis for the DQw1, DR1; DQw2, DR3; DQw3.2, DR4; and DQw3.1, DR4 specificities. Differences between the DQw3.1 and 3.2 specificities (DQβ) indicate that amino acids 13, 26, 45 and/or 57 are important for IDDM susceptibility and resistance. Oligonucleotide probes have been designed around DQβ amino acid coding region 26, 37–38 and 57 (Todd *et al.*, 1987; D. Owerbach *et al.*, unpublished) and DQα residues 46–57 (Saiki *et al.*, 1986; D. Owerbach *et al.*, unpublished) and used in studies of IDDM (Todd *et al.*, 1987; D. Owerbach *et al.*, unpublished). The basis of the assay is genomic amplification of the exon 2 regions using oligonucleotide primers and DNA polymerase, a method described by Saiki *et al.* (1986). The amplification process increases the amount of DNA defined by the primers, to be replicated thousands of times. This DNA is analyzed by dot blot hybridization using 17–19-mer DQβ and DQα oligonucleotide probes. Figure 11.4 shows the dot blot analysis of 8 individuals' amplified DNA using 4 DQα-chain and two β-chain probes. In addition, Table 11.3 shows the DNA haplotypes from Table 11.1 as detected with the various oligonucleotide probes. Finally, the oligonucleotide sequences present in IDDM patients versus non-diabetic siblings are shown in Table 11.4.

Fourteen of the 31 diabetic patients (D.Owerbach *et al.*, unpublished) had both DNA haplotypes 3A(B)/4A, which are detected by hybridization with probes α4, β4A, α3 and β3 but not with probes α1, β1, α7 or β4B. Two additional patients had haplotypes 4A/5B and 4A/8A

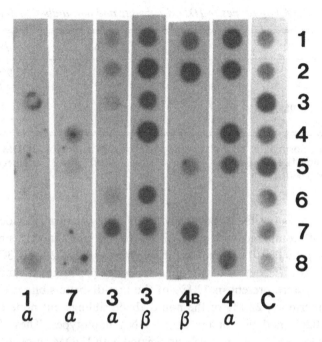

Fig. 11.4. Dot blot analysis on amplified DNA using oligonucleotide probes. Amplified genomic DNA from individuals 1–8 were hybridized with ^{32}P-labeled 19-mer probes α1, α7, α3, β3, β4B and α4. Lane C was hybridized with a genomic probe containing DQβ exon 2 sequences and is a positive control lane. The DNA haplotypes of the individuals are: 1, 3A/4B; 2, 3A/4B; 3, 3A/2A; 4, 4A/7A; 5, 4B/7C; 6, 3A/8A; 7, 3B/5A; and 8, 4A/10. Hybridizations with probes β1 and β4A are not pictured.

respectively, which also hybridized with the α4, β4A, α3 and β3 probes. Another 5 patients had DNA haplotypes 4A/4A or 3A(B)/3A(B) and hybridized with probes α4, β4A or α3, β3 respectively. Therefore, 64.5% of the patients had DQα and DQβ genes that had identity with α4, β4A and/or α3, β3 oligonucleotide sequences. In contrast, only 17.6% of the non-diabetic siblings had these combinations of sequences (Table 11.4). Therefore, the genes associated with these DQα and DQβ sequences are strongly associated with IDDM susceptibility.

An additional 22.6% of the IDDM patients in the study were heterozygous for an IDDM-associated DNA haplotype and DNA haplotypes 1 or 8B, which are recognized by the β1 probe (Table 11.4). HLA-DR1 and 8 have previously been shown to be neutral with respect to IDDM susceptibility (Svejgaard et al., 1986). In total, 87.1% of the IDDM patients had

Table 11.4. *DNA haplotypes in IDDM subjects and non-diabetic siblings*

	+ / +	+ /n	− /x
IDDM	64.5%	22.6%	12.9%
	($n = 20$)	($n = 7$)	($n = 4$)
non-IDDM siblings	17.6%	10.3%	72.1%
	($n = 12$)	($n = 7$)	($n = 49$)

Symbols: +, positive; n, neutral; −, negative; x, any haplotype. IDDM compared with non-IDDM siblings (+ / +, + /n vs. − /x): $\chi^2 = 31.2$, 1 degree of freedom, $p < 0.00001$.

DNA haplotypes positively/positively or positively/neutrally associated with IDDM, compared with only 27.9% of the non-diabetic siblings (Table 11.4; $p < 0.00001$). Other DNA haplotypes (4B, 5A, 2A, 2B, 3C, 4C, 6A, 6B, 7A, 7B, 7C), detected with oligonucleotide probes $\alpha1$, $\alpha7$ and $\beta4B$ (Table 11.3), were present in 72.1% of the non-diabetic siblings (Table 11.4). Furthermore, 22.1% of the non-diabetic siblings but none of the IDDM patients had two of the above DNA haplotypes. These DNA haplotypes appear to be negatively associated with IDDM susceptibility. The DNA haplotypes can thus be sorted into positive, neutral or negative IDDM associations (Table 11.3). The DNA haplotype susceptibilities are even more specifically defined by oligonucleotide sequences of DQα and DQβ genes.. These results, along with studies by Todd *et al.* (1987), demonstrate that alleles of DQβ and DQα determine susceptibility and resistance to IDDM.

Todd *et al.* (1987) further indicate that amino acid residue 57 of the β-chain specifies the autoimmune response in IDDM; i.e. if residue 57 is Asp the β-chain is negatively associated with IDDM, while Ala or other amino acids are risk-associated. This explanation, however, fails to explain why DQβ genes associated with DRw9 and DRw8 which have Asp at residue 57 (Todd *et al.*, 1987) are not negatively associated with IDDM susceptibility (Svejgaard *et al.*, 1986). In addition, their observation fails to explain the excess of HLA-DR 3/4 heterozygotes in IDDM patients (Platz *et al.*, 1981; Wolf *et al.*, 1983); DR3/4 heterozygotes are at a higher risk of developing IDDM than the corresponding homozygotes, suggesting the interaction of at least two complementary genes. These observations suggest a more complex explanation than a simple single amino acid difference in the DQβ chain.

The data of D. Owerbach and co-workers (unpublished) suggest that other amino acid residues, in addition to residue 57 of the DQβ chain, are important for IDDM susceptibility or resistance. For example, the β4B and β1 probes are centered on Tyr^{26} or Gly^{26}, and are negatively or neutrally associated with IDDM susceptibility, respectively, whereas the β3 probe is designed around amino acid residues Ile^{37} and Val^{38} and detects sequences associated with IDDM susceptibility.

We also conclude that the DQα genes play an active role in IDDM susceptibility. Sequences detected by α3 and α4 probes are positively associated with IDDM susceptibility, but α7 sequences are negatively associated. Therefore, the specific interaction of DQα and DQβ genes seems to be of critical importance in IDDM. The excess of HLA-DR3/4 heterozygotes may be a result of transallelic complementation of the DQα and DQβ loci; for example, DQ molecules detected by the α4 probe are associated with DQ molecules detected by the β3 probe. Normally, the products of these genes are encoded on separate haplotypes. However, on DNA haplotype 4D (Tables 11.1, 11.3), a chance genetic recombination seems to have placed both the α4- and β3-detected specificities on the same chromosome in a diabetic patient.

11.6. Summary and conclusions

Population and family studies during the past decade have demonstrated that a significant portion of the genetic susceptibility to IDDM is provided by the HLA region of chromosome 6 in man. The initial observations were with HLA class I loci followed by associations with class III and class II loci, with the class II association being strongest to IDDM. With the advent of molecular cloning techniques, the class II region has been further dissected into 14 (15) different genes. Analysis of these genes by RFLP analysis demonstrates DQβ genes to be the most strongly associated with IDDM. DNA sequence analysis, along with hybridization studies using oligonucleotide probes, further localizes the susceptibility region to specific regions of the amino-terminal domains of DQα and DQβ genes. The specific interaction of DQα and DQβ loci seems to be of critical importance in the MHC susceptibility to IDDM.

The MHC region, while necessary for IDDM, is not sufficient in itself to target, precipitate or perpetuate the disease. In identical twin studies of IDDM, approximately 50% of both twins are concordant for the disease (Barnett et al., 1981). Furthermore, in HLA identical non-diabetic sibs of a diabetic patient, these sibs have a 12–24% risk of IDDM (Platz et al., 1981; Wolf et al., 1983). Finally, in the Caucasoid population, of the HLA-DR3/4 individuals having the high risk DQα and DQβ genes, only about 1 in 30 will ever develop IDDM. Clearly other genes and/or environmental factors are involved.

T-cell receptor α and β genes (Hoover et al., 1986) and the polymorphic region flanking the human insulin gene (Mandrup-Poulsen et al., 1985) have been implicated as playing a role in IDDM. Further studies are needed to clarify the role of these markers in IDDM. Other genes may code for regulatory proteins that interact with class II gene promoter and/or enhancer sequences and control the levels of class II gene expression during IDDM. The promoter and enhancer regions of DQα and DQβ genes and the proteins that bind to them are currently being investigated (Boss & Strominger, 1986; Miwa et al., 1987). Finally, other genes may be found that serve in the targeting (endogenous viruses, islet antigens), destruction (cytotoxins) and/or regeneration (insulin enhancer region) of the pancreatic B-cells. The identification of the specific DQα and DQβ genes providing the MHC risk to IDDM will now facilitate the search for these other genetic and/or environmental factors that interact with the MHC components to precipitate and perpetuate the disease.

This work was in part supported by grants from the NIH and the Juvenile Diabetes Foundation. I thank Dr Kenneth Gabbay for helpful discussions, and acknowledge the technical assistance over the past 8 years of Lissi Aagaard (Copenhagen, Denmark), Cheryl Rich and Shirley Carnegie (Worcester, MA) and Yvette Crawford, George Ty and Linda Wible (Houston, TX).

References

Andersson, G., Larhammar, D., Widmark, E., Servenius, B., Peterson, P.A. & Rask, L. (1987). Class II genes of the human major histocompatibility complex – organization and evolutionary relationship of the DRβ genes. Journal of Biological Chemistry 262, 8748–58.

Arnheim, N., Strange, C. & Erlich, H. (1985). Use of pooled DNA samples to detect linkage disequilibrium of polymorphic restriction fragments and human disease: studies of the HLA class II loci. Proceedings of the National Academy of Sciences of the U.S.A. 82, 6970–4.

Auffray, C., Korman, A.J., Roux-Dosseto, M., Bono, R. & Strominger, J.L. (1982). cDNA clone for the heavy chain of the human B cell alloantigen DC1: strong sequence homology to the HLA-DR heavy chain. Proceedings of the National Academy of Sciences of the U.S.A. 79, 6337–41.

Bach, F.H., Ohta, N., Anichini, A. & Reinsmoen, N.L. (1986). Cellular detection of HLA class II–encoded determinants: subtype polymorphisms of HLA-D. In HLA class II antigens (ed. B.G. Solheim, E. Moller & S. Ferrone), pp. 249–65. Berlin: Springer.

Barnett, A.H., Eff, C., Leslie, R.D. & Pyke, D.A. (1981). Diabetes in identical twins. Diabetologia 20, 87–93.

Benacerraf, B. (1981). Role of MHC gene products in immune regulation. Science 212, 1229–38.

Bodmer, J. & Bodmer, W. (1984). Histocompatibility 1984. Immunology Today 5, 251–4.

Böhme, J., Carlsson, B., Wallin, J., Möller, E., Persson, B., Peterson, P.A. & Rask, L. (1986). Only one DQβ restriction fragment pattern of each DR specificity is associated with insulin-dependent diabetes. Journal of Immunology 137, 941–7.

Boss, J.M. & Strominger, J.L. (1986). Regulation of a transfected human class II major histocompatibility complex gene in human fibroblasts. *Proceedings of the National Academy of Sciences of the U.S.A.* **83**, 9139–43.

Bottazzo, G.F., Dean, B.M., McNally, J.M., MacKay, E.H., Swift, P.G.E. & Ga:nble, D.R. (1985). In situ characterization of autoimmune phenomena and expression of HLA-molecules in the pancreas in diabetic insulitis. *New England Journal of Medicine* **313**, 353–60.

Bottazzo, G.F., Pujol-Borrell, R. & Hanafusa, T. (1983). Role of aberrant HLA-DR expression and antigen presentation in induction of endocrine autoimmunity. *Lancet* **ii**, 1115–18.

Cahill, G.F. Jr & McDevitt, H.O. (1981). Insulin-dependent diabetes mellitus: the initial lesion. *New England Journal of Medicine* **304**, 1454–65.

Cohen, N., Brautbar, C., Font, M.P., Dausset, J. & Cohen, D. (1986). HLA-DR2 associated Dw subtypes correlate with RFLP clusters: most DR2 IDDM patients belong to one of these clusters. *Immunogenetics* **23**, 84–9.

Cohen-Haguenauer, O., Robbins, E., Massart, C., Busson, M., Deschamps, I., Hors, J., Lalouel, J.M., Dausset, J. & Cohen, D. (1985). A systematic study of HLA class II-β DNA restriction fragments in insulin-dependent diabetes mellitus. *Proceedings of the National Academy of Sciences of the U.S.A.* **82**, 3335–9.

Conner, B.J., Reyes, A.A., Morin, C., Itakura, K., Teplitz, R.L. & Wallace, R.B. (1983). Detection of sickle cell beta-S-globin allele by hybridization with synthetic oligonucleotides. *Proceedings of the National Academy of Sciences of the U.S.A.* **80**, 278–82.

DeJongh, B.M., Termijtelen, A., Bruining, G.J., DeVries, R.R.P. & Van Rood, J.J. (1984). Relation of insulin-dependent diabetes mellitus (IDDM) and the HLA-linked SB system. *Tissue Antigens* **23**, 87–93.

Eisenbarth, G.S. (1986). Type I diabetes mellitus, a chronic autoimmune disease. *New England Journal of Medicine* **314**, 1360–8.

Festenstein, H., Awad, J., Hitman, G.A., Cutbush, S., Groves, A.V., Cassell, P., Ollier, W. & Sachs, J.A. (1986). New HLA DNA polymorphisms associated with autoimmune diseases. *Nature* **322**, 64–7.

Foulis, A.K., Liddle, C.N., Farquharson, M.A., Richmond, J.A. & Weir, R.S. (1986). The histopathology of the pancreas in type I (insulin-dependent) diabetes mellitus: a 25-year review of deaths in patients under 20 years of age in the United Kingdom. *Diabetologia* **29**, 267–74.

Freeman, S.M., Saunders, T.L., Madden, M., Segall, M., Bach, F.H. & Wu, S. (1987). Comparison of DRβ1 alleles from diabetic and normal individuals. *Human Immunology* **19**, 1–6.

Hardy, D., Bell, J., Long, E., Linstein, T. & McDevitt, H.O. (1986). Mapping of the class II region of the human major histocompatibility complex by pulse-field gel electrophoresis. *Nature* **323**, 453–5.

Henson, V., Maclaren, N., Riley, W. & Wakeland, E.K. (1987). Polymorphisms of DQβ genes in HLA-DR4 haplotypes from healthy and diabetic individuals. *Immunogenetics* **25**, 152–60.

Hitman, G.A., Sachs, J., Cassell, P., Awad, J., Bottazzo, G.F., Tarn, A.C., Schwartz, G., Monson, J.P. & Festenstein, H. (1986). A DR3-related DXα gene polymorphism strongly associates with insulin-dependent diabetes mellitus. *Immunogenetics* **23**, 47–51.

Hitman, G.A., Sachs, J.A. & Niven, J.M. (1987). HLA class II region α chain polymorphism and the genetic susceptibility to insulin dependent (Type 1) diabetes mellitus and coeliac disease. In *Protides of the biological fluids*, vol. 35, pp. 101–10. New York: Pergamon Press.

Hoover, M.L., Angelini, G., Ball, E., Stastny, P., Marks, J., Rosenstock, J., Raskin, P., Ferrara, G.B., Tosi, R. & Capra, J.D. (1986). HLA-DQ and T-cell receptor genes in insulin-dependent diabetes mellitus. *Cold Spring Harbor Symposia on Quantitative Biology* **51**, 803–9.

Kim, S.J., Holbeck, S.L., Nisperos, B., Hansen, J.A., Maeda, H. & Nepom, G.T. (1985). Identification of a polymorphic variant associated with HLA-DQw3 and characterized by specific restriction sites within the DQ β-chain gene. *Proceedings of the National Academy of Sciences of the U.S.A.* **82**, 8139–43.

Korman, A.J., Boss, J.M., Spies, T., Sorrentino, R., Okada, K. & Strominger, J.L. (1985). Genetic complexity and expression of human class II histocompatibility antigens. *Immunological Reviews* **85**, 45–86.

Lacy, P.E., Davie, J.M. & Finke, E.H. (1981). Transplantation of insulin-producing tissue. *American Journal of Medicine* **70**, 589–94.

Like, A.A., Anthony, M., Guberski, D.L. & Rossini, A.A. (1983). Spontaneous diabetes mellitus in BB/W rat: effects of glucocorticoids, cyclosporin A and antiserum to rat lymphocytes. *Diabetes* **32**, 326–30.

Mandrup-Poulsen, T., Owerbach, D., Nerup, J., Johanson, K., Ingerslev, J. & Hansen, A. (1985). Insulin-gene flanking sequences, diabetes mellitus and atherosclerosis: a review. *Diabetologia* **28**, 556–64.

Michelsen, B. & Lernmark, Å. (1987). Molecular cloning of a polymorphic DNA endonuclease fragment associates insulin-dependent diabetes mellitus with HLA-DQ. *Journal of Clinical Investigation* **79**, 1144–52.

Miwa, K., Doyle, C. & Strominger, J.L. (1987). Sequence-specific interactions of nuclear factors with conserved sequences of human class II major histocompatibility complex genes. *Proceedings of the National Academy of Sciences of the U.S.A.* **84**, 4939–43.

Nepom, B.S., Palmer, J., Kim, S.J., Hansen, J.A., Holbeck, S.L. & Nepom, G.T. (1986). Specific genomic markers for the HLA-DQ subregion discriminate between DR4⁺ insulin-dependent diabetes mellitus and DR4⁺ seropositive juvenile rheumatoid arthritis. *Journal of Experimental Medicine* **164**, 345–50.

Nerup, J., Mandrup-Poulsen, T. & Mølvig, J. (1987). The HLA-IDDM association: implications for etiology and pathogenesis of IDDM. *Diabetes Metabolism Reviews* **3**, 779–802.

Owerbach, D. (1987). DNA markers linked to diabetes. In *Protides of the biological fluids*, vol. 35, pp. 97–100. New York: Pergamon Press.

Owerbach, D., Carnegie, S., Rich, C., Metzger, B.E. & Freinkel, N. (1987a). Gestational diabetes mellitus is associated with HLA-DQβ-chain endonuclease fragments. *Diabetes Research* **6**, 109–12.

Owerbach, D., Hägglöf, B., Lernmark, Å. & Holmgren, G. (1984). Susceptibility to insulin-dependent diabetes defined by restriction enzyme polymorphism of HLA-D region genomic DNA. *Diabetes* **33**, 958–65.

Owerbach, D., Lernmark, Å., Platz, P., Ryder, L.P., Rask, L., Peterson, P.A. & Ludvigsson, J. (1983a). HLA-D region beta-chain DNA endonuclease fragments differ between HLA-DR identical healthy and insulin-dependent diabetic individuals. *Nature* **303**, 815–17.

Owerbach, D., Lernmark, Å., Rask, L., Peterson, P.A., Platz, P. & Svejgaard, A. (1983b). Detection of HLA-D/DR-related DNA polymorphism in HLA-D homozygous typing cells. *Proceedings of the National Academy of Sciences of the U.S.A.* **80**, 3758–61.

Owerbach, D., Rich, C., Carnegie, S. & Taneja, K. (1987b). Molecular biology of the HLA system in insulin-dependent diabetes mellitus. *Diabetes Metabolism Reviews* **3**, 819–33.

Owerbach, D., Rich, C. & Taneja, K. (1986). Characterization of three HLA-DR beta genes isolated from an HLA-DR3/4 insulin-dependent diabetic patient. *Immunogenetics* **24**, 41–6.

Panzram, G. & Zabel-Langhenning, R. (1981). Prognosis of diabetes mellitus in a geographically defined population. *Diabetologia* **20**, 587–91.

Platz, P., Jakobsen, B.K., Morling, N., Ryder, L.P., Svejgaard, A., Thomsen, M., Christy, M., Kromann, H., Benn, J., Nerup, J., Green, A. & Hauge, M. (1981). HLA-D and DR antigens in genetic analysis of insulin-dependent diabetes mellitus. *Diabetologia* 21, 108–15.

Pujol-Borrell, R., Todd, I., Doshi, M., Bottazzo, G.F., Sutton, R., Gray, D., Adolf, G.R. & Feldmann, M. (1987). HLA-class II induction in human islet cells by interferon – γ plus tumor necrosis factor or lymphotoxin. *Nature* 326, 304–6.

Raum, D., Awdeh, Z., Yunis, E.J., Alper, C.A. & Gabbay, K.H. (1984). Extended major histocompatibility complex haplotypes in Type 1 diabetes mellitus. *Journal of Clinical Investigation* 74, 449–54.

Rossini, A.A. (1983). Immunotherapy for insulin-dependent diabetics? *New England Journal of Medicine* 308, 333–5.

Rossini, A.A. & Chick, W.L. (1980). Microvascular pathology in diabetes mellitus. In *Microcirculation*, vol. 3 (ed. G. Kaley & M. Burton), pp. 245–71. Baltimore: University Park Press.

Saiki, R.K., Bugawan, T.L., Horn, G.T., Mullis, K.B. & Erlich, H.A. (1986). Analysis of enzymatically amplified β-globin and HLA-DQα DNA with allele-specific oligonucleotide probes. *Nature* 324, 163–6.

Segall, M., Noreen, H., Schluender, L., Swenson, M., Barbosa, J. & Bach, F.H. (1986). DR2 haplotypes in insulin-dependent diabetes: analysis of DNA restriction fragment length polymorphisms. *Human Immunology* 17, 61–8.

Southern, E.M. (1975). Detection of specific sequences among DNA fragments separated by gel electrophoresis. *Journal of Molecular Biology* 98, 503–17.

Stastny, P., Ball, E.J., Dry, P.J. & Nunez, G. (1983). The human immune response region (HLA-D) and disease susceptibility. *Immunological Reviews* 70, 113–53.

Stetler, D., Grumet, F.C. & Erlich, H.A. (1985). Polymorphic restriction endonuclease sites linked to the HLA-DRα gene: localization and use as genetic markers of insulin-dependent diabetes. *Proceedings of the National Academy of Sciences of the U.S.A.* 82, 8100–4.

Svejgaard, A., Jakobsen, B.K., Platz, P., Ryder, L.P., Nerup, J., Christy, M., Borch-Johnsen, K., Parving, H.H., Deckert, T., Mølsted-Petersen, L., Kühl, C., Buschard, K. & Green, A. (1986). HLA associations in insulin-dependent diabetes: search for heterogeneity in different groups of patients from a homogenous population. *Tissue Antigens* 28, 237–44.

Todd, J.A., Bell, J.I. & McDevitt, H.O. (1987). HLA-DQβ gene contributes to susceptibility and resistance to insulin-dependent diabetes mellitus. *Nature* 329, 599–604.

Wagener, D.K., Sachs, J.M., LaPorte, R.E. & Macgregory, M.J. (1982). The Pittsburgh study of insulin-dependent diabetes mellitus, risk for diabetes among relatives of IDDM. *Diabetes* 31, 136–44.

Walker, A. & Cudworth, A.G. (1980). Type I (insulin-dependent) diabetic multiplex families: mode of genetic transmission. *Diabetes* 29, 1036–9.

Wolf, E., Spencer, K.M. & Cudworth, A.G. (1983). The genetic susceptibility to type I (insulin-dependent) diabetes: analysis of the HLA-DR association. *Diabetologia* 24, 224–30.

12

The role of the insulin gene in diabetes: use of restriction fragment length polymorphisms in diagnosis

STEVEN C. ELBEIN AND M. ALAN PERMUTT

12.1. Introduction

The genetic nature of diabetes has been well documented by studies of familial aggregation, concordance in identical twins and, for insulin-dependent diabetes, HLA analysis in sib pairs (Rotter & Rimoin, 1981). For insulin-dependent (Type 1) diabetes mellitus (IDDM), autoimmune pancreatic destruction clearly occurs and appears to be mediated in part by genetic susceptibility in the HLA class II genes. None the less, considerable evidence suggests involvement of other loci, including the insulin gene (Rotter & Rimoin, 1981; Bell et al., 1984). For non-insulin-dependent diabetes mellitus (Type 2, NIDDM), concordance in identical twins approaches 100%; the familial nature is well documented (Rotter & Rimoin, 1981; Kobberling & Tillil, 1982). The physiological defect remains controversial, but elements of insulin resistance, increased hepatic glucose output, and relative insulin deficiency are all present in the fully developed phenotype (Weir, 1982; Truglia et al., 1985; Ward et al., 1984). Although defects in insulin production could account for part of this spectrum, a role for insulin gene defects has been described only in a few pedigrees with mild glucose intolerance or diabetes and hyperinsulinemia or hyperproinsulinemia (see below).

12.2. Insulin gene structure

The cloning of the human insulin gene (Bell et al., 1980; Ullrich et al., 1980) permitted a new approach to the role of insulin gene mutations in IDDM and NIDDM. Insulin is encoded by a single gene of 1430 base pairs (bp). Both somatic cell hybrids (Owerbach et al., 1980) and in situ hybridization (Zabel et al., 1985) have localized the gene to the short arm of chromosome 11 (11p). The gene consists of three exons (coding regions) representing the 5'-non-coding region (exon 1), the prepeptide, B-chain and first half of the C-peptide (exon 2), and the second half of the C-peptide, A-chain and 3'-non-coding region (exon 3). The three exons are separated by introns of 179 bp (exons 1 and 2) and 786 bp (exons 2 and 3) (Ullrich et al., 1980; Bell et al., 1980). This 1430 base pair DNA is transcribed and processed by splicing out of the two introns to give a preproinsulin mRNA of 446 bases with an additional 100–200 polyadeno-sine residues. Single nucleotide alterations were present in two of the alleles originally sequenced (Ullrich et al., 1980). These consisted of one mutation in the first intron, one in the second intron, and two in the 3'-non-coding region. Only one of these alters a restriction site and results in a new PstI site (Bell et al., 1980; Ullrich et al., 1980). These mutations do not appear to be significant for insulin gene expression. The new PstI site does not appear to be common in several populations, and thus is not a useful polymorphism for clinical studies (S.C.E. & M.A.P., unpublished data).

The insulin gene has the expected regulatory sequences in the 5'-flanking region. These include a transcriptional promoter (TATAAA) located 24 base pairs 5' from the transcription start site. In the 3' region is the expected polyadenylation signal (AATAAA). In addition, a region 168 to 258 bp upstream (5') to the transcription start site appears to be important as an enhancer element, and thus may be essential to full insulin gene expression (Walker et al., 1983).

12.3. Restriction fragment length polymorphisms of the insulin gene

No useful polymorphisms have been detected in the coding, non-coding or intronic regions of the insulin gene. A single site mutation in the 3' region of the gene has been described (Elbein et al., 1985a) but has not been useful in subsequent studies. In contrast, variations in the 5'-flanking region have been extensively studied in diabetes. Approximately 500 nucleotides from the transcription initiation site is a region which appears to represent a common insertion of variable length (Bell et al., 1981). Initial reports suggested two classes of insertions: those 0–600 bp and those 1600–2200 bp in length. Sequence analysis of this region from alleles of different sizes demonstrated that the apparent insertion actually represents a variable number of 14–15 bp repeating units (Bell et al., 1982; Ullrich et al., 1982; Owerbach & Aagaard, 1984; Rotwein et al., 1986). Such restriction fragment length polymorphisms (RFLPs) are now known to be common in the human genome, and have been distinguished from single base pair mutations by the term 'variable number tandem repeats' (VNTR). For the insulin gene, the number of repeating units varies from 30 to 139 among common alleles, but over 540 repeats are present in some individuals. These repeats are easily detected in DNA extracted from individuals and digested with a number of different restriction endonucleases (see below) using the technique of Southern transfer (Southern, 1975) and cloned insulin gene fragments which detect the tandem repeats. The amount of variation in this region depends on the length of the restriction fragment examined. Thus, digests generating large restriction fragments initially suggested that only two sizes were present, whereas the use of a probe specific for the repeating unit and enzymes which cut very close to this region of repeats demonstrates considerably more heterogeneity (Bell et al., 1981; Elbein et al., 1985a).

In addition to the region of tandem repeats (hereafter referred to as VNTR), two common site mutations have been noted in the far 5'-flanking region of the gene (Elbein et al., 1985a). The RFLPs are randomly associated with each other and the VNTR; thus, when all three RFLPs are used together, this locus becomes highly informative for family studies, with most individuals having two distinguishable insulin alleles (heterozygosity near 100% (Elbein et al., 1985a)). Additionally, the insulin gene has been mapped close to the c-Harvey ras oncogene (White et al., 1985; Zabel et al., 1985). Many studies have utilized another VNTR polymorphism in the 3' region of that gene to further increase the heterozygosity (number of distinguishable alleles) for pedigree studies, although recombination between these loci occurs at a rate of 3% (White et al., 1985). Finally, the IGF-II gene lies close to the insulin gene (Bell et al., 1985), although RFLPs of this gene have not been utilized in published studies.

12.4. Insulin gene RFLPs and association studies with diabetes

The earliest studies of the insulin gene in diabetes have involved the region of tandem repeats (VNTR) 5' to the insulin gene, variously known as the insulin gene polymorphism and the 5'-flanking polymorphism, and referred to in this discussion as the INS VNTR, to distinguish it from other insulin gene RFLPs. Analysis of these RFLPs utilizes the technique of Southern blotting (Southern, 1975). This technique has been in use for most contemporary genetic studies, and thus has been described in more detail elsewhere. Briefly, DNA is extracted from peripheral leukocytes and purified by well-described protocols. This DNA is digested with restriction endonucleases which recognize specific DNA sequences. In the case of the INS VNTR, several enzymes which recognize sites on either side of the tandem repeats may be used. The DNA is separated by size through electrophoresis on an agarose gel, is denatured to render it single-stranded, and is transferred by capillary action (or alternatively by electroblotting) to nylon membranes. DNA is bound to these membranes, and can subsequently be hybridized to specific DNA fragments labeled with $[^{32}P]$nucleotides. After hybridization for 24 h the blots are washed to remove non-specifically bound radioactive probe and are exposed to X-ray film. Since all DNA fragments are present, one blot can be hybridized with probes from several genes providing the RFLP is detected by that restriction enzyme.

Initial reports from several groups (Rotwein et al., 1981, 1983; Bell et al., 1984; Owerbach & Nerup, 1982) suggested an association of a large allele (1600 bp of insertion) with NIDDM. Additionally, early reports suggested a weak association of a smaller allele (0–600) with IDDM. These studies utilized restriction enzymes which generated fragments of 7000 bp or larger. Consequently, these association studies left an impression that only three types of allele were present: small, or Class 1 alleles; large, or Class 3 alleles; and intermediate sized alleles, or Class 2 (Bell et al., 1984). Racial differences in the frequencies of these alleles were apparent, with Class 2 alleles restricted to black populations (Bell et al., 1984; Elbein et al., 1985c), and Class 3 alleles rare in Japanese populations (Takeda et al., 1986). Despite racial differences in frequencies, initial reports tended to treat all racial groups together. Subsequent work in larger and racially uniform populations suggested only a weak association of Class 3 alleles and NIDDM in Caucasians ($p = 0.025$) (Bell et al., 1984) and no association of this allele with NIDDM in either blacks (Elbein et al., 1985c), Pima Indians (Knowler et al., 1984), Nauruans (Serjeantson et al., 1983) or Japanese (Takeda et al., 1986).

Associations of the Class 1 allele with IDDM were found to be significant in Caucasians even when all available data from several groups were combined (Bell et al., 1984); the results were confirmed by one other group (Ferns et al., 1986). In addition, several groups have reported associations of the INS VNTR with hypertriglyceridemia (Jowett et al., 1984a, b) and atherosclerosis (Mandrup-Poulsen et al., 1986), although these findings have yet to be confirmed in larger association studies or linkage studies.

12.5. Linkage analyses with the insulin gene RFLPs

Linkage analyses of the insulin gene and diabetes have been largely restricted to NIDDM. These studies have utilized the VNTR polymorphism, but have varied in the amount of polymorphism detected at this RFLP depending on the enzyme used. Some investigators have used the nearby HRAS locus (the c-Ha-*ras* oncogene) and some have combined this with the two 5'-flanking site polymorphisms (Elbein *et al.*, 1985*b*, 1988*a*, *b*). The best-studied pedigrees at this locus are those of the autosomal dominant subtype of NIDDM known as maturity onset diabetes of the young, or MODY (Tattersall & Fajans, 1975). Since many of these pedigrees are hypoinsulinemic (Fajans, 1982), insulin gene mutations were thought to be a possibility. Studies of several pedigrees have demonstrated lack of linkage, however, suggesting that this locus is not involved in the pathogenesis of MODY (Bell *et al.*, 1983; Andreone *et al.*, 1985; Owerbach *et al.*, 1983; Johnston *et al.*, 1984).

Studies of more typical NIDDM pedigrees have been more limited, and interpretation is made difficult by the unknown mode of inheritance and the possibility of multiple genes predisposing to diabetes, such that insulin gene defects could be present in some family members but the INS locus not linked to diabetes. Hitman *et al.* (1984) reported on large pedigrees in which lack of linkage was demonstrated despite scoring of the INS VNTR as only two alleles (large and small). Dobs *et al.* (1986) similarly studied 7 small pedigrees (5 Caucasian and 2 black), but their small pedigrees and ability to distinguish only two alleles really demonstrated only lack of segregation of the Class 3 alleles with NIDDM. We have recently studied a large number of Caucasian pedigrees from Utah, of which 12 were large enough to be informative (Elbein *et al.*, 1988*b*). Our analysis utilized the INS VNTR with an enzyme which generated multiple alleles, the two 5' site polymorphisms, and the HRAS polymorphism. No linkage was demonstrated in 6 pedigrees separately analysed, and no pedigree demonstrated statistically significant segregation of NIDDM and the insulin gene locus. Examination of sharing of identical alleles among siblings and of pooled analysis of the pedigrees under several models likewise failed to demonstrate significant involvement of the insulin gene. We have demonstrated lack of linkage in several other Caucasian pedigrees (Elbein *et al.*, 1988*a* and unpublished data). When the pedigrees studied by us are combined with that studied by Hitman (Hitman *et al.*, 1984), 11 pedigrees demonstrate no linkage, a number which suggests that insulin gene defects are not a major cause of NIDDM.

The positive association of the insulin gene polymorphism with IDDM has suggested a role for INS as a second locus after the HLA region. Two published studies have failed to find linkage of the insulin gene and IDDM (Ferns *et al.*, 1986; Cox *et al.*, 1988). A large collaborative study is in progress to evaluate the role of this locus in IDDM. Preliminary results from our studies also suggest lack of linkage, however (M.A. Permutt *et al.*, unpublished data). The reason for the discrepancy between association studies and linkage studies is unknown.

12.6. Does the insulin gene VNTR play a role in diabetes?

Studies have failed to uniformly demonstrate an association of the INS VNTR with NIDDM, but studies with IDDM demonstrate an association despite apparent lack of linkage in other studies. If this RFLP is indeed a marker for diabetes, it must in itself effect insulin gene expression, or it must be non-randomly associated with another mutation of the insulin gene or a surrounding gene in the population. Both possibilities have been evaluated. In studies of the 5' regulatory elements and enhancer sequence, Walker *et al.* (1983) could not demonstrate an effect of the VNTR region on insulin gene expression in cultured cells. Studies of insulin secretion physiology in humans using a sensitive method to determine pancreatic reserve also failed to demonstrate an association of the larger (Class 3) insertions with diminished insulin secretion (Permutt *et al.*, 1985).

To examine the possibility that association studies were the result not of a direct effect of the VNTR polymorphism, but rather of non-random association of that region with the predisposing gene, relationships between the VNTR and other RFLPs, both 5' and 3', were examined. Surprisingly, despite a relatively short distance between these RFLPs, the associations were nearly random (Chakravarti *et al.*, 1986). These data suggested that the VNTR region promotes recombination, and thus would be less likely to associate with a mutation of the insulin gene or a surrounding gene in a population. Thus several independent lines of evidence have failed to confirm a role for the INS VNTR in diabetes.

12.7. Insulin gene mutations

While a common role for insulin gene mutations in diabetes remains questionable, insulin gene defects were well described before the cloning of the insulin gene (Gabbay *et al.*, 1976). Several pedigrees have been described. They are characterized by autosomal dominant inheritance of hyperinsulinemia or hyperproinsulinemia and mild glucose intolerance to relatively mild NIDDM (Haneda *et al.*, 1984). All cases described thus far are heterozygous for insulin gene defects. The defects essentially fall into two classes: those which block cleavage of the C-peptide and thus result in high circulating levels of proinsulin or partially cleaved intermediates, and those which result in high circulating levels of normally cleaved insulin with diminished activity. Other classes of mutations which might give low circulating insulin levels, such as promoter or enhancer mutations or nonsense mutations leading to rapid protein degradation, have not been described despite the occurrence of such defects in other genes.

Familial hyperproinsulinemia was the first defect described by Gabbay (Gabbay et al., 1976) and later by Kanazawa (Kanazawa et al., 1978). The former defect was originally reported to be a defect in cleavage of the B-chain C-peptide linkage (Gabbay et al., 1979), but this was later found to be incorrect. The defect in both pedigrees appears to be a mutation of Arg^{65}, one of two dibasic amino acids required for cleavage of the A-chain from the C-peptide (Robbins et al., 1981). Cloning and sequencing of the defective gene from the second pedigree has demonstrated a G to A transition which results in a His for Arg^{65} substitution (Shibasaki et al., 1985). A third pedigree presented a similar clinical syndrome but appeared to have normal circulating proinsulin, and was thus thought to represent a defect in conversion (Gruppuso et al., 1984). To examine the question of whether the conversion defect was linked to the insulin gene, the insulin gene RFLPs described above were used to examine affected and unaffected members of the pedigree. This study demonstrated that the defect was linked to the insulin gene (Elbein et al., 1985b), but did not prove that an insulin gene mutation was present. Subsequent cloning and sequencing of this gene demonstrated a histidine for aspartic acid substitution at position 10 of the B-chain in the appropriate allele (Chan et al., 1987). The reason for the defective processing of this molecule remains speculative.

The first case of hyperinsulinemia from a mutant insulin was described by Tager and associates (Tager et al., 1979; Given et al., 1980) in 1979 in a lean 51 year old man with a 10 year history of NIDDM. Subsequently, a leucine for phenylalanine substitution at position B25 was suggested as a result of high-pressure liquid chromatography (Tager et al., 1979). The mutation was characterized by sequencing as a C to G transversion and has been called 'insulin Chicago' (Kwok et al., 1983). A unique feature of this mutation is that it falls in a restriction endonuclease recognition site. Thus, the mutation may be detected as the loss of an MboII site or a new RsaI site on Southern transfers. A second mutant, known as 'insulin Los Angeles', results from serine for phenylalanine substitution at position B24; the mutation is also in the MboII recognition site (Shoelson et al., 1983). Thus both of these mutations may be detected by restriction digests. Sanz et al. (1986) screened 213 unrelated individuals (426 alleles) with

NIDDM for mutations resulting in loss of this MboII site, but found only one mutation. Further examination showed an apparently normal insulin by high-pressure liquid chromatography, thus suggesting a silent mutation (Sanz et al., 1986). Thus these mutations do not appear to be common in NIDDM. The effect of B24 and B25 substitutions is apparently to alter the region of the insulin molecule which binds to the receptor.

A third mutation has been described in a 56 year old Japanese woman and has been called 'insulin Wakayama' (Nanjo et al., 1986). This patient had polyuria, ketonuria and weight loss; her condition was thus more severe than the mutations discussed above. The mutation results from a G to T transversion at the codon for valine at position 3 in the A-chain, resulting in a leucine at that position. This same mutation has been identified in one other unrelated Japanese pedigree (Sakura et al., 1986). Like the B24 and B25 mutations, this A-chain mutation is thought to alter the conformation of the receptor binding region.

12.8. Summary

The cloning of the insulin gene and identification of RFLPs at this locus has permitted an evaluation of the role of insulin gene mutations in diabetes. While these RFLPs provide the tools to rapidly evaluate families with possible insulin gene defects for linkage to the insulin locus, available data have not confirmed the initially exciting results from association studies. Thus, insulin gene defects appear to be unusual in Caucasians with NIDDM. Other racial groups have not been extensively analyzed by linkage studies, and may have such defects. In IDDM, association studies suggest a role for the insulin locus while linkage studies have thus far failed to confirm this finding. This discrepancy remains to be resolved, but since IDDM must involve at least two loci, current linkage studies may be inadequate to assess this role. The insulin gene VNTR does not appear to have a role in insulin production or secretion, however. Furthermore, the variable spectrum of normal glucose intolerance to overt diabetes in individuals with heterozygous insulin gene mutations raises questions of whether these mutations are the sole cause of NIDDM in these pedigrees. At the current time insulin gene RFLPs do not have a routine diagnostic role in diabetes, although they could be useful in the analysis of some pedigrees with possible insulin gene defects.

References

Andreone, T., Fajans, S.S., Rotwein, P., Skolnick, M. & Permutt, M.A. (1985). Insulin gene analysis in a family with maturity-onset diabetes of the young. Diabetes 34, 108–14.

Bell, G.I., Gerhard, D.S., Fong, N.M., Sanchez-Pescador, R. & Rall, L.B. (1985). Isolation of the human insulin-like growth factor genes: insulin-like growth factor II and insulin genes are contiguous. *Proceedings of the National Academy of Sciences of the U.S.A.* **82**, 6450–4.

Bell, G.I., Horita, S. & Karam, J.H. (1984). A polymorphic locus near the human insulin gene is associated with insulin-dependent diabetes mellitus. *Diabetes* **33**, 176–83.

Bell, G.I., Karam, J.H. & Rutter, W.J. (1981). Polymorphic DNA region adjacent to the 5' end of the human insulin gene. *Proceedings of the National Academy of Sciences of the U.S.A.* **78**, 5759–63.

Bell, G.I., Pictet, R.L., Rutter, W.J., Cordell, B., Tischer, E. & Goodman, H.M. (1980). Sequence of the human insulin gene. *Nature* **284**, 26–32.

Bell, G.I., Selby, M.J. & Rutter, W.J. (1982). The highly polymorphic region near the human insulin gene is composed of simple tandemly repeating sequences. *Nature* **295**, 31–5.

Bell, J.I., Wainscoat, J.S., Old, J.M., Chlouverakis, C., Keen, H., Turner, R.C. & Weatherall, D.J. (1983). Maturity onset diabetes of the young is not linked to the insulin gene. *British Medical Journal* **286**, 590–2.

Chakravarti, A., Elbein, S.C. & Permutt, M.A. (1986). Evidence for increased recombination near the human insulin gene: implication for disease association studies. *Proceedings of the National Academy of Sciences of the U.S.A.* **83**, 1045–9.

Chan, S.J., Seino, S., Gruppuso, P.A., Schwartz, R. & Steiner, D.F. (1987). A mutation in the B chain coding region is associated with impaired proinsulin conversion in a family with hyperproinsulinemia. *Proceedings of the National Academy of Sciences of the U.S.A.* **84**, 2194–7.

Cox, N.J., Baker, L. & Spielman, R.S. (1988). Insulin-gene sharing in sib pairs with insulin-dependent diabetes mellitus: no evidence for linkage. *American Journal of Human Genetics* **42**, 167–72.

Dobs, A.S., Phillips, J.A. III, Mallonee, R.L., Saudek, C.D. & Ney, R.L. (1986). Pedigree analysis of the 5' flanking region of the insulin gene in familial diabetes mellitus. *Metabolism* **35**, 13–17.

Elbein, S.C., Corsetti, L., Goldgar, D., Skolnick, M. & Permutt, M.A. (1988a). Insulin gene linkage in familial NIDDM: lack of linkage in Utah Mormon pedigrees. *Diabetes* **37**, 569–76.

Elbein, S.C., Corsetti, L. & Permutt, M.A. (1985a). New polymorphisms at the insulin locus increase its usefulness as a genetic marker. *Diabetes* **34**, 1139–44.

Elbein, S.C., Gruppuso, P., Schwartz, R., Skolnick, M. & Permutt, M.A. (1985b). Hyperproinsulinemia in a family with a proposed defect in conversion is linked to the insulin gene. *Diabetes* **34**, 821–4.

Elbein, S.C., Rotwein, P., Permutt, M.A., Bell, G.I., Sanz, N. & Karam, J.H. (1985c). Lack of association of the polymorphic locus in the 5'-flanking region of the human insulin gene and diabetes in American Blacks. *Diabetes* **34**, 433–9.

Elbein, S.C., Ward, W.K., Beard, J.C. & Permutt, M.A. (1988b). Familial NIDDM: molecular-genetic analysis and assessment of insulin action and pancreatic β-cell function. *Diabetes* **37**, 377–82.

Fajans, S.S. (1982). Heterogeneity between various families with non-insulin dependent diabetes of the MODY type. In *The genetics of diabetes mellitus* (ed. J. Kobberling & R. Tattersall), pp. 251–60. London and New York: Academic Press.

Ferns, G.A.A., Hitman, G.A., Trembath, R., Williams, L., Tarn, A., Gale, E.A. & Galton, D.J. (1986). DNA polymorphic haplotypes on the short arm of chromosome 11 and the inheritance of type I diabetes mellitus. *Journal of Medical Genetics* **23**, 210–16.

Gabbay, K.H., Bergenstal, R.M., Wolff, J., Mako, M.E. & Rubenstein, A.H. (1979). Familial hyperproinsulinemia: partial characterization of circulating proinsulin-like material. *Proceedings of the National Academy of Sciences of the U.S.A.* **76**, 2881–5.

Gabbay, K.H., DeLuca, K., Fisher, J.N. Jr, Mako, M.E. & Rubenstein, A.H. (1976). Familial hyperproinsulinemia: an autosomal dominant defect. *New England Journal of Medicine* **294**, 911–15.

Given, B.D., Mako, M.E., Tager, H.S., Baldwin, D., Markese, J., Rubenstein, A.H., Olefsky, J., Kobayashi, M., Kolterman, O. & Poucher, R. (1980). Diabetes due to secretion of an abnormal insulin. *New England Journal of Medicine* **302**, 129–35.

Gruppuso, P.A., Gorden, P., Kahn, C.R., Cornblath, M., Zeller, W.P. & Schwartz, R. (1984). Familial hyperproinsulinemia due to a proposed defect in conversion of proinsulin to insulin. *New England Journal of Medicine* **311**, 629–34.

Haneda, M., Polonsky, K.S., Bergenstal, R.M., Jaspan, J.B., Shoelson, S.E., Blix, P.M., Chan, S.J., Kwok, S.C.M., Wishner, W.B., Zeidler, A., Olefsky, J.M., Freidenberg, G., Tager, H.S., Steiner, D.F. & Rubenstein, A.H. (1984). Familial hyperinsulinemia due to a structurally abnormal insulin: definition of an emerging new clinical syndrome. *New England Journal of Medicine* **310**, 1288–94.

Hitman, G.A., Jowett, N.I., Williams, L.G., Humphries, S., Winter, R.M. & Galton, D.J. (1984). Polymorphisms in the 5'-flanking region of the insulin gene and non-insulin-dependent diabetes. *Clinical Science* **66**, 383–8.

Johnston, C., Owerbach, D., Leslie, R.D.G., Pyke, D.A. & Nerup, J. (1984). Mason-type diabetes and DNA insertion polymorphism. *Lancet* i, 280.

Jowett, N.I., Rees, A., Williams, L.G., Stocks, J., Vella, M.A., Hitman, G.A., Katz, J. & Galton, D.J. (1984a). Insulin and apolipoprotein A-1/C-III gene polymorphisms relating to hypertriglyceridaemia and diabetes mellitus. *Diabetologia* **27**, 180–3.

Jowett, N.I., Williams, L.G., Hitman, G.A. & Galton, D.J. (1984b). Diabetic hypertriglyceridaemia and related 5' flanking polymorphism of the human insulin gene. *British Medical Journal* **288**, 96–8.

Kanazawa, Y., Hayashi, M., Ikeuchi, M., Kasuga, M., Oka, Y., Sato, H., Hiramatsu, K. & Kosaka, K. (1978). Familial proinsulinemia: a rare disorder of insulin biosynthesis. In *Proinsulin, insulin, C-peptide* (International Congress Series 468) (ed. S. Baba, T. Kaneko & N. Yanaihara), pp. 262–9. Amsterdam: Excerpta Medica.

Knowler, W.C., Pettitt, D.J., Vasquez, B., Rotwein, P.S., Andreone, T.L. & Permutt, M.A. (1984). Polymorphism in the 5' flanking region of the human insulin gene: relationships with non-insulin-dependent diabetes mellitus, glucose and insulin concentrations, and diabetes treatment in the Pima Indians. *Journal of Clinical Investigation* **74**, 2129–35.

Kobberling, J. & Tillil, H. (1982). Empirical risk figures for first degree relatives of non-insulin dependent diabetics. In *The genetics of diabetes mellitus* (ed. J. Kobberling & R. Tattersall), pp. 201–9. London and New York: Academic Press.

Kwok, S.C.M., Steiner, D.F., Rubenstein, A.H. & Tager, H.S. (1983). Identification of a point mutation in the human insulin gene giving rise to a structurally abnormal insulin gene (Insulin Chicago). *Diabetes* **32**, 872–5.

Mandrup-Poulsen, T., Owerbach, D., Nerup, J., Johansen, K. & Hansen, A.T. (1986). Diabetes mellitus, atherosclerosis, and the 5' flanking polymorphism of the human insulin gene. *Journal of Inherited Metabolic Disease* **9** (suppl. 1), 98–110.

Nanjo, K., Sanke, T., Miyano, M., Okai, K., Sowa, R., Kondo, M., Nishimura, S., Iwo, K., Miyamura, K., Given, B.D., Chan, S.J., Tager, H.S., Steiner, D.F. & Rubenstein, A.H. (1986). Diabetes due to secretion of a structurally abnormal insulin (Insulin Wakayama): clinical and functional characteristics of [Leu^{A3}]insulin. *Journal of Clinical Investigation* **77**, 514–19.

Owerbach, D. & Aagaad, L. (1984). Analysis of a 1963-bp polymorphic region flanking the human insulin gene. *Gene* **32**, 475–9.

Owerbach, D., Bell, G.I., Rutter, W.J. & Shows, T.B. (1980). The insulin gene is located on chromosome 11 in humans. *Nature* **286**, 82–4.

Owerbach, D. & Nerup, J. (1982). Restriction fragment length polymorphism of the insulin gene in diabetes mellitus. *Diabetes* **31**, 275–7.

Owerbach, D., Thomsen, B., Johansen, K., Lamm, L.U. & Nerup, J. (1983). DNA insertion sequences near the insulin gene are not associated with maturity-onset diabetes of young people. *Diabetologia* **25**, 18–20.

Permutt, M.A., Rotwein, P., Andreone, T., Ward, W.K. & Porte, D. Jr (1985). Islet β-cell function and polymorphism in the 5′-flanking region of the human insulin gene. *Diabetes* **34**, 311–14.

Robbins, D.C., Blix, P.M., Rubenstein, A.H., Kanazawa, Y., Kosaka, K. & Tager, H.S. (1981). A human proinsulin variant at arginine 65. *Nature* **291**, 679–81.

Rotter, J.I. & Rimoin, D.L. (1981). The genetics of the glucose intolerance disorders. *American Journal of Medicine* **70**, 116–26.

Rotwein, P.S., Chirgwin, J., Province, M., Knowler, W.C., Pettitt, D.J., Cordell, B., Goodman, H.M. & Permutt, M.A. (1983). Polymorphism in the 5′ flanking region of the human insulin gene: a genetic marker for non-insulin-dependent diabetes. *New England Journal of Medicine* **308**, 65–71.

Rotwein, P., Chyn, R., Chirgwin, J., Cordell, B., Goodman, H.M. & Permutt, M.A. (1981). Polymorphism in the 5′-flanking region of the human insulin gene and its possible relation to type 2 diabetes. *Science* **213**, 1117–20.

Rotwein, P., Yokoyama, S., Didier, D.K. & Chirgwin, J.M. (1986). Genetic analysis of the hypervariable region flanking the human insulin gene. *American Journal of Human Genetics* **39**, 291–9.

Sakura, H., Iwamoto, Y., Sakamoto, Y., Kuzuya, T. & Hirata, H. (1986). Structurally abnormal insulin in a diabetic patient: characterization of the mutant insulin A3(val→leu) isolated from the pancreas. *Journal of Clinical Investigation* **78**, 1666–72.

Sanz, N., Karam, J.H., Horita, S. & Bell, G.I. (1986). Prevalence of insulin-gene mutations in non-insulin-dependent diabetes mellitus. *New England Journal of Medicine* **314**, 1322–3.

Serjeantson, S.W., Owerbach, D., Zimmet, P., Nerup, J. & Thoma, K. (1983). Genetics of diabetes in Nauru: effects of foreign admixture, HLA antigens and the insulin-gene-linked polymorphism. *Diabetologia* **25**, 13–17.

Shibasaki, Y., Kawakami, T., Kanazawa, Y., Akanuma, Y. & Takaku, F. (1985). Posttranslational cleavage of proinsulin is blocked by a point mutation in familial hyperproinsulinemia. *Journal of Clinical Investigation* **76**, 378–80.

Shoelson, S., Haneda, M., Blix, P., Nanjo, A., Sanke, T., Inouye, K., Steiner, D., Rubenstein, A. & Tager, H. (1983). Three mutant insulins in man. *Nature* **302**, 540–3.

Southern, E.M. (1975). Detection of specific sequences among DNA fragments separated by gel electrophoresis. *Journal of Molecular Biology* **98**, 503–17.

Tager, H., Given, B., Baldwin, D., Mako, M., Markese, J., Rubenstein, A., Olefsky, J., Kobayashi, M., Kolterman, O. & Poucher, R. (1979). A structurally abnormal insulin causing human diabetes. *Nature* **281**, 122–5.

Takeda, J., Seino, Y., Fukumoto, H., Koh, G., Otsuka, A., Ikeda, M., Kuno, S., Yawata, M., Moridera, K., Morita, T., Tsuda, K. & Imura, H. (1986). The polymorphism linked to the human insulin gene: its lack of association with either IDDM or NIDDM in Japanese. *Acta Endocrinologica* **113**, 268–71.

Tattersall, R.B. & Fajans, S.S. (1975). A difference between the inheritance of classical juvenile-onset and maturity-onset type diabetes of young people. *Diabetes* **24**, 44–53.

Truglia, J.A., Livingston, J.N. & Lockwood, D.H. (1985). Insulin resistance: receptor and post-binding defects in human obesity and non-insulin-dependent diabetes mellitus. *American Journal of Medicine* **79** (suppl. 2B), 13–22.

Ullrich, A., Dull, T.J. & Gray, A. (1980). Genetic variation in the human insulin gene. *Science* **209**, 612–15.

Ullrich, A., Dull, T.J., Gray, A., Philips, J.A. III & Peter, S. (1982). Variation in the sequence and modification state of the human insulin gene flanking regions. *Nucleic Acids Research* **10**, 2225–40.

Walker, M.D., Edlund, T., Boulet, A.M. & Rutter, W.J. (1983). Cell-specific expression controlled by the 5'-flanking region of insulin and chymotrypsin genes. *Nature* **306**, 557–61.

Ward, W.K., Beard, J.C., Halter, J.B., Pfeifer, M.A. & Porte, D. Jr (1984). Pathophysiology of insulin secretion in non-insulin-dependent diabetes mellitus. *Diabetes Care* **7**, 491–502.

Weir, G.C. (1982). Non-insulin dependent diabetes mellitus: interplay between B-cell inadequacy and insulin resistance. *American Journal of Medicine* **73**, 461–4.

White, R., Leppert, M., Bishop, D.T., Barker, D., Berkowitz, J., Brown, C., Callahan, P., Holm, T. & Jerominski, L. (1985). Construction of linkage maps with DNA markers for human chromosomes. *Nature* **313**, 101–5.

Zabel, B.U., Kronenberg, H.M., Bell, G.I. & Shows, T.B. (1985). Chromosome mapping of genes on the short arm of human chromosome 11: parathyroid hormone gene is at 11p15 together with the genes for insulin, c-Harvey-*ras* 1, and β-hemoglobin. *Cytogenetics and Cell Genetics* **39**, 200–5.

13

Abnormal products of the human insulin gene

HOWARD S. TAGER

13.1. Introduction

As has been amply documented by preceding chapters, the structures of hormone genes, related mRNAs and precursor polypeptides are critical to determining the nature of peptides secreted from the pancreatic islet. Whereas transcription of the human insulin gene can be expected usually to result in the formation of normal insulin in normal amounts, five classes of insulin gene mutations might affect the complex processes of insulin biogenesis. As illustrated by Fig. 13.1, mutations might alter (a) the rate of control of insulin gene transcription (by affecting promoter, enhancer, termination or other elements); (b) the efficiency or correctness of processing of the corresponding mRNA precursor (by interfering with excision of intervening sequences, 5'-capping or 3'-polyadenylation); (c) the rate or control of translation of mature mRNA (by altering ribosome binding sites, regulatory regions or mRNA stability); (d) the length of the product of insulin mRNA translation (by changing the position of the first stop codon encountered during the process); or (e) the amino acid sequences of peptides resulting from the accurate and controlled translation of insulin mRNA. The last group of mutations could involve (a) amino acid replacements in the signal peptide region in preproinsulin (leading to failure of the precursor to be translocated from the cytoplasm to the cisternum of the rough endoplasmic reticulum, or to failure of the precursor to be sorted to the Golgi apparatus and secretion granules); (b) replacements at critical processing sites in proinsulin (leading to failure of the hormone precursor to be converted enzymatically to insulin); or (c) replacements in the regions of the insulin A- and B-chains (leading ultimately to the formation of products with altered structures). Multiple articles concerning the subject

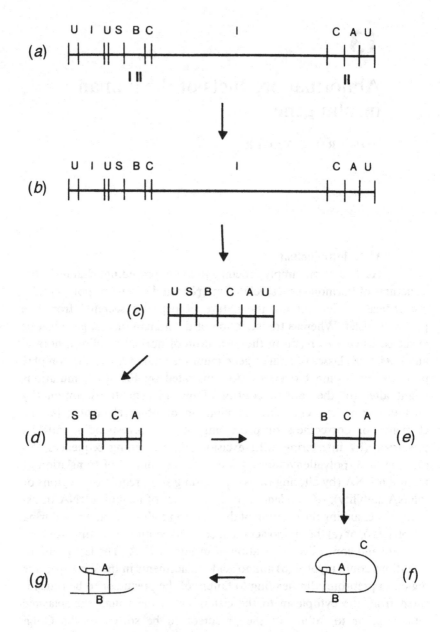

Fig. 13.1. Diagrammatic representation of the process of insulin biosynthesis. The figure illustrates the structures of (*a*) the insulin gene, (*b*) the insulin mRNA precursor (resulting from transcription of the gene), (*c*) the mature insulin mRNA (resulting from excision of intervening sequences, 5′ capping and 3′ polyadenylation), (*d*) preproinsulin (the NH$_2$-terminal signal peptide of which is removed

of insulin biogenesis have appeared (Steiner *et al.*, 1969, 1980; Steiner, 1977; Bell *et al.*, 1980; Ullrich *et al.*, 1980; Given *et al.*, 1985).

Whereas mutations leading to incorrect excision of intervening segments from the insulin mRNA precursor and those leading to incorrect translational termination would certainly cause major changes in the structures of insulin gene products, and whereas mutations in various other regions could have important effects in altering the rate or control of product formation, known mutations in the human insulin gene concern simply the exchange of one codon for another in structural regions of the gene. All involve single base changes, and all result in single amino acid replacements. The consequences of these replacements, however, can extend well beyond what might have been expected. This chapter will consider separately the matters of (*a*) relationships between insulin gene mutations and diabetes, (*b*) insulin gene mutations leading to the formation of abnormal 5 kDa insulins, (*c*) insulin gene mutations affecting proinsulin conversion to insulin, and (*d*) structure–activity relationships and implications of insulin gene mutations in understanding the evolved structure of insulin.

13.2. Insulin gene mutations and human diabetes

Although the details of specific insulin gene mutations will be provided later, it may be helpful at this stage to identify the clinical characteristics of patients with insulin gene mutations and to clarify the relationship between these mutations and diabetes. In general, patients subsequently identified to have a mutant insulin gene and to secrete an abnormal insulin gene product can be characterized as being hyperinsulinemic (as judged by simple measurement of serum insulin by radioimmunoassay), as having an increased serum insulin:C-peptide molar ratio (because of the hyperinsulinemia), as lacking any evidence for the existence of high levels of contra-insulin hormones, as exhibiting normal levels of insulin receptors on isolated monocytes, as showing mild to moderate glucose intolerance, and as responding normally to the administration of exogenous insulin during an insulin tolerance test (Given *et al.*,

Caption to Fig. 13.1 (*cont.*)

cotranslationally), (*e*) linear proinsulin, (*f*) folded proinsulin, and (*g*) insulin. Throughout: U, untranslated region; I, intervening sequence; S, signal peptide; B, insulin B-chain; C, proinsulin C-peptide; A, insulin A-chain. Bars under the illustration in (*a*) indicate sites of known mutation in the human insulin gene.

1980; Haneda *et al.*, 1984; Nanjo *et al.*, 1986*b*; Iwamoto *et al.*, 1986*b*). In its simplest expression, a patient with diabetes associated with secretion of an abnormal insulin shows higher than normal levels of serum immunoreactive insulin, seems not to respond to elevated levels of endogenous hormone as expected, and responds to exogenous insulin much as a normal subject would. A logical extension of these findings suggests that the insulin from an affected subject somehow differs from normal insulin in a way that decreases its biological potency. Additional information identifying that the insulin-immunoreactive material from the serum of affected subjects (when purified by immunoaffinity chromatography or other means) exhibits decreased receptor binding and biological potency *in vitro* and that the related abnormality of hyperinsulinemia is inherited with a kindred leads to what now seems an obvious conclusion: patients with the characteristics described possess a mutant insulin gene, the expression of which results in the formation and secretion of a low-activity, abnormal insulin. It should be noted, however, that the picture is not always so clear as the general example just given.

Questions arising from the clinical presentation of patients with insulin gene mutations involve consideration of the biological potencies of the relevant abnormal gene products, the causes of hyperinsulinemia, the genetics and inheritance of mutant insulin genes, and the association of insulin gene mutations with diabetes. The matters of the biological potencies of abnormal insulin gene products and of hyperinsulinemia are actually tightly interrelated. Measurements of the biological potencies of the abnormal insulin gene products identified to date (determined, in part, as the result of analysis of peptides prepared by semisynthetic methods and assays conducted *in vitro*) show them, in the aggregate, to have only about 1–10% of the potency of normal insulin (Tager *et al.*, 1980; Wollmer *et al.*, 1981; Inouye *et al.*, 1981; Kobayashi *et al.*, 1982, 1984*b*, 1986; Shoelson *et al.*, 1983*a*; Haneda *et al.*, 1985; Peavy *et al.*, 1985). Most important, decrements in the affinity of the insulin receptor for the abnormal gene products and decrements in the ability of the abnormal hormones to effect typical cellular responses to insulin (for example, stimulation of 2-deoxyglucose uptake or glucose oxidation) are usually parallel and equivalent. It is thus the case that the low biological potencies of the abnormal gene products can be attributed specifically to their poor abilities to combine with insulin receptors. Under most circumstances the abnormal gene products (notwithstanding their high levels in the circulation, usually about 100 μunits/ml) would be expected to contribute only marginally, or not at all, to normal glucose homeostasis. In fact, the hyperinsulinemia associated with expression of a mutant insulin gene and the high levels of abnormal insulin gene products seen in affected individuals arise from the low affinities of the abnormal products for the insulin receptor and from the consequently decreased degradation of the abnormal products by the usually applicable receptor-mediated mechanisms (Terris & Steiner, 1975, 1976). The abnormal insulin gene products do not efficiently bind to insulin receptors, do not exert significant biological activity, and are not efficiently removed from the circulation (Shoelson et al., 1984).

All individuals studied to date who express abnormal insulin genes have been found to be heterozygous. That is, analysis of gene structure (by molecular cloning or by restriction fragment mapping and Southern analysis) (Kwok et al., 1981, 1983; Shoelson et al., 1983b; Haneda et al., 1983; Nanjo et al., 1986b) has identified in each case a normal insulin allele and an abnormal allele which has undergone a single nucleotide replacement. Parallel analysis of serum insulin (most often by study of immunopurified material and by use of high performance liquid chromatography coupled with radioimmunoassay (Shoelson et al., 1983b) has identified the existence of both normal and abnormal hormone forms in the circulation. Fig. 13.2 illustrates what might be called an HPLC pedigree developed for the kindred expressing the abnormal insulin Insulin Los Angeles (Shoelson et al., 1984). As can be seen, (a) the abnormal insulin is readily detected in the serum of the propositus by HPLC analysis; (b) the serum of the propositus (like that of affected relatives) contains small amounts of normal insulin as well as larger amounts of the abnormal form; (c) the abnormal insulin allele (as reflected by the appearance of the abnormal insulin in serum) can be traced to the father of the propositus; (d) all three siblings of the propositus can be seen to express the abnormal insulin allele (and can thus be regarded to have inherited the allele from the paternal source); and (e) a single member of the third generation also appears to have inherited and to express the abnormal insulin allele. It is thus clear that the abnormal insulin allele is inherited in an autosomal fashion in these heterozygous subjects, and that, in each case, the majority of serum insulin is represented by the abnormal insulin which arises from the mutant insulin allele.

Although it is clear that individuals secreting insulin Los Angeles are heterozygous for the mutant allele, the assessment of the relative expression of normal and abnormal alleles in this kindred has not been possible. In two other examples of abnormal human insulins (one involving insulin Chicago and the other insulin Wakayama), however, the question of

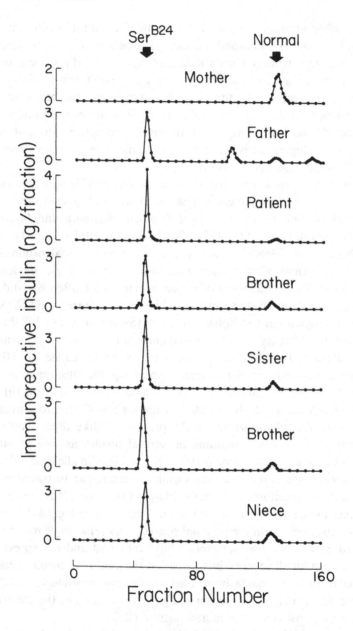

Fig. 13.2. Analysis of serum insulin from the kindred expressing insulin Los Angeles by high performance liquid chromatography. The panels show elution profiles (assayed by radioimmunoassay) for the serum insulin derived from the patient (a 27 year old woman), her mother, her father, sister, each of two brothers, and a niece. The earlier eluting peak (a peak representing the vast majority of serum insulin in

allelic expression has been addressed. In each of two cases where it has been possible to measure the content of normal and abnormal insulin in pancreatic biopsies from individuals expressing abnormal insulin alleles in the heterozygous state (Tager *et al.*, 1979; Iwamoto *et al.*, 1986*a*, *b*), it has been found that normal and abnormal forms of the hormone are stored in pancreatic tissue in approximately equimolar amounts. It thus appears that normal and abnormal alleles are codominantly expressed, and that pancreatic secretions from affected subjects most likely contain equimolar amounts of both hormonal forms. To see whether the equimolar secretion of normal and abnormal insulins (a situation resulting from the codominant expression of normal and abnormal alleles) could actually account for the predominance of the abnormal form in the circulation, we infused equimolar amounts of normal human insulin and of insulin Los Angeles into the hepatic circulation of anesthetized dogs (Shoelson *et al.*, 1984). Results indeed showed that the equimolar infusion of the two insulins resulted in the appearance in the periphery of an insulin mixture in which the abnormal insulin predominated and which mimicked closely the mixture seen in the peripheral blood of affected human subjects.

From results presented above, it would seem that (*a*) the autosomal inheritance of an abnormal insulin allele in heterozygous subjects, (*b*) the codominant expression of normal and abnormal alleles in affected individuals, (*c*) the equimolar secretion of both forms into the pancreatic vein, and (*d*) the low receptor binding potency of the abnormal form and its concomitantly decreased rate of metabolic clearance (Shoelson *et al.*, 1984; Kobayashi *et al.*, 1984*a*; Haneda *et al.*, 1985) could, in sum, account for the predominance of the abnormal gene product in the peripheral circulation. As noted earlier, it is the product of the abnormal insulin allele which causes the hyperinsulinemia typical of individuals expressing abnormal insulin genes; the level of normal insulin (a hormone form resulting from the expression of the normal allele) is most often either at the lower level of that seen in normal individuals or is lower (Shoelson *et al.*, 1983*b*, 1984) (Fig. 13.2).

Given that identified subjects expressing abnormal insulin genes are invariably heterozygous, and that normal and abnormal alleles are

Caption to Fig. 13.2 (*cont.*)

affected subjects) corresponds to human [Ser^{B24}]insulin (insulin Los Angeles). The position taken by normal human insulin (a component present in the serum of all subjects in much lower amount) is indicated by the word 'normal'. Only the patient and her father are diabetic. The figure is modified from one presented by Shoelson *et al.* (1984).

apparently codominantly expressed, the association of abnormal insulin gene expression with glucose intolerance or with diabetes is not always clear. That is, affected subjects would obviously be at risk (since only half of their secreted insulin would have significant biological activity), but a 50% deficit in the synthesis of biologically active insulin *per se* would not necessarily be expected to cause clinical symptoms. Although the matter was not identified earlier, it is important to note that only two members of the kindred expressing insulin Los Angeles (the propositus and her father) (see Fig. 13.2) are diabetic; other affected members are either asymptomatic or only slightly glucose-intolerant (Shoelson *et al.*, 1984). The explanation, as it turns out, is that expression of a mutant insulin gene does not itself cause diabetes. Glucose intolerance, in any but the most minor degree of severity, appears to arise only as the result of a perturbing secondary defect. In the example of insulin Los Angeles, the propositus appears to suffer a defect in the ability to secrete insulin as the result of a glucose stimulus: whereas serum insulin (or C-peptide levels) increase only slightly if at all after oral glucose in the case of the propositus, her essentially asymptomatic siblings respond to oral glucose with the robust and sustained secretion of insulin to account for the decreased average potency of the secreted products (Shoelson *et al.*, 1984). Perhaps, to oversimplify, the propositus suffers a compound impairment of the insulin system, whereas her siblings are able to compensate (at least at this time) by what would ordinarily be considered as exaggerated degrees of insulin secretion; whether the currently asymptomatic affected members of the kindred will retain their compensating ability over long periods, however, remains to be seen.

It can be expected that secondary factors potentially contributing to glucose intolerance or to diabetes in subjects expressing abnormal insulin alleles will include, as well as blunted secretory responses to oral glucose, varying degrees of apparent insulin resistance. Preliminary evidence for the contribution of insulin resistance to the glucose intolerance of some individuals expressing abnormal insulin genes has already been obtained. In these cases, however, affected subjects will not necessarily respond normally to the administration of insulin during insulin tolerance tests, and hypotheses as to the direct or secondary causes of glucose intolerance will be more difficult to assess. All considered, it is now clear that insulin gene mutations do not actually cause diabetes (at least in heterozygous subjects), and that the association of insulin gene mutations with severe glucose intolerance seems necessarily to require a secondary defect in the processes by which insulin is secreted from the endocrine pancreas or by which insulin acts at peripheral sites. Thus, the diagnosis of glucose intolerance or overt diabetes as the result of insulin gene mutation ultimately requires analysis of insulin gene structure or analysis of the structures of secreted gene products.

13.3. Mutations leading to abnormal insulins

The first clear evidence for the existence of an insulin gene mutation in man and for the secretion of an abnormal insulin associated with human diabetes was obtained about 10 years ago (Tager et al., 1979). While the clinical findings of the relevant patient were in accord with the criteria for insulin gene mutation identified above (Given et al., 1980), proof for the existence of an abnormal insulin relied on the analysis of insulin purified from a pancreatic biopsy derived from the affected subject. Initial experiments involving determination of the amino acid composition of the isolated insulin suggested (a) the occurrence of a Leu for Phe replacement at position B24 or B25 in the patient's abnormal insulin, and (b) the storage of normal and abnormal insulins in islet B-cells in approximately equimolar amounts (Tager et al., 1979). Although restriction enzyme digestion and Southern analysis of leukocyte DNA confirmed the existence of a nucleotide change in the codon serving for Phe^{B24} or Phe^{B25} (normally TTC in both cases) (Kwok et al., 1981), insufficient amounts of the isolated insulin were available for sequence analysis. Identification of the site of mutation and of amino acid replacement therefore had to rely on semisynthesis of the corresponding two possible abnormal insulins ($[Leu^{B24}]$insulin and $[Leu^{B25}]$insulin) and analysis of the natural insulin and the semisynthetic analogs by use of reverse-phase HPLC (Shoelson et al., 1983b); subsequent studies involved molecular cloning and determination of the nucleotide sequence of abnormal insulin alleles (Kwok et al., 1983). It turns out that the first identified abnormal insulin (insulin Chicago) corresponded to human $[Leu^{B25}]$insulin and arose from a single nucleotide change in which G replaced C in the phenylalanine codon for position B25 (TTC→TTG). Insulin Chicago was thus identified as human insulin B25 (Phe→Leu) (Fig. 13.3 and Table 13.1).

The second and third examples of abnormal insulins arising from insulin gene mutations were first identified by use of HPLC applied to immunoaffinity-purified serum insulin derived from affected subjects (Shoelson et al., 1983b). In each of these examples, the serum insulin migrated on reverse-phase columns to a position different from that taken by normal insulin. In the second example (insulin Los Angeles), the

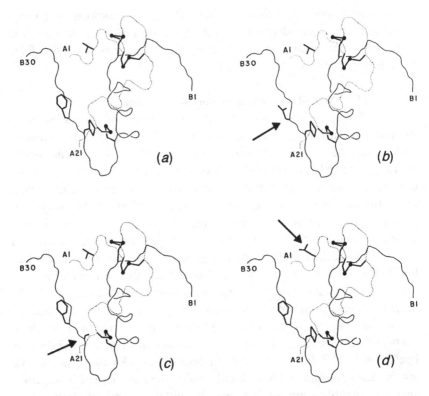

Fig. 13.3. Diagrammatic structures of the abnormal insulins. The diagrams show the basic structure of the insulin monomer (viewed in a direction perpendicular to the 3-fold axis) in the 2-zinc insulin hexamer, as described by Blundell *et al.* (1972). Side chains are shown explicitly only for residues at positions A3, B24 and B25 (Val, Phe and Phe respectively in normal insulin). Arrows show sites of amino acid replacement. Solid lines, the insulin B-chain; dotted lines, the insulin A-chain. (*a*) Normal insulin; (*b*) insulin Chicago (Phe^{B25}→Leu); (*c*) insulin Los Angeles (Phe^{B24}→Ser); (*d*) insulin Wakayama (Val^{A3}→Leu).

elution position of the abnormal insulin indicated that it possessed characteristics more hydrophilic than those of normal insulin or insulin Chicago, whereas in the third example (insulin Wakayama), the abnormal insulin possessed characteristics consistent with its being more hydrophobic than normal human insulin or insulin Chicago. In neither case was a pancreatic biopsy available for extraction of insulin or for analysis of the structure of the abnormal hormone. Nevertheless, restriction mapping of the insulin gene from the patient expressing insulin Los Angeles (by use of the enzyme *Mbo*II, an enzyme that cleaves DNA at the sequence TCTTC)

Table 13.1. *Abnormal products of the human insulin gene*

name	amino acid change	codon change
insulin Chicago	$Phe^{B25} \rightarrow Leu$	$TTC \rightarrow TTG$
insulin Los Angeles	$Phe^{B24} \rightarrow Ser$	$TTC \rightarrow TCC$
insulin Wakayama	$Val^{A3} \rightarrow Leu$	$GTG \rightarrow TTG$
proinsulin Tokyo	$Arg^{65} \rightarrow His$	$CGC \rightarrow CAC$
proinsulin Providence	$His^{B10} \rightarrow Asp$	$CAC \rightarrow GAC$

indicated that the abnormal allele (like that from the patient expressing insulin Chicago) had undergone a nucleotide substitution in the codon corresponding to residue Phe^{B24} or Phe^{B25} (Shoelson *et al.*, 1983*b*). Reference to the genetic code suggested that a single nucleotide change resulting in replacement of Phe^{B24} or Phe^{B25} by a hydrophilic residue would most likely entail replacement of Phe by Ser. Two methods were applied to the molecular description of insulin Los Angeles. Semisynthesis of [Ser^{B24}]insulin and [Ser^{B25}]insulin (one or the other of which most likely corresponded to insulin Los Angeles) coupled with HPLC analysis (Shoelson *et al.*, 1983*a*), and sequence determination of the patient's abnormal insulin gene (Haneda *et al.*, 1983), both identified that Ser had replaced Phe at position B24. The corresponding change in the phenylalanine codon was $TTC \rightarrow TCC$. Insulin Los Angeles could therefore be identified as human insulin B24 (Phe→Ser) (Fig. 13.3 and Table 13.1).

In the third example of insulin gene mutation resulting in the formation of an abnormal 5 kDa insulin (insulin Wakayama), clues as to the nature of the relevant nucleotide and amino acid replacements were lacking. That is, the increased apparent hydrophobicity of the abnormal insulin could have been consistent with any of dozens of potential amino acid substitutions, insufficient amounts of material were available for chemical analysis, and no apparent changes in restriction enzyme cleavage sites were noted in the patient's insulin gene. Molecular cloning and sequence analysis of the abnormal allele, however, identified the relevant amino acid replacement as occurring at position A3. A single nucleotide change resulting in the codon GTG being altered to TTG was shown to account for the replacement of Val^{A3} by Leu. Insulin Wakayama could therefore be identified as human insulin A3 (Val→Leu) (Fig. 13.3 and Table 13.1). Interestingly, three examples of insulin Wakayama have been identified within apparently unrelated kindreds in the Japanese population (Nanjo *et al.*, 1986*a*, *b*; Iwamoto *et al.*, 1986*a*, *b*). Notwithstanding the absence of screening of broad populations by HPLC analysis of serum insulin or by clinical testing, nor the preliminary description of a few additional examples of abnormal insulins (Seino et al., 1985), it would appear that the frequency of insulin gene mutations within the human population is low. The screening of leukocyte DNA from several hundred diabetic patients by use of selected restriction enzyme systems in fact identified only a single change (Sanz et al., 1985). Nevertheless, subsequent analysis showed that the relevant nucleotide replacement represented a so-called silent mutation in which one phenylalanine codon replaced another. The corresponding insulin gene product therefore had normal structure.

13.4.   Mutations leading to impaired proinsulin processing

As noted above, mutations in the human insulin gene have the potential for affecting the conversion of proinsulin to insulin, as well as for altering the amino acid sequence of the usual B-cell product insulin. In fact, the description of the syndrome known as familial hyperproinsulinemia (one subsequently shown to arise from insulin gene mutation) actually predated the identification of the first abnormal human insulin (Gabbay et al., 1986). The clinical and laboratory findings in familial hyperproinsulinemia are little different from those associated with expression of mutant insulin genes yielding abnormal insulins. Nevertheless, individuals with familial hyperproinsulinemia (a) in general exhibit normal or near-normal glucose tolerance; and (b) have circulating, as the cause of apparent hyperinsulinemia, insulin-immunoreactive material having a molecular mass of about 9 kDa (Gabbay et al., 1976). Early studies identified the insulin-immunoreactive material in serum, in the case of proinsulin Boston, as a two-chained intermediate of proinsulin conversion rather than as single-chained proinsulin (Gabbay et al., 1976). Subsequent analysis of related material by a combination of immunochemical and chemical methods identified the intermediates in both proinsulin Boston (Robbins et al., 1984) and proinsulin Tokyo (Robbins et al., 1981, 1984) as derivatives of des-Arg^{31},Arg^{32}-proinsulin; that is, these peptides were identified as intermediates of proinsulin conversion in which the carboxyl terminus of the C-peptide remained joined to the amino terminus of the insulin A-chain, and in which the paired dibasic amino acids linking the carboxyl terminus of the insulin B-chain to the amino terminus of the C-peptide had been removed (presumably by normal processing enzymes within the islet B-cell). Further analysis indicated that an amino acid replacement had occurred at proinsulin residue 65, and that the usual

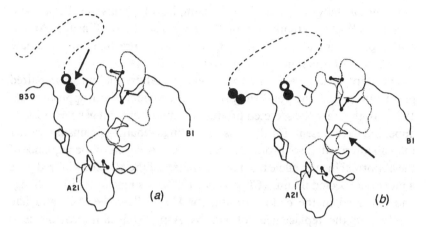

Fig. 13.4. Diagrammatic structures of the abnormal proinsulins or processing intermediates. The basic structure shown is that determined for insulin by crystallographic methods (see legend to Fig. 13.3). Lysine and arginine residues present in the dibasic amino acid pairs joining the C-peptide to the A- and B-chains are illustrated by open and filled circles, respectively. Solid lines, the insulin B-chain; dotted lines, the insulin A-chain; dashed lines, the C-peptide. Since the three-dimensional structure of proinsulin has not been determined, the diagrams should be regarded as being approximate. Arrows show sites of amino acid replacement. (a) Proinsulin Tokyo/Boston (Arg^{65}→His, shown as the intermediate of proinsulin processing present in serum); (b) proinsulin Providence (His^{B10}→Asp).

dibasic pair Lys^{64}–Arg^{65} had been modified in some way to yield a linking region that did not serve as substrate for the usual processing enzyme.

Although the absence of pancreatic tissue (from which the proposed abnormal proinsulin conversion intermediate might have been isolated) hindered significantly the identification of the proposed abnormal conversion intermediate, molecular cloning and gene sequence analysis identified for proinsulin Tokyo that Arg^{65} had been replaced by His; the causative nucleotide change at the codon for Arg^{65} (CGC→CAC) thus resulted in the synthesis and secretion of the abnormal proinsulin conversion intermediate des-Arg^{31},Arg^{32}-[His^{65}]proinsulin (Shibasaki et al., 1985). Proinsulin Tokyo (and most probably proinsulin Boston) could thus be identified as human proinsulin 65 (Arg→His) (Fig. 13.4 and Table 13.1). The finding that the Arg–His sequence was not cleaved by enzymes that process proinsulin to insulin in vivo emphasized the importance of paired dibasic amino acid residues in the recognition of substrate by the usual proinsulin converting enzymes. Note, however, that the abnormal conversion intermediates secreted in the examples of proinsulin Tokyo and proinsulin Boston actually contain the normal structure of insulin. Amino acid replacement at the processing site, alone, precludes the ultimate release of the normal hormone product.

The third example of an insulin gene mutation leading to impaired proinsulin processing concerns proinsulin Providence (Gruppuso et al., 1984). In this case the secreted product of mutant insulin gene expression appeared to represent, for the most part, single-chained proinsulin rather than an intermediate of proinsulin conversion. Analysis of the sequence of the abnormal insulin allele from two members of the affected kindred gave a most unexpected finding (Chan et al., 1987). A single nucleotide change had occurred in the codon serving for His^{B10} (CAC→GAC) and had resulted in the replacement of His by Asp. Proinsulin Providence is therefore human proinsulin 10 (His→Asp) (Fig. 13.4 and Table 13.1). It was later determined that [Asp^{B10}]insulin (a possible product of the abnormal insulin allele, but one which has actually yet to be identified among the insulin-related peptides in the sera of affected subjects) exhibits receptor binding and biological potencies in vitro several times greater than the corresponding potencies determined for normal insulin (Schwartz et al., 1987). Questions concerning how an amino acid replacement so distant from both proinsulin processing sites (Arg^{31}–Arg^{32} and Lys^{64}–Arg^{65}) could affect proinsulin conversion to insulin, why the effect on processing seems to be felt equivalently at both sites, and whether the impairment in processing alters the nature of products of the normal insulin allele as well as that of the abnormal allele, all remain to be answered. It is clear, however, that seemingly minor changes in proinsulin structure can have major consequences in determining the course of conversion of the hormone precursor to insulin. Matters related to the co-segregation of hormone precursor and processing enzymes in the Golgi apparatus and secretion granules (Orci et al., 1985), as well as those related to the inherent structure of the precursor, could be important to understanding why proinsulin Providence is apparently not converted to [Asp^{B10}]insulin in vivo.

13.5. Structure–activity and evolutionary implications of abnormal insulin gene products

Sites of amino acid replacement in the abnormal human insulins are known to be important to insulin–receptor interactions. These sites have remained invariant (or nearly so) in the many animal insulins for which sequences have been determined, and they identify loci where structural change might be expected to have significant functional consequences. Insulin gene mutations resulting in amino acid replacements at additional critical sites (for example, those corresponding to Gly^{A1}, Ile^{A2}, Glu^{A4}, Gln^{A5}, Leu^{A16}, Tyr^{A19}, Leu^{B6}, Gly^{B8}, Leu^{B11}, Val^{B12}, Leu^{B15}, Tyr^{B16} and Gly^{B23}) would also be expected to cause the formation of insulins with greatly decreased biological activity, and to have the potential for association with glucose intolerance or diabetes in affected subjects. A variety of mutations resulting in amino acid replacements at less critical sites, however, might be expected to go unnoticed. Replacements involving position B30 at the B-chain carboxy-terminus (a position that is notable for interspecific variation) represent extreme cases. For example, a single nucleotide change in the codon for Thr^{B30} in the human insulin gene (ACC→GCC) could result in the replacement of Thr by Ala. Affected subjects would synthesize and secrete human $[Ala^{B30}]$insulin, a hormone identical to porcine insulin, but would be asymptomatic since human and porcine insulins have identically high biological activities.

Overall, only about 43% of the 51-residue sequence of typical vertebrate insulins has remained invariant throughout evolution. When cysteine residues (residues of obvious importance to maintaining insulin's tertiary structure through the formation of disulfide bonds) are omitted from consideration, the value for invariance decreases to 35%. For mutations resulting in amino acid replacements outside of the insulin A- and B-chains, one must consider both the critical importance of the dibasic amino acid pairs that normally join the insulin A- and B-chains to the C-peptide in the hormone precursor (and that represents sites for proinsulin processing), and the significant interspecific variability of sequences in proinsulin C-peptides (Steiner, 1977); it is probable that most amino acid replacements resulting from insulin gene mutations in the C-peptide region would have little biochemical or physiological consequence. We know that some rather significant changes apparently can be accepted in the primary structure of insulin (at least when they occur in ways which apply to the natural hormones), but that particular changes (including those that apply to the abnormal human insulins) can have very deleterious effects.

Sites of amino acid replacement in insulins Chicago and Los Angeles (positions B25 and B24, respectively) occur within a sequence in the COOH-terminal B-chain domain (Gly^{B23}–Phe^{B24}–Phe^{B25}–Tyr^{B26}) that has remained nearly invariant throughout insulin evolution. This region, like the domain (Gly^{A1}–Ile^{A2}–Val^{A3}–Glu^{A4}) associated with the replacement identified at position A3 in insulin Wakayama, has long been known to play an important role in insulin–receptor interactions (Blundell *et al.*, 1972; DeMeyts *et al.*, 1978; Gammeltoft, 1984). Knowledge of the crystallographic structure of insulin has been critical to understanding how replacements at positions B24, B25 and A3 might decrease the affinity of the insulin receptor nearly 100-fold for the respective, naturally occurring insulin analogs (Blundell *et al.*, 1972; Dodson *et al.*, 1979, 1983). It turns out that the side chain of Phe^{B24} occurs at the surface of a hydrophobic pocket in the central region of the insulin molecule (Dodson *et al.*, 1983). Replacement of the residue with either hydrophobic or hydrophilic amino acids appears to alter the orientation of the COOH-terminal domain of the B-chain relative to the rest of insulin, and to cause propagated structural changes that can be observed by measurements of circular dichroic spectra or other methods (Wollmer *et al.*, 1981; Assoian *et al.*, 1982). Whereas the role of the Phe^{B24} side chain in insulin-receptor interactions may not yet be completely understood, it is clear that B24-substituted insulins (such as insulin Los Angeles) actually maintain overall structures in solution rather different from the structure taken by normal insulin.

In contrast to the disposition of the side chain of insulin residue B24, the side chain of Phe^{B25} (the site of amino acid replacement in insulin Chicago) occurs on the hormone surface and in a position suitable for a direct side chain contact with receptors (Blundell *et al.*, 1972; Wollmer *et al.*, 1981; Dodson *et al.*, 1983). Most important, replacements at position B25 do not appear to alter the secondary or tertiary structures of insulin, and seem to affect insulin–receptor interactions by mechanisms different from those that apply to insulins bearing replacements at position B24 (Wollmer *et al.*, 1981; Assoian *et al.*, 1982). As the result of studies involving the synthesis and analysis of many insulin analogs containing amino acid replacements at position B25, it now appears that (*a*) normal interaction of insulin with its receptor requires at position B25 a residue with a β-aromatic ring; (*b*) the importance of the aromatic ring arises from consideration of steric rather than electronic effects; and (*c*) the Phe^{B25} side chain functions to direct concerted conformational changes, in insulin and receptor, which occur during high-affinity hormone binding (Nakagawa & Tager, 1986, 1987). Since these conformational changes are necessary to reverse a potentially negative interaction involving the COOH-terminal B-chain domain and the insulin receptor (Nakagawa & Tager, 1986, 1987), amino acid replacements at position B25 are actually better tolerated in insulin analogs in which residues B26–B30 (Tyr–Thr–Pro–Lys–Thr, residues apparently not necessary for high affinity binding of insulin to receptor) have been deleted. Taken together, results obtained by study of insulin Chicago and related analogs suggest that the side chain of Phe^{B25} is important because of its role in altering the structure of insulin at the receptor, rather than because of its potential for directly conferring binding energy on insulin–receptor interactions.

In two examples of animal insulins (those of the coypu and dogfish), the seemingly critical residue Phe^{B25} has been deleted, apparently by deletion of the entire relevant codon. Nevertheless, activity is retained, most probably because of the resulting appearance of Tyr (a residue with a β-aromatic side chain) at position B25; the sequence in the COOH-terminal B-chain domain Gly–Phe–Phe–Tyr thus becomes Gly–Phe–Tyr. It seems that the sequence of amino acids COOH-terminal to residue B25 is not actually critical, and that all necessary criteria for effective insulin–receptor interactions are met, even in these unusual insulins. Replacement of Phe^{B25} by Tyr may actually increase somewhat the affinity of the receptor for related insulin analogs (Nakagawa & Tager, 1986). Much less is known about the effects of replacing Val^{A3} by Leu in insulin Wakayama. It is the case that the NH_2-terminal region of the insulin A-chain is important to maintaining (a) the position of the Gly^{A1} amino terminus, (b) the conformation of the helical domain in the region spanning residues A4–A9, and (c) the orientation of the A chain amino terminus relative to the rest of the molecule (through interactions between the side chains of Ile^{A2} and Tyr^{A19}) (Blundell et al., 1972). Interestingly, the replacement of Val^{A3} by Leu (the replacement occurring in insulin Wakayama) has also been seen to occur in examples of insulin from fish. Nevertheless, multiple amino acid replacements have occurred in some of these fish insulins, and complementary structural changes could well have compensated for local and even propagated structural changes that might have occurred as the result of any single amino acid replacement.

As noted earlier, the finding that expression of proinsulin Tokyo leads to the secretion of human des-Arg^{31},Arg^{32}-[His^{65}]proinsulin is consistent with the importance of paired dibasic amino acids at sites for proinsulin processing to insulin. The failure of endogenous converting enzymes to cleave the sequence Lys^{64}–His^{65} (a sequence corresponding to Lys^{64}–Arg^{65} in normal proinsulin), under circumstances where they cleave normally the sequence Arg^{31}–Arg^{32} in the same proinsulin molecule, identifies the specificity of the processing events. In fact, very many peptide hormone and other precursors are known to be converted to products by selective and often multiple proteolytic events that occur at dibasic sequences corresponding to Arg–Arg, Lys–Arg, Lys–Lys or Arg–Lys. Notwith-

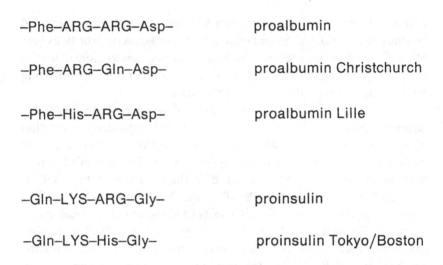

–Phe–ARG–ARG–Asp–   proalbumin

–Phe–ARG–Gln–Asp–   proalbumin Christchurch

–Phe–His–ARG–Asp–   proalbumin Lille

–Gln–LYS–ARG–Gly–   proinsulin

–Gln–LYS–His–Gly–   proinsulin Tokyo/Boston

Fig. 13.5. Normal and modified processing sites in proalbumin and proinsulin. Amino acid sequences surrounding the relevant pairs of dibasic amino acids at precursor processing sites are shown in each case. Dibasic residues are indicated in capital letters. Whereas the paired dibasic sequences are cleaved in the normal precursors, the sequences in the abnormal precursors (wherein one or the other dibasic amino acid has been replaced) are not. (See text for further details.)

standing the facts that (a) not all paired dibasic sites are necessarily acceptable sites for processing (even in peptide hormone precursors); (b) each of the four potential dibasic pairs seems to have its own relative propensity for cleavage by relevant processing enzymes (Schwartz et al., 1983); and (c) the Lys–Arg and Arg–Arg sequences in proinsulin may be cleaved by different converting enzymes within the pancreatic B-cell (Davidson et al., 1988), it appears that for most prohormones paired dibasic sites are necessary, if not sufficient, for directing normal processing of precursor to product.

Although defects related to prohormone processing are currently known only for insulin, related defects have also been observed for the processing of abnormal proalbumins that arise from mutations within the human albumin gene. In one example (Brennan & Carrell, 1978), the normal processing sequence Arg–Arg (the cleavage of which liberates serum albumin and a short, NH_2-terminal propeptide) is changed to Arg–Gln (proalbumin Christchurch), whereas in the other (Abdo et al., 1981) it is changed to His–Arg (proalbumin Lille) (Fig. 13.5). Whether considering proinsulins Tokyo or Boston, or proalbumins Christchurch hypothetical ancestral sequence

```
      20   21   22   23   24   25   26   27   28   29   30   31   32   33   34
1..  -gly-glu-ARG-gly-phe-phe-tyr-thr-pro-LYS-thr-ARG-gln-glu-ala-
     ggg  gaa  CGA  ggc  ttc  ttc  tac  aca  ccc  AAG  acc  CGC  cag  gag  gca

2.   -gly-LYS-ARG-gly-phe-phe-tyr-thr-pro-LYS-thr-ARG-gln-glu-ala-
     ggg  AAA  CGA  ggc  ttc  ttc  tac  aca  ccc  AAG  acc  CGC  cag  gag  gca

3.   -gly-glu-ARG-ARG-phe-phe-tyr-thr-pro-LYS-thr-ARG-gln-glu-ala-
     ggg  gaa  CGA  CGC  ttc  ttc  tac  aca  ccc  AAG  acc  CGC  cag  gag  gca

4.   -gly-glu-ARG-gly-phe-phe-tyr-thr-ARG-LYS-thr-ARG-gln-glu-ala-
     ggg  gaa  CGA  ggc  ttc  ttc  tac  aca  CGC  AAG  acc  CGC  cag  gag  gca

5.   -gly-glu-ARG-gly-phe-phe-tyr-thr-pro-LYS-thr-ARG-ARG-glu-ala-
     ggg  gaa  CGA  ggc  ttc  ttc  tac  aca  ccc  AAG  acc  CGC  CGG  gag  gca
```

proinsulin

Fig. 13.6. Illustration of a hypothetical course whereby mutations in an ancestral proinsulin-related sequence could give rise to a protein subject to prohormone processing. The amino acid sequences correspond to residues 20–34 of human proinsulin; the nucleotide sequences correspond to the human insulin gene (Bell *et al.*, 1980). Dibasic amino acid residues and their corresponding codons are shown by capital letters. Sequence 1, a hypothetical ancestral sequence which is identical to that of proinsulin with the exception that a single nucleotide change has occurred in the codon for Arg^{32} leading to replacement of Arg by Gln; the protein sequence contains no paired dibasic sites for potential precursor processing. Sequences 2–5, sequences related to the sequence of 1 which might have arisen due to single nucleotide changes; in each case, the resulting amino acid replacement would result in the appearance of a paired dibasic amino acid site that might be subject to precursor processing. The sequence shown in 5 is actually that of human proinsulin. (See text for further explanation.)

or Lille, it appears that the alteration of one or the other residue in a dibasic amino acid pair can prevent conversion of precursor to product and can result in the secretion of the corresponding uncleaved precursor.

Whereas the above account identifies clearly the consequences of genetic mutation in preventing the conversion of prohormone precursors to normal products (by single nucleotide changes which lead to single amino acid replacements), it also identifies how prohormone conversion sites and small peptide hormones might have evolved. Fig. 13.6 illustrates at the top a hypothetical 15-residue sequence which might have occurred in an ancestral proinsulin. (The sequence corresponds to the COOH-terminal B-chain domain and to the very NH$_2$-terminal region of the C-peptide). The region contains three single dibasic residues, but no paired dibasic residues that might serve as processing sites for cleavage by modern proinsulin-converting enzymes. Sequences 2–5 in Fig. 13.6 document how single nucleotide changes in the hypothetical, ancestral gene could have led to the replacement of amino acid residues and to the generation of potential processing sites at sequences corresponding to Lys–Arg, Arg–Arg or Arg–Lys. With the assumptions that normal processing would occur at an existing Lys^{64}–Arg^{65} site joining the C-peptide to the insulin A-chain, and that a dibasic amino acid-specific exopeptidase would remove any COOH-terminal basic residues (as it does in the normal conversion of proinsulin to insulin), the complete processing of precursors 2–5 would yield des-decapeptide(B21–B30)-insulin, des-nonapeptide(B22–B30)-insulin, des-tripeptide(B28–B30)-insulin, and normal insulin, respectively. Potencies of the first and second of these hypothetical products for interaction with the modern insulin receptor would be only about 0.1% of that of insulin; the potency of the third would most probably approach that of normal insulin (cf. Nakagawa & Tager, 1986). Nevertheless, the dibasic amino acid sequence Arg–Lys, a sequence potentially appearing in peptide 4 of Fig. 13.6, appears not to be nearly so good a potential conversion site as Arg–Arg or Lys–Arg; the relevant hypothetical precursor might not be expected to undergo efficient processing.

The arguments provided above are flawed, of course, by failure to consider three important matters. First, ancestral processing enzymes may well have had specificities different from those that apply today. Second, the ancestral insulin receptor could have had structural requirements for ligand–receptor interactions quite different from those of the modern receptor. Third, it is not obvious why proinsulin (the unprocessed precursor) was not itself accepted during evolution as a peptide effector in the absence of any proteolytic processing. It should be noted in the last regard that the related peptide hormones insulin-like growth factors I and II show homology to insulin, but retain a single-chained disulfide-bonded structure similar to that of proinsulin (Blundell et al., 1978). In any case, the formation of a paired dibasic amino acid sequence could represent a very large step in the evolution of a peptide effector, especially if the newly formed sequence were to be cleaved by an existing proteolytic enzyme. Under this set of circumstances, even a single nucleotide change in a gene for a peptide effector could change in very substantial ways the size, structure, character and activity of the ultimate gene product.

All considered, it is clear that the study of abnormal human insulins has provided us with important clues about the association of insulin gene mutations with a relatively rare type of diabetes, and about both insulin structure–activity relationships and the evolution of peptide hormone effectors. In many ways, abnormal products of the human insulin gene represent important, naturally occurring examples of interesting peptide hormone and prohormone analogs. Further study of the abnormal human insulins and of potentially abnormal products of additional peptide hormone genes will be critical to our eventual understanding of clinical, molecular and physiological aspects of endocrine function in man and of the multiple facets of human endocrine disease.

Work reported from this laboratory was supported by Grants DK 18347 and DK 20595 from the National Institutes of Health.

References

Abdo, Y., Rousseau, J. & Dautrevaux, M. (1981). Proalbumin Lille, a new variant of human serum albumin. *FEBS Letters* **131**, 286–8.

Assoian, R.K., Thomas, N.E., Kaiser, E.T. & Tager, H.S. (1982). [Leu^{B25}] insulin and [Ala^{B24}]insulin: Altered structures and cellular processing of B24-substituted insulin analogs. *Proceedings of the National Academy of Sciences of the U.S.A.* **79**, 5147–51.

Bell, G.I., Pictet, R.L., Rutter, W.J., Cordell, B., Tischer, E. & Goodman, H.M. (1980). Sequence of the human insulin gene. *Nature* **284**, 26–32.

Blundell, T.L., Bedarkar, S., Rinderknecht, E. & Humbel, R.E. (1978). Insulin-like growth factor: a model for tertiary structure accounting for immunoreactivity and receptor binding. *Proceedings of the National Academy of Sciences of the U.S.A.* **75**, 180–4.

Blundell, T., Dodson, G., Hodgkin, D. & Mercola, D. (1972). Insulin: the structure in the crystal and its reflection in chemistry and biology. *Advances in Protein Chemistry* **26**, 279–402.

Brennan, S.O. & Carrell, R.W. (1978). A circulating variant of human proalbumin. *Nature* **274**, 908–9.

Chan, S.J., Seino, S., Gruppuso, P.A., Schwartz, R. & Steiner, D.F. (1987). A mutation in the B chain coding region is associated with impaired proinsulin conversion in a family with hyperproinsulinemia. *Proceedings of the National Academy of Sciences of the U.S.A.* **84**, 2194–7.

Davidson, H.W., Rhodes, C.J. & Hutton, J.C. (1988). Intraorganellar calcium and pH control proinsulin cleavage in the pancreatic β cell via two distinct site-specific endopeptidases. *Nature* **333**, 93–6.

DeMeyts, P., Van Obberghen, E., Roth, J., Wollmer, A. & Brandenburg, D. (1978). Mapping of the residues responsible for the negative cooperativity of the receptor-binding region of insulin. *Nature* **273**, 504–9.

Dodson, E.J., Dodson, G.G., Hodgkin, D.C. & Reynolds, C.D. (1979). Structural relationships in the two-zinc insulin hexamer. *Canadian Journal of Biochemistry* **57**, 469–79.

Dodson, E.J., Dodson, G.G., Hubbard, R.E. & Reynolds, C.D. (1983). Insulin's structural behavior and its relation to activity. *Biopolymers* **22**, 281–91.

Gabbay, K.H., Bergenstal, R.M., Wolff, J., Mako, M.E. & Rubenstein, A.H. (1979). Familial hyperproinsulinemia: partial characterization of circulating proinsulin-like material. *Proceedings of the National Academy of Sciences of the U.S.A.* **76**, 2882–5.

Gabbay, K.H., DeLuca, K., Fisher, J.N. Jr, Mako, M.E. & Rubenstein, A.H. (1976). Familial hyperproinsulinemia: an autosomal dominant defect. *New England Journal of Medicine* **294**, 911–15.

Gammeltoft, S. (1984). Insulin structure and function. *Physiological Reviews* **64**, 1321–78.

Given, B.D., Cohen, R.M., Shoelson, S.E., Frank, B.H., Rubenstein, A.H. & Tager, H.S. (1985). Biochemical and clinical implications of proinsulin conversion intermediates. *Journal of Clinical Investigation* **76**, 1398–405.

Given, B.D., Mako, M.E., Tager, H.S., Baldwin, D., Markese, J., Rubenstein, A.H., Olefsky, J., Kobayashi, M., Kolterman, O. & Poucher, R. (1980). Diabetes due to secretion of an abnormal insulin. *New England Journal of Medicine* **302**, 129–35.

Gruppuso, P.A., Gorden, P., Kahn, C.R., Cornblath, M., Zeller, W.P. & Schwartz, R. (1984). Familial hyperproinsulinemia due to a proposed defect in conversion of proinsulin to insulin. *New England Journal of Medicine* **311**, 629–34.

Haneda, M., Chan, S.J., Kwok, S.C.M., Rubenstein, A.H. & Steiner, D.F. (1983). Studies on mutant insulin genes: identification and sequence analysis of a gene encoding [Ser^{B24}]insulin. *Proceedings of the National Academy of Sciences of the U.S.A.* **80**, 6366–70.

Haneda, M., Kobayashi, M., Maegawa, H., Watanabe, N., Takata, Y., Ishibashi, O., Shigeta, Y. & Inouye, K. (1985). Decreased biological activity and degradation of human [Ser^{B24}]insulin, a second mutant insulin. *Diabetes* **34**, 568–73.

Haneda, M., Polonsky, K.S., Bergenstal, R.M., Jaspan, J.B., Shoelson, S.E., Blix, P.M., Chan, S.J., Kwok, S.C.M., Wishner, W.B., Zeidler, A., Olefsky, J.M., Freidenberg, G., Tager, H.S., Steiner, D.F. & Rubenstein, A.H. (1984). Familial hyperinsulinemia due to a structurally abnormal insulin: definition of an emerging new clinical syndrome. *New England Journal of Medicine* **310**, 1288–94.

Inouye, K., Watanabe, K., Tochino, Y., Kobayashi, M. & Shigeta, Y. (1981). Semisynthesis and properties of some insulin analogs. *Biopolymers* **20**, 1845–58.

Iwamoto, Y., Sakura, H., Yui, R., Fujita, T., Sakamoto, Y., Matsuda, A. & Kuzuya, T. (1986a). Identification and characterization of a mutant insulin isolated from the pancreas of a patient with abnormal insulinemia. *Diabetes* **35** (suppl. 1), 77A.

Iwamoto, Y., Sakura, H., Ishii, Y., Yamamoto, R., Kumakura, S., Sakamoto, Y., Masuda, A. & Kuzuya,T. (1986b). Radioreceptor assay for serum insulin as a useful method for detection of abnormal insulin with a description of a new family of abnormal insulinemia. *Diabetes* **35**, 1237–42.

Kobayashi, M., Haneda, M., Ishibashi, O., Takata, Y., Maegawa, H., Watanabe, N. & Shigeta, Y. (1984a). Prolonged disappearance rate of a structurally abnormal mutant insulin from the blood. *Diabetes* **33** (suppl. 1), 17A.

Kobayashi, M., Haneda, M., Maegawa, H., Watanabe, N., Takada, Y., Shigeta, Y. & Inouye, K. (1984b). Receptor binding and biological activity of [Ser^{B24}]insulin, an abnormal mutant insulin. *Biochemical and Biophysical Research Communications* **119**, 49–57.

Kobayashi, M., Ohgaku, S., Iwasaki, M., Maegawa, H., Shigeta, Y. & Inouye, K. (1982). Characterization of [Leu^{B24}]- and [Leu^{B25}]insulin analogues. *Biochemical Journal* **206**, 597–603.

Kobayashi, M., Takata, Y., Ishibashi, O., Sasaoka, T., Iwasaki, M., Shigeta, Y. & Inouye, K. (1986). Receptor binding and negative cooperativity of a mutant insulin [Leu^{A3}]insulin. *Biochemical and Biophysical Research Communications* **137**, 250–7.

Kwok, S.C.M., Chan, S.J., Rubenstein, A.H., Poucher, R. & Steiner, D.F. (1981). Loss of a restriction endonuclease cleavage site in the gene of a structurally abnormal insulin. *Biochemical and Biophysical Research Communications* **98**, 844–9.

Kwok, S.C.M., Steiner, D.F., Rubenstein, A.H. & Tager, H.S. (1983). Identification of the mutation giving rise to Insulin Chicago. *Diabetes* **32**, 872–5.

Nakagawa, S.H. & Tager, H.S. (1986). Role of the phenylalanine B25 side chain in directing insulin interaction with its receptor. *Journal of Biological Chemistry* **261**, 7332–41.

Nakagawa, S.H. & Tager, H.S. (1987). Role of the COOH-terminal B-chain domain in insulin-receptor interactions. *Journal of Biological Chemistry* **262**, 12054–8.

Nanjo, K., Given, B., Sanke, T., Kondo, M., Miyano, M., Okai, K., Miyama, K., Chan, S., Tager, H., Steiner, D., Polonsky, K & Rubenstein, A. (1986a). Pancreatic function in the mutant insulin syndrome. *Diabetes* **35** (suppl. 1), 77A.

Nanjo, K., Sanke, T., Miyano, M., Okai, K., Sowa, R., Kondo, M., Nishimura, S., Iwo, K., Miyamura, K., Given, B.D., Chan, S.J., Tager, H.S., Steiner, D.F. & Rubenstein, A.H. (1986b). Diabetes due to secretion of a structurally abnormal insulin (Insulin Wakayama). *Journal of Clinical Investigation* **77**, 514–19.

Orci, L., Ravazzola, M., Amherdt, M., Madsen, O., Vassalli, J.-D. & Perrelet, A. (1985). Direct identification of prohormone conversion site in insulin-secreting cells. *Cell* **42**, 671–81.

Peavy, D.E., Brunner, M.R., Duckworth, W.C., Hooker, C.S. & Frank, B.H. (1985). Receptor binding and biological potency of several split forms (conversion intermediates) of human proinsulin. *Journal of Biological Chemistry* **260**, 13989–94.

Robbins, D.C., Blix, P.M., Rubenstein, A.H., Kanazawa, Y., Kosaka, K. & Tager, H.S. (1981). A human proinsulin variant at arginine 65. *Nature* **291**, 679–81.

Robbins, D.C., Shoelson, S.E., Rubenstein, A.H. & Tager, H.S. (1984). Familial hyper-proinsulinemia: two cohorts secreting indistinguishable type II intermediates of proinsulin conversion. *Journal of Clinical Investigation* **73**, 714–19.

Sanz, N., Karam, J.H., Horiat, S. & Bell, G.I. (1985). DNA screening for insulin gene mutations in non-insulin-dependent diabetes mellitus (NIDDM). *Diabetes* **34** (suppl. 1), 85A.

Schwartz, G.P., Burke, G.T. & Katsoyannis, P.G. (1987). A superactive insulin: [B10–aspartic acid]insulin (human). *Proceedings of the National Academy of Sciences of the U.S.A.* **84**, 6408–11.

Schwartz, T.W., Wittels, B. & Tager, H.S. (1983). Hormone precursor processing in the pancreatic islet. In *Peptides: structure and function* (ed. V.J. Hruby & D.H. Rich), pp. 229–38. Rockford, Illinois: Pierce Chemical Company.

Seino, S., Funakoshi, A., Fu, Z.Z. & Vinik, A. (1985). Identification of insulin variants in patients with hyperinsulinemia by reversed-phase, high performance liquid chromatography. *Diabetes* **34**, 1–7.

Shibasaki, Y., Kawakami, T., Kanazawa, Y., Akanuma, Y. & Takaku, F. (1985). Posttrans-lational cleavage of proinsulin is blocked by a point mutation in familial hyperproinsuline-mia. *Journal of Clinical Investigation* **76**, 378–80.

Shoelson, S., Fickova, M., Haneda, M., Nahum, A., Musso, G., Kaiser, E.T., Rubenstein, A.H. & Tager, H.S. (1983a). Identification of a mutant insulin predicted to contain a serine-for-phenylalanine substitution. *Proceedings of the National Academy of Sciences of the U.S.A.* **80**, 7390–4.

Shoelson, S., Haneda, M., Blix, P., Nanjo, K., Sanke, T., Inouye, K., Steiner, D., Rubenstein, A. & Tager, H.S. (1983b). Three mutant insulins in man. *Nature* **302**, 540–3.

Shoelson, S.E., Polonsky, K.S., Zeidler, A., Rubenstein, A.H. & Tager, H.S. (1984). Human insulin (Phe→Ser): secretion and metabolic clearance of the abnormal insulin in man and in a dog model. *Journal of Clinical Investigation* **73**, 1351–8.

Steiner, D.F. (1977). Insulin today. *Diabetes* **26**, 322–40.

Steiner, D.F., Clark, J.L., Nolan, C., Rubenstein, A.H., Margoliash, E., Aten, B. & Oyer, P.E. (1969). Proinsulin and the biosynthesis of insulin. *Recent Progress in Hormone Research* **25**, 207–82.

Steiner, D.F., Quinn, P.S., Chan, S.J., Marsh, J. & Tager, H.S. (1980). Processing mechanisms in the biosynthesis of proteins. *Annals of the New York Academy of Sciences* **343**, 1–16.

Tager, H., Given, B., Baldwin, D., Mako, M., Markese, J., Rubenstein, A.H., Olefsky, J., Kobayashi, M., Kolterman, O. & Poucher, R. (1979). A structurally abnormal insulin causing human diabetes. *Nature* **281**, 122–5.

Tager, H., Thomas, N., Assoian, R., Rubenstein, A., Saekow, M., Olefsky, J. & Kaiser, E.T. (1980). Semisynthesis and biological activity of porcine [Leu^{B24}]insulin and [Leu^{B25}]insulin. *Proceedings of the National Academy of Sciences of the U.S.A.* **77**, 3181–5.

Terris, S. & Steiner, D.F. (1975). Binding and degradation of ^{125}I-insulin by rat hepatocytes. *Journal of Biological Chemistry* **250**, 8389–98.

Terris, S. & Steiner, D.F. (1976). Retention and degradation of ^{125}I-insulin by perfused livers from diabetic rats. *Journal of Clinical Investigation* **57**, 885–96.

Ullrich, A., Dull, T.J., Gray, A., Brosius, J. & Sures, I. (1980). Genetic variation in the human insulin gene. *Science* **209**, 612–15.

Wollmer, A., Strassburger, W., Glatler, V., Dodson, G.G., McCall, M., Danho, W., Brandenburg, D., Gattner, H.-G. & Rittel, W. (1981). Two mutant forms of human insulin: structural consequences of the substitution of invariant B24 or B25 by leucine. *Hoppe-Seyler's Zeitschrift für Physiologische Chemie* **362**, 581–92.

14

A novel gene, *rig*, activated in insulinomas

SHIN TAKASAWA, CHIYOKO INOUE, KIYOTO
SHIGA AND MOTOO KITAGAWA

14.1. Introduction

As discussed in Chapter 10, insulinomas can be induced in
experimental animals by diabetogenic agents such as streptozotocin and
alloxan (Lazarow, 1952; Rakieten *et al.*, 1971; Yamagami *et al.*, 1985) and
by viruses (Uchida *et al.*, 1979). A unifying concept was proposed for the
diabetogenic and oncogenic effects of the B-cytotoxins, in which the
principal action of streptozotocin and alloxan was considered to be the
induction of lesions in the DNA of B-cells of islets of Langerhans (see Fig.
10.5). To elucidate the molecular mechanisms of B-cell oncogenesis, we
have looked for genes whose expression is altered in insulinomas. We
identified a novel gene named *rig* (rat insulinoma gene), which is activated
in chemically induced rat insulinomas (Takasawa *et al.*, 1986). The gene
was also activated in virus-induced hamster insulinomas and in spontaneously occurring human insulinomas (Inoue *et al.*, 1987).

14.2. Isolation of *rig* from insulinomas

The combined administration of streptozotocin or alloxan with
poly(ADP-ribose) synthetase inhibitors such as nicotinamide to rats
induces a high incidence of insulinomas (Yamagami *et al.*, 1985). We
constructed a cDNA library from poly(A)$^+$ RNA of rat insulinomas
induced by the combined administration of streptozotocin and nicotinamide (Takasawa *et al.*, 1986). From this library, independent clones were
duplicated on nitrocellulose filters for differential screening. One set of
filters was hybridized to ^{32}P-labeled cDNA that was reverse-transcribed
from poly(A)$^+$ RNA of insulinomas with [^{32}P]deoxycytidine triphosphate
(dCTP) and oligo(dT) as primer. The other set of filters was hybridized to
^{32}P-labeled cDNA synthesized from poly(A)$^+$ RNA of normal islets. In

```
                                                 -1   1                                                           60
Rat       TTTTTCCCAGCAGCCGCCAAG ATG GCC GAA GTG GAG CAG AAG AAA AAG CGA ACC TTC CGC AAG TTC ACC TAC CGT GGC GTG
Hamster   TTTTTTACGAGAAGCCGTCAAA                 A                        G   G         G                      C   C
Human     AAAGCGATCTCTTCTGAGGATCCGGCAAG         A                            G             C                  C       A
Mouse     CTTTTTTCCTTTCCGAGTAACCGCCAAG          A                        G               C                         C
          Met Ala Glu Val Glu Gln Lys Lys Lys Arg Thr Phe Arg Lys Phe Thr Tyr Arg Gly Val
           1                           10                                      20


                                                                                                          144
Rat       GAC CTC GAC CAG CTG GAC ATG TCC TAT GAG CTG CAG TTC TAC AGC GCC CAG CGG AGA CGG CTG AAC CGA GGC
Hamster   G                       A                   C                   G        G   G         G        C
Human       A       C                             C       C    C  G   A                               T   T
Mouse                                           A                                    C  G A       T        T
          Asp Leu Asp Gln Leu Asp Met Ser Tyr Glu Gln Leu Tyr Ser Ala Arg Gln Arg Arg Leu Asn Arg Gly
                          30                                      40


                                                                                                          228
Rat       CTC CGG AGG AAG CAG CAC TCA CTC CTC AAG CGC TTG AGG GAC ATG ATC CTC GAG AAG AAG GAG GCG CCA CCC ATG GAG CCG AAG GAG GTC GTG
Hamster       C   C           C           G       C           G   C       C   A                            G    G          C       A   G
Human       T   C                         A           G            A   C   C   A               A                          T           G
Mouse                                                                     G                    A                          A
          Leu Arg Arg Lys Gln His Ser Leu Leu Lys Arg Leu Arg Asp Met Ile Leu Glu Lys Lys Glu Ala Pro Pro Met Glu Pro Lys Glu Val Val
                          50                                      60                                      70


                                                                                                          312
Rat       AAG ACC CTT AGG GAC ATG ATT CTG CCC GAG ATC GGC ATG GTG TAC AGC ATG GGT GTG TAC AAC GGC AAG ACC TTC AAC CAG
Hamster   G   G   G                   C   C   A                       G   G   C                             G
Human     G   G   C                   C   C   A   G                   G   G   C                             G
Mouse     T                           G           G                       T   G                             G
          Lys Thr His Leu Arg Asp Met Ile Ile Leu Pro Glu Met Val Gly Ser Met Val Gly Val Tyr Asn Gly Lys Thr Phe Asn Gln
                          80                                      90                                      100
```

```
Rat       CTG GAG ATC AAA CCC GAG ATG ATC GGC CAC TAC CTG GGC GAG TTC TCC ATC ACC TAC AAG CCT GTG AAG CAC GGC CGG CCC CCC GGC  396
Hamster                   G   T                                                                                    G
Human                     G                                                   C       A       T
Mouse                         A                                               C
          Val Glu Ile Lys Pro Glu Met Ile Gly His Tyr Leu Gly Glu Phe Ser Ile Thr Tyr Lys Pro Val Lys His Gly Arg Pro Gly
                       110                          120                                              130


Rat       ATT GGT GCC ACC TCC TCC CGA TTT ATC CCC CTC AAG  435  TAGTGGGACAATAAAGACTCGTTTTCAGCC(A)n
Hamster       C   T               G   C   G                      ACTGCCCAATAAAGACTCTAGAGTCCG(A)n
Human         C   G                   C   T                      ATGGCTCAGCTAATAAAGGCGCACACTCACTCC(A)n
Mouse                                 C                          CCGAGGCCAATAAAGACTGGTTTTGGTCCCTG(A)n
          Ile Gly Ala Thr His Ser Ser Arg Phe Ile Pro Leu Lys End
                           140                      145
```

Fig. 14.1. Nucleotide and deduced amino acid sequences of rat, hamster, human and mouse *rig* (Takasawa *et al.*, 1986; Inoue *et al.*, 1987). Nucleotide residues are numbered in the 5' to 3' direction, beginning with the first residue of the ATG triplet encoding the initiator methionine; the nucleotides on the 5' side of residue 1 are indicated by negative numbers. The upper nucleotide sequence shows the rat *rig* cDNA. Nucleotide differences found in the coding region of hamster, human and mouse sequences are displayed beneath the rat sequence. The deduced amino acids are given below the nucleotide sequence and numbered beginning with the initiator methionine.

the differential screening of 5000 clones, one clone consistently hybridized preferentially to ^{32}P-labeled cDNA from insulinomas. As shown in Fig. 14.1, the cDNA stretches for 487 nucleotides plus the poly(A) tail and has one open reading frame, coding for a protein of 145 amino acids (on the assumption that ATG at nucleotide 1–3 is the start codon and TAG at nucleotide 436–438 is the stop codon). The nucleotide sequence of CCAAGATGG around the methionine codon agrees well with the initiation sequence characteristic of many eukaryotic mRNAs (Kozak, 1984). The calculated relative molecular mass (17 040) of the deduced protein is in good agreement with that of the *in vitro* translation product of RNA that was synthesized from the cloned cDNA. The nucleotide and deduced amino acid sequences of the cDNA were screened for possible relationships to other genes and proteins stored in the nucleic acid and protein data banks of the European Molecular Biology Laboratory (Heidelberg, F.R.G.), Los Alamos National Laboratory (Los Alamos, U.S.A.), and the National Biomedical Research Foundation (Washington, D.C., U.S.A.) with a rapid similarity search algorithm (Wilbur & Lipman, 1983). This screening indicated that there are neither genes nor gene products that are identical or highly homologous to the cDNA.

The cDNA was labeled by nick translation with [α-^{32}P]dCTP and utilized to probe for levels of the corresponding mRNA species in rat insulinomas. As shown in Fig. 14.2, the cDNA probe hybridized to an RNA of 0.7 kilobases, which was clearly present at a much higher level in each insulinoma induced by streptozotocin and nicotinamide (lanes 2–4) than in normal islets of Langerhans (lane 1); the sequence of the normal counterpart isolated from a rat pancreatic islet cDNA library was identical to that of insulinomas (Takasawa *et al.*, 1986). We next analyzed RNAs from insulinomas induced by alloxan and nicotinamide and found that the level of the 0.7 kilobase mRNA was also increased in all the alloxan–nicotinamide-induced insulinomas examined (lanes 5–7). On the other hand, the level of the mRNA was low in regenerating islets (lane 8) as well as in liver, kidney and brain (lanes 9–11). The novel gene identified was activated in both streptozotocin–nicotinamide-induced insulinomas and alloxan–nicotinamide-induced insulinomas of the rat, but not in normal pancreatic islets or in regenerating islets. We designated the novel gene *rig* (rat insulinoma gene).

To see whether the gene is also activated in insulinomas of other mammalian species, we analyzed RNAs from BK virus-induced hamster insulinoma cells (Uchida *et al.*, 1979; Yamamoto *et al.*, 1980) and from

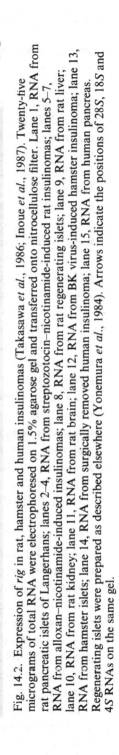

Fig. 14.2. Expression of *rig* in rat, hamster and human insulinomas (Takasawa *et al.*, 1986; Inoue *et al.*, 1987). Twenty-five micrograms of total RNA were electrophoresed on 1.5% agarose gel and transferred onto nitrocellulose filter. Lane 1, RNA from rat pancreatic islets of Langerhans; lanes 2–4, RNA from streptozotocin–nicotinamide-induced rat insulinomas; lanes 5–7, RNA from alloxan–nicotinamide-induced insulinomas; lane 8, RNA from rat regenerating islets; lane 9, RNA from rat liver; lane 10, RNA from rat kidney; lane 11, RNA from rat brain; lane 12, RNA from BK virus-induced hamster insulinoma; lane 13, RNA from hamster islets; lane 14, RNA from surgically removed human insulinoma; lane 15, RNA from human pancreas. Regenerating islets were prepared as described elsewhere (Yonemura *et al.*, 1984). Arrows indicate the positions of 28S, 18S and 4S RNAs on the same gel.

surgically removed human insulinomas, using *rig* cDNA as probe. As shown in Fig. 14.2, an RNA species that hybridized to *rig* cDNA was clearly present in hamster insulinoma cells (lane 12) and in human insulinoma (lane 14). The RNA was 0.7 kb long, the same as *rig* mRNA (lanes 2–7). The level of the 0.7 kb RNA was much higher in both hamster and human insulinomas than in normal tissues (lanes 13, 15). The results indicate that *rig* homologs are present and activated in hamster and human insulinomas. The hamster insulinoma was induced by BK virus (Uchida *et al.*, 1979); the human insulinoma was considered to have occurred spontaneously. The rat insulinomas in which *rig* was initially identified had been induced by DNA-damaging agents such as streptozotocin and alloxan (Yamamoto *et al.*, 1981; Yamagami *et al.*, 1985). Examination of chemically induced, virus-induced, and spontaneously occurring insulinomas indicates that the activation of *rig* is a general feature of pancreatic B-cell transformation.

14.3. Evolutionary conservation of *rig*

As mentioned above, *rig* homologs were present in hamster and human insulinomas. The *rig* homolog has also been detected in mouse (NIH 3T3) cells. We isolated hamster, human and mouse homologs to *rig* from hamster and human insulinoma-derived cDNA libraries and from a mouse NIH 3T3 cell cDNA library by screening with synthetic oligodeoxyribonucleotide corresponding to the first 60 nucleotides of the coding region of *rig* (see Fig. 14.1). As shown in Fig. 14.1, the hamster, human and mouse cDNAs were 486, 498 and 500 nucleotides long, respectively, plus poly(A) tails, and all the three cDNAs had the same large open reading frame of 435 nucleotides, assuming that ATG at nucleotides 1–3 is the start codon and TAA (hamster and human) and TAG (mouse) at nucleotides 436–438 are the stop codon. A polyadenylation signal (AATAAA) (Proudfoot & Brownlee, 1976) was present 20, 21 and 24 nucleotides upstream from the poly(A) tail in hamster, human and mouse cDNAs, respectively. We then compared the nucleotide and deduced amino acid sequences of hamster, human and mouse cDNAs with those of *rig* cDNA. The coding region of hamster, human and mouse cDNAs was exactly the same in length as that of rat *rig* cDNA and exhibited a high degree of homology with the rat sequence (93% between hamster and rat; 91% between human and rat; 92% between mouse and rat). It should be emphasized that most of the nucleotide substitutions were in the third position of the triplet and that the deduced 145 amino acid sequences of the hamster, human, mouse, and rat were absolutely identical. This suggested that *rig* has evolved under extraordinarily strong selective constraints, which eliminate any base substitutions resulting in amino acid changes. Recently, the *rig* homolog has also been detected in chicken (*Gallus gallus domesticus*), frog (*Xenopus laevis*), carp (*Cyprinus carpio*) and crab (*Sesarma intermedia*) cells by Northern blot hybridization. Several copies of *rig* pseudogenes have recently been identified in rat, human and hamster genomes (K. Shiga, I. Kato, J. Aruga, S. Takasawa, H. Yamamoto & H. Okamoto, unpublished data). The occurrence of *rig* in species from the crustacean to the higher vertebrates, and existence of the pseudogenes, suggest that *rig* is a gene that has an ancient evolutionary origin.

14.4. Possible function of *rig*

The hydrophilicity analysis of *rig* protein, according to Kyte & Doolittle (1982), showed that the *rig* protein of 145 amino acids can be divided into a highly hydrophilic amino-terminal domain (amino acid residues 1–83) and a less hydrophilic carboxy-terminal domain (amino acid residues 84–145). The amino-terminal domain is rich in basic amino acids and contains a putative nuclear location signal (Smith *et al.*, 1985) consisting of a stretch of predominantly basic amino acids with two prolines at the carboxyl side (amino acid residues 61–69). The carboxy-terminal domain contains the sequence Gly–Val–Try–Asn–Gly–Lys–Thr–Phe–Asn–Gln–Val at amino acid residues 95–105, which is similar to the consensus sequence (Gly/Ala–(Xaa)$_3$–Gly–(Xaa)$_5$–Val/Ile) conserved in a number of DNA-binding proteins and in some nuclear oncogene products (Fig. 14.3). In addition, a computerized secondary structure prediction of the *rig* protein indicated that the 11-amino acid sequence can form an α-helix–β-turn–α-helix structure, a characteristic structural motif shared by the DNA-binding proteins (Sauer *et al.*, 1982; Laughon & Scott, 1984; Vogt *et al.*, 1987). The two α-helices have been reported to make contact with the major groove of the B-form of the DNA helix (Pabo & Sauer, 1984). These results strongly suggest that the *rig* protein is transferred to the nucleus and interacts with DNA.

Based upon the sequence data of hamster *rig* cDNA (Fig. 14.1), the 21-base oligodeoxyribonucleotide complementary to the 5′ portion of hamster *rig* mRNA (nucleotides − 6 to 15) was synthesized and injected into the hamster insulinoma cells (Inoue *et al.*, 1987). The length of the antisense oligodeoxyribonucleotide was determined so that the formation of DNA–RNA hybrids in the injected cells would proceed at the maximal rate at the culture temperature (37° C). Microinjection of the antisense *rig*

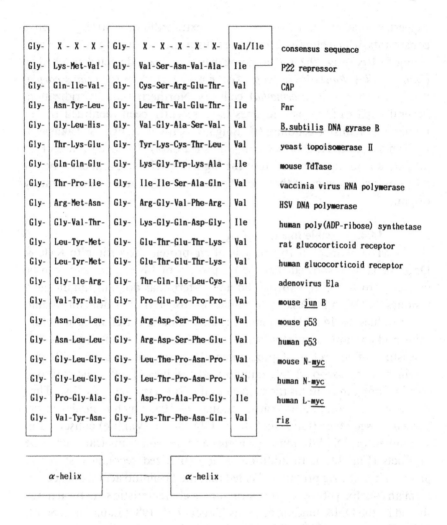

Gly-	X - X - X -	Gly-	X - X - X - X - X-	Val/Ile	consensus sequence
Gly-	Lys-Met-Val-	Gly-	Val-Ser-Asn-Val-Ala-	Ile	P22 repressor
Gly-	Gln-Ile-Val-	Gly-	Cys-Ser-Arg-Glu-Thr-	Val	CAP
Gly-	Asn-Tyr-Leu-	Gly-	Leu-Thr-Val-Glu-Thr-	Ile	Fnr
Gly-	Gly-Leu-His-	Gly-	Val-Gly-Ala-Ser-Val-	Val	B.subtilis DNA gyrase B
Gly-	Thr-Lys-Glu-	Gly-	Tyr-Lys-Cys-Thr-Leu-	Val	yeast topoisomerase II
Gly-	Gln-Gln-Glu-	Gly-	Lys-Gly-Trp-Lys-Ala-	Ile	mouse TdTase
Gly-	Thr-Pro-Ile-	Gly-	Ile-Ile-Ser-Ala-Gln-	Val	vaccinia virus RNA polymerase
Gly-	Arg-Met-Asn-	Gly-	Arg-Gly-Val-Phe-Arg-	Val	HSV DNA polymerase
Gly-	Gly-Val-Thr-	Gly-	Lys-Gly-Gln-Asp-Gly-	Ile	human poly(ADP-ribose) synthetase
Gly-	Leu-Tyr-Met-	Gly-	Glu-Thr-Glu-Thr-Lys-	Val	rat glucocorticoid receptor
Gly-	Leu-Tyr-Met-	Gly-	Glu-Thr-Glu-Thr-Lys-	Val	human glucocorticoid receptor
Gly-	Gly-Ile-Arg-	Gly-	Thr-Gln-Ile-Leu-Cys-	Val	adenovirus E1a
Gly-	Val-Tyr-Ala-	Gly-	Pro-Glu-Pro-Pro-Pro-	Val	mouse jun B
Gly-	Asn-Leu-Leu-	Gly-	Arg-Asp-Ser-Phe-Glu-	Val	mouse p53
Gly-	Asn-Leu-Leu-	Gly-	Arg-Asp-Ser-Phe-Glu-	Val	human p53
Gly-	Gly-Leu-Gly-	Gly-	Leu-The-Pro-Asn-Pro-	Val	mouse N-myc
Gly-	Gly-Leu-Gly-	Gly-	Leu-Thr-Pro-Asn-Pro-	Val	human N-myc
Gly-	Pro-Gly-Ala-	Gly-	Asp-Pro-Ala-Pro-Gly-	Ile	human L-myc
Gly-	Val-Tyr-Asn-	Gly-	Lys-Thr-Phe-Asn-Gln-	Val	rig

α-helix ──────── α-helix

Fig. 14.3. Putative DNA-binding domain within *rig* protein. The consensus sequence (Gly–X–X–X–Gly–X–X–X–X–X–Val/Ile) is conserved in a number of DNA-binding proteins, including P22 repressor (Pabo *et al.*, 1982), CAP (catabolite gene activator protein) (Weber *et al.*, 1982), Fnr (regulatory gene for fumarate and nitrate reduction) protein (Shaw & Guest, 1982), *Bacillus subtilis* DNA gyrase B (Moriyama *et al.*, 1985), yeast DNA topoisomerase II (Giaever *et al.*, 1986), mouse TdTase (terminal deoxynucleotidyltransferase) (Koiwai *et al.*, 1986), vaccinia virus RNA polymerase 147 kDa subunit (Broyles & Moss, 1986), herpes simplex virus (HSV) DNA polymerase (Gibbs *et al.*, 1985; Quinn & McGeoch, 1985), human poly(ADP-ribose) synthetase (Cherney *et al.*, 1987; Kurosaki *et al.*, 1987), rat glucocorticoid receptor (Miesfeld *et al.*, 1986), human glucocorticoid receptor (Hollenberg *et al.*, 1985), adenovirus E1a protein (Ralston &

oligodeoxyribonucleotide into the hamster insulinoma cells resulted in a significant inhibition of DNA synthesis. The 20 base oligodeoxyribonucleotide complementary to another portion of hamster *rig* mRNA (nucleotides 29–48) also inhibited DNA synthesis in the insulinoma cells. On the other hand, neither the antisense oligodeoxyribonucleotide complementary to the 5' portion of hamster insulin mRNA (Bell & Sanchez-Pescador, 1984) nor the sense *rig* oligodeoxyribonucleotide had any inhibitory effect. This result suggests that the *rig* products have a role in the events leading to or occurring during DNA synthesis in the insulinoma cells.

Recently we have found that *rig* is transiently expressed in the rat regenerating liver and synchronously cultured hepatocytes (Inoue *et al.*, 1988). In synchronously cultured rat hepatocytes, DNA synthesis started from 45 h after plating, rose steeply to a peak at 55 h and subsided by 90 h. The *rig* mRNA level was found to have increased from 20 h after plating, to reach a peak at 40 h, and to decline when DNA synthesis was maximal, indicating that *rig* mRNA level was elevated at the mid-to-late G_1 phase of the cell cycle (Inoue *et al.*, 1988). As shown in Fig. 14.4, the intracellular localization of the *rig* protein around the cell cycle was investigated by indirect immunofluorescence using antiserum against the *rig* protein. Hepatocytes in the G_1 phase exhibited weak fluorescence staining in the cytoplasm. As the cells progressed through the S phase, a strong staining pattern was observed in the nucleus; fluorescent granules were distributed throughout the nucleoplasm except in the nucleoli. After entering into the G_2 phase, the granules decreased in number and their fluorescence intensity was reduced. No fluorescence staining was observed immediately after plating. These results indicated that the *rig* protein accumulated in the nucleus during the S phase of the cell cycle, and that the cytoplasmic fluorescence in the G_1 phase represented the *de novo* synthesized *rig* protein.

14.5. Conclusions

By screening a cDNA library constructed from streptozotocin-nicotinamide-induced rat insulinomas, we have isolated a novel gene. The

Caption to Fig. 14.3 (*cont.*)

Bishop, 1983), mouse *jun*-B protein (Ryder *et al.*, 1988), mouse p53 (Jenkins *et al.*, 1984; Pennica *et al.*, 1984), human p53 (Harlow *et al.*, 1985), mouse N-*myc* protein (dePinho *et al.*, 1986), human N-*myc* protein (Kohl *et al.*, 1986; Stanton *et al.*, 1986) and human L-*myc* protein (Kaye *et al.*, 1988).

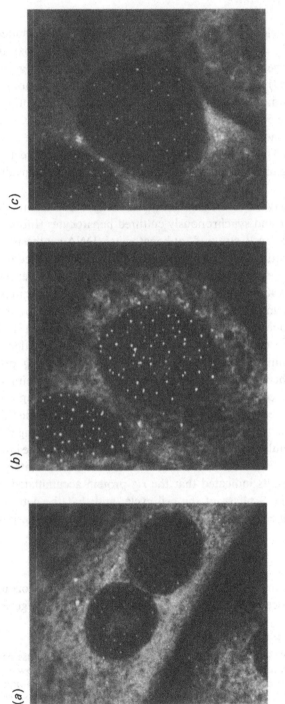

Fig. 14.4. Immunofluorescent staining of *rig* protein during cell cycle in primary cultured hepatocytes (Inoue *et al.*, 1987). (*a*) Cells in G$_1$ phase (30 h after plating); (*b*) cells in S phase (55 h after plating); (*c*) cells in G$_2$ phase (90 h after plating). Magnifications × 250.

gene, designated *rig* (rat insulinoma gene), is expressed in chemically induced rat insulinomas, a virus-induced hamster insulinoma and spontaneously occurring human insulinomas. The deduced 145 amino acid sequence remained invariant in the rat, hamster, human and mouse, suggesting that *rig* has evolved under extraordinarily strong selective constraints. The *rig* protein is supposed to be a nuclear protein interacting with DNA. The gene is also expressed in a variety of human tumors, such as esophageal cancers and colon cancers, and in many cultured cells such as NB-1 (human neuroblastoma cells), NIH 3T3 cells and rat embryo fibroblasts (K. Shiga *et al.*, unpublished data). Transient expression of *rig* was observed at the proliferative phase of regenerating liver and cycling hepatocytes. Thus *rig* is thought to be involved in a more general way in cell growth and replication. More recently, we isolated the genomic *rig* from a human esophageal carcinoma and determined its complete nucleotide sequence. The gene is composed of about 3000 nucleotides and divided into four exons separated by three introns. The 5'-flanking region of the gene has no apparent TATA-box or CAAT-box sequence. However, two GC-boxes are found at -235 and -76 base pairs upstream from the first ATG codon and 5 GC-boxes also in introns 1 and 2, suggesting that *rig* expression can be regulated by a transcription factor Sp1. Furthermore, the gene is bounded in the 5' region by CpG islands, regions of DNA with a high GC content and a high frequency of CpG dinucleotides relative to the bulk genome. These results indicate that *rig* belongs to the class of 'housekeeping' genes, whose products are necessary for the growth of all cell types (K. Shiga, H. Yamamoto & H. Okamoto, submitted for publication).

References

Bell, G.I. & Sanchez-Pescador, R. (1984). Sequence of a cDNA encoding Syrian hamster preproinsulin. *Diabetes* **33**, 297–300.

Broyles, S.S. & Moss, B. (1986). Homology between RNA polymerases of poxviruses, prokaryotes, and eukaryotes: nucleotide sequence and transcriptional analysis of vaccinia virus genes encoding 147-kDa and 22-kDa subunits. *Proceedings of the National Academy of Sciences of the U.S.A.* **83**, 3141–5.

Cherney, B.W., McBride, O.W., Chen, D., Alkhatib, H., Bhatia, K., Hensley, P. & Smulson, M.E. (1987). cDNA sequence, protein structure, and chromosomal location of the human gene for poly(ADP-ribose) polymerase. *Proceedings of the National Academy of Sciences of the U.S.A.* **84**, 8370–4.

DePinho, R.A., Legouy, E., Feldman, L.B., Kohl, N.E., Yancopoulos, G.D. & Alt, F.W. (1986). Structure and expression of the murine N-*myc* gene. *Proceedings of the National Academy of Sciences of the U.S.A.* **83**, 1827–31.

Giaever, G., Lynn, R., Goto, T. & Wang, J.C. (1986). The complete nucleotide sequence of the structural gene TOP2 of yeast DNA topoisomerase II. *Journal of Biological Chemistry* **261**, 12448–54.

Gibbs, J.S., Chiou, H.C., Hall, J.D., Mount, D.W., Retondo, M.J., Weller, S.K. & Coen, D.M. (1985). Sequence and mapping analyses of the herpes simplex virus DNA polymerase gene predict a C-terminal substrate binding domain. *Proceedings of the National Academy of Sciences of the U.S.A.* **82**, 7969–73.

Harlow, E., Williamson, N.M., Ralston, R., Helfman, D.M. & Adams, T.E. (1985). Molecular cloning and in vitro expression of a cDNA clone for human cellular tumor antigen p53. *Molecular and Cellular Biology* **5**, 1601–10.

Hollenberg, S.M., Weinberger, C., Ong, E.S., Cerelli, G., Oro, A., Lebo, R., Thompson, E.B., Rosenfeld, M.G. & Evans, R.M. (1985). Primary structure and expression of a functional human glucocorticoid receptor cDNA. *Nature* **318**, 635–41.

Inoue, C., Igarashi, K., Kitagawa, M., Terazono, K., Takasawa, S., Obata, K., Iwata, K., Yamamoto, H. & Okamoto, H. (1988). Expression of the insulinoma gene *rig* during liver regeneration and in primary cultured hepatocytes. *Biochemical and Biophysical Research Communications* **150**, 1302–8.

Inoue, C., Shiga, K., Takasawa, S., Kitagawa, M., Yamamoto, H. & Okamoto, H. (1987). Evolutionary conservation of the insulinoma gene *rig* and its possible function. *Proceedings of the National Academy of Sciences of the U.S.A.* **84**, 6659–62.

Jenkins, J.R., Rudge, K., Redmond, S. & Wade-Evans, A. (1984). Cloning and expression analysis of full length mouse cDNA sequences encoding the transformation associated protein p53. *Nucleic Acids Research* **12**, 5609–26.

Kaye, F., Battey, J., Nau, M., Brooks, B., Seifter, E., De Greve, J., Birrer, M., Sausville, E. & Minna, J. (1988). Structure and expression of the human L-*myc* gene reveal a complex pattern of alternative mRNA processing. *Molecular and Cellular Biology* **8**, 186–95.

Kohl, N.E., Legouy, E., DePinho, R.A., Nisen, P.D., Smith, R.K., Gee, C.E. & Alt, F.W. (1986). Human N-*myc* is closely related in organization and nucleotide sequence to c-*myc*. *Nature* **319**, 73–7.

Koiwai, O., Yokota, T., Kageyama, T., Hirose, T., Yoshida, S. & Arai, K. (1986). Isolation and characterization of bovine and mouse terminal deoxynucleotidytransferase cDNAs expressible in mammalian cells. *Nucleic Acids Research* **14**, 5777–92.

Kozak, M. (1984). Compilation and analysis of sequences upstream from the translational start site in eukaryotic mRNAs. *Nucleic Acids Research* **12**, 857–72.

Kurosaki, T., Ushiro, H., Mitsuuchi, Y., Suzuki, S., Matsuda, M., Matsuda, Y., Katunuma, N., Kangawa, K., Matsuo, H., Hirose, T., Inayama, S. & Shizuta, Y. (1987). Primary structure of human poly(ADP-ribose) synthetase as deduced from cDNA sequence. *Journal of Biological Chemistry* **262**, 15990–7.

Kyte, J. & Doolittle, R.F. (1982). A simple method for displaying the hydropathic character of a protein. *Journal of Molecular Biology* **157**, 105–32.

Laughon, A. & Scott, M.P. (1984). Sequence of a *Drosophila* segmentation gene: protein structure homology with DNA-binding proteins. *Nature* **310**, 25–31.

Lazarow, A. (1952). Spontaneous recovery from alloxan diabetes in the rat. *Diabetes* **1**, 363–72.

Miesfeld, R., Rusconi, S., Godowski, P.J., Maler, B.A., Okret, S., Wikstöm, A-C., Gustafsson, J-Å. & Yamamoto, K.R. (1986). Genetic complementation of a glucocorticoid receptor deficiency by expression of cloned receptor cDNA. *Cell* **46**, 389–99.

Moriyama, S., Ogasawara, N. & Yoshikawa, H. (1985). Structure and function of the region of the replication origin of the *Bacillus subtilis* chromosome. III. Nucleotide sequence of some 10,000 base pairs in the origin region. *Nucleic Acids Research* **13**, 2251–65.

Pabo, C.O., Krovatin, W., Jeffrey, A. & Sauer, R.T. (1982). The N-terminal arms of λ repressor wrap around the operator DNA. *Nature* **298**, 441–3.

Pabo, C.O. & Sauer, R.T. (1984). Protein–DNA recognition. *Annual Review of Biochemistry* **53**, 293–321.

Pennica, D., Goeddel, D.V., Hayflick, J.S., Reich, N.C., Anderson, C.W. & Levine, A.J.

(1984). The amino acid sequence of murine p53 determined from a c-DNA clone. *Virology* **134**, 477–82.

Proudfoot, N.J. & Brownlee, G.G. (1976). 3'Non-coding region sequences in eukaryotic messenger RNA. *Nature* **263**, 211–14.

Quinn, J.P. & McGeoch, D.J. (1985). DNA sequence of the region in the genome of herpes simplex virus type 1 containing the genes for DNA polymerase and the major DNA binding protein. *Nucleic Acids Research* **13**, 8143–63.

Rakieten, N., Gordon, B.S., Beaty, A., Cooney, D.A., Davis, R.D. & Schein, P.S. (1971). Pancreatic islet cell tumors produced by the combined action of streptozotocin and nicotinamide (35561). *Proceedings of the Society for Experimental Biology and Medicine* **137**, 280–3.

Ralston, R. & Bishop, J.M. (1983). The protein products of the *myc* and *myb* oncogenes and adenovirus Ela are structurally related. *Nature* **306**, 803–6.

Ryder, K., Lau, L.F. & Nathans, D. (1988). A gene activated by growth factors is related to the oncogene v-*jun*. *Proceedings of the National Academy of Sciences of the U.S.A.* **85**, 1487–91.

Sauer, R.T., Yocum, R.R., Doolittle, R.F., Lewis, M. & Pabo, C.O. (1982). Homology among DNA-binding proteins suggests use of a conserved super-secondary structure. *Nature* **298**, 447–51.

Shaw, D.J. & Guest, J.R. (1982). Nucleotide sequence of the *fnr* gene and primary structure of the Fnr protein of *Escherichia coli*. *Nucleic Acids Research* **10**, 6119–30.

Smith, A.E., Kalderon, D., Roberts, B.L., Colledge, W.H., Edge, M., Gillett, P., Markham, A., Paucha, E. & Richardson, W.D. (1985). The nuclear location signal. *Proceedings of the Royal Society of London* **B226**, 43–58.

Stanton, L.W., Schwab, M. & Bishop, M. (1986). Nucleotide sequence of the human N-*myc* gene. *Proceedings of the National Academy of Sciences of the U.S.A.* **83**, 1772–6.

Takasawa, S., Yamamoto, H., Terazono, K. & Okamoto, H. (1986). Novel gene activated in rat insulinomas. *Diabetes* **35**, 1178–80.

Uchida, S., Watanabe, S., Aizawa, T., Furuno, A. & Muto, T. (1979). Polyoncogenicity and insulinoma-inducing ability of BK virus, a human papovavirus, in Syrian golden hamsters. *Journal of the National Cancer Institute* **63**, 119–26.

Vogt, P.K., Bos, T.J. & Doolittle, R.F. (1987). Homology between the DNA-binding domain of the GCN4 regulatory protein of yeast and the carboxyl-terminal region of a protein coded for by the oncogene *jun*. *Proceedings of the National Academy of Sciences of the U.S.A.* **84**, 3316–19.

Weber, I.T., McKay, D.B. & Steitz, T.A. (1982). Two helix DNA binding motif of CAP found in *lac* repressor and *gal* repressor. *Nucleic Acids Research* **10**, 5085–102.

Wilbur, W.J. & Lipman, D.J. (1983). Rapid similarity searches of nucleic acid and protein data banks. *Proceedings of the National Academy of Sciences of the U.S.A.* **80**, 726–30.

Yamagami, T., Miwa, A., Takasawa, S., Yamamoto, H. & Okamoto, H. (1985). Induction of rat pancreatic B-cell tumors by the combined administration of streptozotocin or alloxan and poly(adenosine diphosphate ribose) synthetase inhibitors. *Cancer Research* **45**, 1845–9.

Yamamoto, H., Nose, K., Itoh, N. & Okamoto, H. (1980). Quantitation of proinsulin mRNA sequences in hamster insulinoma cells in culture by molecular hybridization. *Experientia* **36**, 187–8.

Yamamoto, H., Uchigata, Y. & Okamoto, H. (1981). Streptozotocin and alloxan induce DNA strand breaks and poly(ADP-ribose) synthetase in pancreatic islets. *Nature* **294**, 284–6.

Yonemura, Y., Takashima, T., Miwa, K., Miyazaki, I., Yamamoto, H. & Okamoto, H. (1984). Amelioration of diabetes mellitus in partially depancreatized rats by poly(ADP-ribose) synthetase inhibitors: evidence of islet B-cell regeneration. *Diabetes* **33**, 401–4.

15

A novel gene, *reg*, expressed in regenerating islets

KIMIO TERAZONO, TAKUO WATANABE AND
YUTAKA YONEMURA

15.1. Introduction

B-cells of the islets of Langerhans, the site of insulin production, have only a limited capacity for regeneration. In adult mouse and rat islets the population of replicating B-cells was estimated to be only about 1% of total B-cells (Logothetopoulos, 1972). However, certain substances, such as glucose (Andersson, 1975), essential amino acids (Swenne *et al.*, 1980), insulin (Rabinovitch *et al.*, 1982), growth hormone (Rabinovitch *et al.*, 1983) and epicatechin (Hii & Howell, 1984), were reported to stimulate some B-cell replication in fetal, newborn, or adult islets; nevertheless, the percentage of replicating B-cells was only about twice as high in B-cells treated with these substances than in control cells. Hyperplastic changes in the islets of Langerhans involving B-cells were observed in the pancreatic remnants of 90% depancreatized rats in the first few weeks after partial pancreatectomy, but thereafter extensive B-cell degeneration and degranulation were noticeable and the rats became severely diabetic (Martin & Lacy, 1963; Clark *et al.*, 1982). This low capacity for regeneration in pancreatic B-cells creates a predisposition for the development of diabetes mellitus (Hellerström *et al.*, 1976).

In 1984, Yonemura *et al.* demonstrated that administration of poly(ADP-ribose) synthetase inhibitors such as nicotinamide and 3-aminobenzamide to 90% depancreatized rats induces continuous regeneration of pancreatic B-cells, thereby preventing the development of diabetes in the partly depancreatized rats (Yonemura *et al.*, 1984; Okamoto, 1985; Okamoto *et al.*, 1985). Recently, Terazono *et al.* (1988) have isolated a novel gene, *reg*, i.e. *reg*enerating gene, which is activated in the regenerating islets, as well as a human homolog to *reg*.

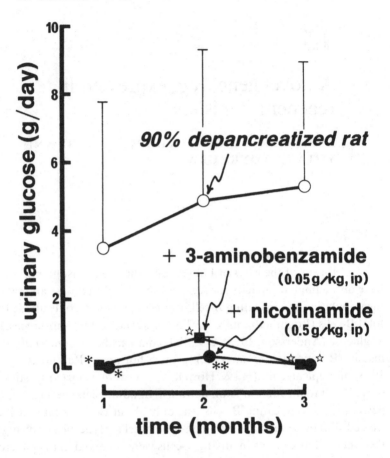

Fig. 15.1. Urinary glucose excretion in partly depancreatized rats with or without poly(ADP-ribose) synthetase inhibitor injections. Rats were given intraperitoneal injections of 3-aminobenzamide (0.05 g/kg) or nicotinamide (0.5 g/kg). The statistical significance of differences between rats treated with and without poly(ADP-ribose) synthetase inhibitors was analyzed using Student's *t*-test. Each point is the mean for five different rats; vertical bars (larger than the symbol indicating the mean value) show standard deviation. Symbols: *, $p < 0.10$; **, $p < 0.05$; ☆, $p < 0.025$ (versus control rats). The time after the partial pancreatectomy is shown on the abscissa.

15.2. Islet regeneration in 90% depancreatized and
poly(ADP-ribose) synthetase inhibitor-administered rats

Rats on which Foglia (1944) had performed 90% pancreatectomy exhibited glucosuria 1–3 months after the operation. The islets in the remaining pancreases of 90% depancreatized rats were relatively less numerous and small in size, and frequently exhibited fibrotic degeneration and degranulation (Martin & Lacy, 1963; Clark *et al.*, 1982). However, Yonemura *et al.* (1984) demonstrated that 90% depancreatized rats administered daily with poly(ADP-ribose) synthetase inhibitors, such as nicotinamide and 3-aminobenzamide, did not develop diabetes. In rats receiving nicotinamide or 3-aminobenzamide, the urinary glucose excretion level decreased markedly (Fig. 15.1). Plasma glucose levels in rats receiving poly(ADP-ribose) synthetase inhibitors were also significantly lower than those in control 90% depancreatized rats. Three months after the partial pancreatectomy, the islets in the remaining pancreases of rats which had received the nicotinamide or 3-aminobenzamide injections were very much larger than islets in the controls (Fig. 15.2). The number of islets per square centimetre was also greater in nicotinamide- or 3-aminobenzamide-injected rats (Yonemura *et al.*, 1984; Okamoto *et al.*, 1985). Since the size of each cell in these islets was normal, the increase in islet size in partly depancreatized rats treated with poly(ADP-ribose) synthetase inhibitors was attributed to an increase in the number of cells. When the remaining pancreases were immunohistochemically stained, almost the entire area of the enlarged islets in nicotinamide- or 3-aminobenzamide-treated rats stained densely for insulin. On the other hand, cells staining for glucagon (A-cells) were localized on the peripheries of the enlarged islets in the remaining pancreases of rats treated with nicotinamide or 3-aminobenzamide, and were almost the same in number as the islets of normal rats, as were cells staining for somatostatin (D-cells). These immunohistochemical findings indicated that it was specifically the B-cell population that increased in the islets of the remaining pancreases of poly(ADP-ribose) synthetase inhibitor-treated rats. The induction of islet regeneration in 90% depancreatized rats by poly(ADP-ribose) synthetase inhibitors through an increase occurring specifically in the B-cell population is the principal conclusion to be drawn from these results (Yonemura *et al.*, 1984; Okamoto, 1985; Okamoto *et al.*, 1985).

15.3. Cloning of regenerating islet-specific transcript

Terazono *et al.* (1988) isolated regenerating islets from the remaining pancreases of 90% depancreatized rats which had received

A

B

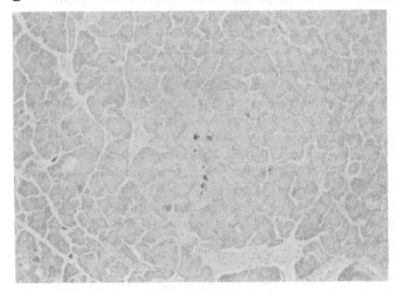

Fig. 15.2. Pancreatic islets stained for insulin. (*A*) Pancreas from normal rat; (*B*) remaining pancreas from control 90% depancreatized rat; (*C*) remaining pancreas from 90% depancreatized and

C

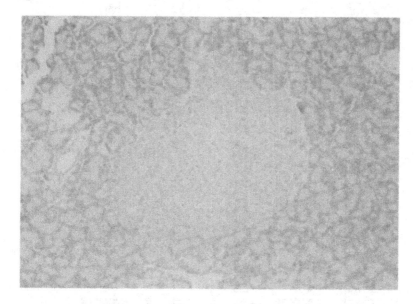

D

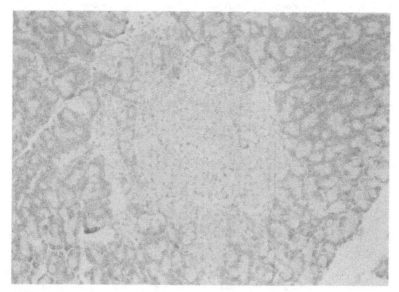

nicotinamide-treated rat; (*D*) remaining pancreas from 90%
depancreatized and 3-aminobenzamide-treated rat.

Rat AGCCTCCAGAGATTGTTGACTTGCCATCCTAAGCAGAAGACAGTTCTGCTCATC
        AA   T   C T   AG   TTCAT   A   G
Human
       -50   -40   -30   -20   -10

                                         1
                                         Met Thr Arg Asn Lys Tyr Phe Ile Leu Leu Ser Cys Leu Met Val
Rat   ATG ACT CGC AAC AAA TAT TTC ATT CTG CTT TCA TGC CTG ATG GTC
Human   ---   C   G   A   C   I T
                                         ATG GCT CAG AC   G   T T
                                         Met Ala Gln Thr Ser Ser   Ile   Phe
                                         -1 -1   10   20   30   40
                                                        Putative Signal Sequence

                                                                                                    40
       Leu Ser Pro Ser Gln Gly Gln Glu Ala Glu Glu Asp Leu Pro   Ser Ala Arg Ile Thr Cys Pro Glu Gly Ser Gly Asn Ala Tyr Ser Ser Tyr
Rat   CTT TCT CCA AGC CAG GGC CAG GAA GCT GAA GAA GAT CTA CCA   TCT GCC AGG ATC ACT TGT CCA GAA GGT TCC AAT GCC TAC AGT TCC TAC
Human G   AG   A   C C G T G   C   C   GC C   A   T   C C
                      Gln   C C G AC   Ser   Thr   Arg
                      50   Gln Thr Glu   Gln
                                      60   70   80   90   100   110   120   130

       Cys Tyr Tyr Phe Met Glu Asp His Leu Ser Trp Ala Glu Ala   Asp Phe Cys Gln Asn Met Asn Gly Tyr Leu Val Val Ser Val Leu
Rat   TGT TAC TAC TTC ATG GAA GAC CAT TTA TCT TGG GCT GAG GCA   GAT TTT TGC CAG AAC ATG AAC TCA GGC TAC TTG GTG TCA GTG CTC
Human   C   T AT   G GAG A C   C A   G   A C   T
                                Asn   Arg Glu Thr   Tyr   Asn
                   140   150   160   170   180   190   200   210   220

       Ser Gln Ala Glu Glu Asn Phe Gly Thr Gly Thr Ala Ala Asn   Val Trp Ile Gly Leu His Asp Pro Lys
Rat   AGC CAG GCT GAG GAG AAC TTT GGT ACT ACA GCT GCC AAT GTC   TGG ATT GGC CTC CAT GAT CCC AAA
Human C   G   T GC   AC TT   C
       Thr   Ala   Asp Asp Phe
                   230   240   250   260   270   280   290   300   310

                                                                                                    130
       Asn Asn Arg Arg Trp His Trp Ser Ser Gly Ser Leu Phe Leu   Tyr Lys Ser Trp Asp Thr Gly Tyr Pro Asn Ser Asn Ala Arg Gly Tyr
Rat   AAT AAT CGC CGC TGG CAC TGG AGC AGT GGG TCT CTG TTT CTC   TAC AAA TCC TGG GAC ACT GGG TAI CCT AAC TCC AAT CGT GGC TAC
Human G C   C TC   G C TC   G T A GCC   G T   C A G GTT   C
       Lys   Val Ser   Gly Ile   Ala   Ser Val   Pro
                   320   330   340   350   360   370   380   390   400

```
                    140                              150                              160                    165
Rat    Cys Val Ser Val Thr Ser Asn Ser Gly Tyr Lys Lys Trp Arg Asp Asn Ser Cys Asp Ala Gln Leu Ser Phe Val Cys Lys Phe Ala
       TGT GTA TCT GTG ACT TCA AAC TCA GGA TAC AAG AAA TGG AGA GAT AAC AGT TGT GAT GCC CAA TTA TCA TTT GTC TGC AAG TTC AAA GCC
Human    G AGC C    C        G A            T C            A   A  G   C    C                                                AA
                   Leu       Ser Thr                   Phe Gln                 Glu Asp Lys Phe                              Asn
         410         420            430          440          450       460          470       480          490

       End
Rat    TGA AATCATCTGAAAAAATAGTCATACAGAGCCAGACAAGAAAATACTATGGAGTCAAAACTACTAGACCATCTATCAAAAGCAAAGTCAACCCCCTCTTCCTAGACAAACATT
Human   AG  GG  G    T C TGTC  G ACTGATCCAGCA TT C ACGGA TCAA    TTAA CCGG CCATC    CCA CTCAACTCAA CTGGA ACTC  T CTCTGCTG G
       500         510         520         530         540         550          560         570         580         590         600          610

Rat    CTTGCCTCACTCGCCTATGGTATTTTATCTCCATTATTGTATGTAACCCTGCACATTTAAATAAAAATACCTTCACAATPolyA
Human  T GC T GTTAAT T CAATAG    ACCTA   CCAG CT TG A C TTAAATA  AA    C  GTT C  TPolyA
       620         630         640          650          660         670           680         690
```

Fig. 15.3. Nucleotide and deduced amino acid sequences of rat *reg* cDNA and the human homolog. The upper two lines show deduced amino acid and nucleotide sequences of rat *reg* cDNA. Nucleotide residues in rat *reg* cDNA are numbered in the 5' to 3' direction, beginning with the first residue of ATG triplet encoding initiator methionine; nucleotides on the 5' side of residue 1 are indicated by negative numbers. Amino acid residues are numbered beginning with initiator methionine in the rat sequence. Nucleotide and amino acid differences found in human sequences are displayed beneath rat sequences. Putative signal sequence and polyadenylation signal are underlined.

nicotinamide for 3 months, and constructed a cDNA library of approximately 2.8×10^6 recombinants from the islet poly(A)$^+$ RNA. From the cDNA library, about 2000 recombinants were duplicated on nitrocellulose filters for differential screening. One set of filters was hybridized to ^{32}P-labeled cDNA that had been reverse-transcribed from the poly(A)$^+$ RNA of regenerating islets. The other set of filters was hybridized to ^{32}P-labeled cDNA synthesized from the poly(A)$^+$ RNA of normal islets. In the differential screening, a clone consistently preferentially hybridizing to ^{32}P-labeled cDNA from regenerating islets was isolated. As shown in Fig. 15.3, the nucleotide sequence of the cloned cDNA was determined: the cDNA was 748 nucleotides long plus poly(A) and had one large open reading frame encoding a 165 amino acid protein, assuming that ATG at nucleotides 1–3 is the start codon and TGA at nucleotides 496–498 is the stop codon. At the amino terminus of the deduced 165 amino acid protein, there was a hydrophobic region similar to the signal sequence of many secretory proteins (Blobel & Dobberstein, 1975). The calculated relative molecular mass (18 656) of the deduced protein was in good agreement with that of the in vitro translation product of RNA which had been synthesized from the cloned cDNA. The nucleotide and deduced amino acid sequences of the cDNA were screened for possible relationships to other genes and proteins stored in the nucleic acid and protein data banks of the European Molecular Biology Laboratory (Heidelberg, F.R.G.), GenBank (Cambridge, Massachusetts, U.S.A.), and the National Biomedical Research Foundation (Washington, D.C., U.S.A.) with a rapid similarity search algorithm (Wilbur & Lipman, 1983). This screening indicated that there are neither genes nor gene products that are identical or statistically significantly homologous to the cDNA. The novel gene was named reg (regenerating gene) (Terazono et al., 1988).

An antisense RNA was synthesized from the PstI–DraI fragment (nucleotides − 51 to 672) of the reg cDNA (Fig. 15.3) and utilized to probe for levels of the reg mRNA in rat regenerating islets. As shown in Fig. 15.4, reg was expressed in regenerating islets but not detected in normal islets, liver, kidney, brain, insulinoma, or regenerating liver. The reg expression during islet regeneration in 90% depancreatized and nicotinamide-administered rats increased 1 month after the partial pancreatectomy, reached a peak after 3 months, and then decreased to an almost undetectable level one year after the operation (Terazono et al., 1988). Terazono et al. (1988) also found that reg was expressed in the hyperplastic islets of aurothioglucose ([1-thio-D-glucopyranosato]gold)-treated NON mice (Fig. 15.4). The standard NON mouse spontaneously

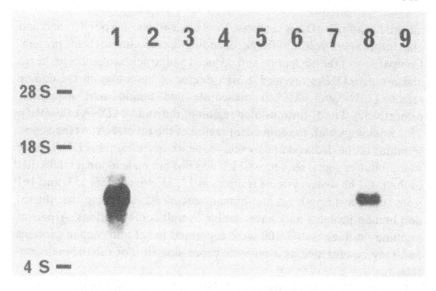

Fig. 15.4. Northern blot analysis of rat *reg* mRNA. Aliquots of 4 μg
RNA were electrophoresed. Bars indicate 28, 18, and 4 *S* RNAs on the
same gel. Lane 1, RNA from rat regenerating islets. Lane 2, RNA
from islets of normal untreated rats. Lanes 3–5, RNA from liver,
kidney, and brain of normal rats. Lane 6, RNA from rat insulinoma.
Lane 7, RNA from rat regenerating liver. Insulinoma was induced by
the combined administration of streptozotocin and nicotinamide
(Yamagami *et al.*, 1985). Regenerating liver was prepared as described
before (Higgins & Anderson, 1931). Lane 8, RNA from hyperplastic
islets of NON mice. NON mice, a strain derived from ICR mice,
received intraperitoneal injections of aurothioglucose at a dose of
0.25 g/kg of body mass on the 42nd and 56th days after birth. Lane 9,
RNA from islets of untreated ICR mice.

displayed impaired glucose tolerance (Tochino *et al.*, 1983; Ohgaku *et al.*,
1987). Aurothioglucose (Debons *et al.*, 1977) administration to NON mice
caused islet hyperplasia, thereby ameliorating the glucose intolerance;
immunohistochemical examination showed that the hyperplastic islets
consisted predominantly of B-cells (Y. Tochino *et al.*, unpublished data).

15.4. Human homolog to *reg*

Terazono *et al.* (1988) also isolated a human homolog to *reg* from
a human pancreas-derived cDNA library. About 10^6 recombinants from
the library were screened with a synthetic 60 base complement of
nucleotide residues 76–135 of *reg* cDNA (Fig. 15.3) as probe, yielding 70
hybridization-positive clones. As shown in Fig. 15.3, the nucleotide
sequence of the human clone with the longest cDNA insert was deter-

mined: the human cDNA comprised 749 nucleotides plus poly(A) and had one large open reading frame encoding a 166 amino acid protein. Comparison of the nucleotide and deduced amino acid sequences in the rat and human cDNAs revealed a high degree of homology in the coding region (75% and 68% in nucleotide and amino acid sequences, respectively). The 5'-untranslated region of the human cDNA exhibited a 75% homology with the equivalent region of the rat cDNA. At the amino terminus of the deduced amino acid sequence of the human cDNA, there was a putative signal sequence which was one amino acid longer than that of the rat. The seven cysteine residues at 12, 35, 46, 63, 136, 153, and 161 were common to both rat and human sequences, suggesting that the rat and human proteins may have similar overall conformations. A pair of arginine residues at 108–109 were conserved in rat and human proteins and may be regarded as a possible processing site for dibasic endopeptidases.

More recently, Watanabe *et al.* (1989) isolated the genomic *reg* from a human leukocyte genomic DNA library and determined the nucleotide sequence. As shown in Fig. 15.5, the gene spanned about 3.0 kilobase pairs, and comparison of the genomic sequence with the cDNA sequence revealed that the coding region was divided into six exons separated by five introns: the nucleotide sequence of the coding region of the gene was identical to that of the cDNA, and all exon–intron junctions conformed to the 'GT–AG' rule (Padgett *et al.*, 1986). Exon 1 encoded the 5'-untranslated region of the mRNA. Exon 2 encoded the remainder of the 5'-untranslated region and the putative signal sequence. Exons 3, 4, 5 and 6 encoded the remainder of the *reg* protein. The 3'-untranslated region was encoded in exon 6. There were several TATA box sequences and three CAAT box sequences in the 1.1 kilobase pairs of the 5'-flanking region (Watanabe *et al.*, 1989).

15.5. Discussion

In this chapter, we demonstrated the existence of a novel gene, *reg*, in regenerating islets. This gene was expressed in regenerating islets but not in normal islets, liver, kidney, or brain. A high expression of *reg* was also found in mouse hyperplastic islets. The regeneration of rat islets was induced by 90% pancreatectomy and nicotinamide administration. The hyperplasia of mouse islets was induced by aurothioglucose administration. Examination of regenerating islets and hyperplastic islets indicated that the activation of *reg* may be a general feature of pancreatic B-cell regeneration. However, *reg* was not expressed in insulinomas or

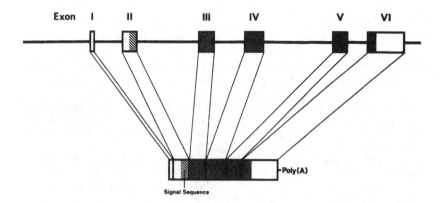

Fig. 15.5. Organization of human genomic *reg*. Exons are indicated as boxes; the coding region is indicated by stippling. Human *reg* mRNA is shown below the genomic *reg*.

regenerating liver, suggesting that the expression of *reg* is highly specific to the normal replication of pancreatic B-cells and not related either to the malignant growth of B-cells or to the regeneration of other tissues. Thus, the expression of *reg* in regenerating islets and hyperplastic islets suggested potential roles for *reg* in the normal replication, growth, and maturation of islet B-cells. The human homolog to *reg* and the genomic *reg* were also isolated and their structures were determined (Terazono *et al.*, 1988; Watanabe *et al.*, 1989). One region of human *reg* protein (residues 35–53) was similar in amino acid sequence to a region of human pancreatic secretory trypsin inhibitor (PSTI) (residues 24–42) (Yamamoto *et al.*, 1985). PSTI, a 56 amino acid secretory protein, was reported to stimulate DNA synthesis in human fibroblasts (Ogawa *et al.*, 1985) and to be identical to endothelial cell growth factor 2a (McKeehan *et al.*, 1986). The sequence similarity between human *reg* protein and PSTI implies that the *reg* protein may have functions similar to those of PSTI.

Recent immunohistochemical studies have demonstrated the presence of the *reg* protein not only in regenerating islet B-cells but also in pancreatic exocrine cells (K. Terazono *et al.*, in preparation). More recently, the amino acid sequence encoded by the human *reg* gene has proved to contain the 144 amino acid sequence of 'pancreatic stone protein' (PSP) (DeCaro *et al.*, 1989) and the partly determined 45 amino acid sequence of 'pancreatic thread protein' (PTP) (Gross *et al.*, 1985). As PSP or PTP is a component of the pancreatic juice, the *reg* gene products may have roles in pancreatic exocrine functions as well as in islet B-cell regeneration.

References

Andersson, A. (1975). Synthesis of DNA in isolated pancreatic islets maintained in tissue culture. *Endocrinology* **96**, 1051–4.

Blobel, G. & Dobberstein, B. (1975). Transfer of proteins across membranes: II. Reconstitution of functional rough microsomes from heterologous components. *Journal of Cell Biology* **67**, 852–62.

Clark, A., Bown, E., King, T., Vanhegan, R.I. & Turner, R.C. (1982). Islet changes induced by hyperglycemia in rats: effect of insulin or chlorpropamide therapy. *Diabetes* **31**, 319–25.

Debons, A.F., Krimsky, I., Maayan, M.L., Fani, K. & Jimenez, F.A. (1977). Gold thioglucose obesity syndrome. *Federation Proceedings* **36**, 143–7.

DeCaro, A.M., Adrich, Z., Fournet, B., Capon, C., Bonicel, J.J., DeCaro, J.D. & Rovery, M. (1989). N-terminal sequence extension in the glycosylated forms of human pancreatic stone protein: the 5-oxoproline N-terminal chain is O-glycosylated on the 5th amino acid residue. *Biochimica et Biophysica Acta* **994**, 281–4.

Foglia, V.G. (1944). Caracteristicas de la diabetes en la rata. *Revista de la Sociedad Argentina de Biologia* **20**, 21–37.

Gross, J., Carlson, R.I., Brauer, A.W., Margolies, M.N., Warshaw, A.L. & Wands, J.R. (1985). Isolation, characterization, and distribution of an unusual pancreatic human secretory protein. *Journal of Clinical Investigation* **76**, 2115–26.

Hellerström, C., Andersson, A. & Gunnarsson, R. (1976). Regeneration of islet cells. *Acta Endocrinologica* **83** (suppl. 205), 145–60.

Higgins, G.M. & Anderson, R.M. (1931). Experimental pathology of the liver: I. Restoration of the liver of the white rat following partial surgical removal. *Archives of Pathology* **12**, 186–202.

Hii, C.S.T. & Howell, S.L. (1984). Effects of epicatechin on rat islets of Langerhans. *Diabetes* **33**, 291–6.

Logothetopoulos, J. (1972). Islet cell regeneration and neogenesis. In *Handbook of Physiology*, sect. 7, vol. 1 (*Endocrine Pancreas*) (ed. D.F. Steiner & N. Freinkel), pp. 67–76. Washington: American Physiological Society.

Martin, J.M. & Lacy, P.E. (1963). The prediabetic period in partially pancreatectomized rats. *Diabetes* **12**, 238–42.

McKeehan, W.L., Sakagami, Y., Hoshi, H. & McKeehan, K.A. (1986). Two apparent human endothelial cell growth factors from human hepatoma cells are tumor-associated proteinase inhibitors. *Journal of Biological Chemistry* **261**, 5378–83.

Ogawa, M., Tsushima, T., Ohba, Y., Ogawa, N., Tanaka, S., Ishida, M. & Mori, T. (1985). Stimulation of DNA synthesis in human fibroblasts by human pancreatic secretory trypsin inhibitor. *Research Communications in Chemical Pathology and Pharmacology* **50**, 155–8.

Ohgaku, S., Morioka, H., Sawa, T., Yano, S., Yamamoto, H., Okamoto, H. & Tochino, Y. (1987). Restriction fragment length polymorphism (RFLP) at the locus of insulin gene in 'NON mice': a new animal model for non-obese NIDDM. In *Best approach to the ideal therapy of diabetes mellitus*, International Congress Series 754 (ed. Y. Shigeta, H.E. Lebovitz, J.E. Gerich & W.J. Malaisse), pp. 133–6. Amsterdam: Excerpta Medica.

Okamoto, H. (1985). Molecular basis of experimental diabetes: degeneration, oncogenesis and regeneration of pancreatic B-cells of islets of Langerhans. *BioEssays* **2**, 15–21.

Okamoto, H., Yamamoto, H. & Yonemura, Y. (1985). Poly(ADP-ribose) synthetase inhibitors induce islet B-cell regeneration in partially depancreatized rats. In *ADP-ribosylation of proteins* (ed. F.R. Althaus, H. Hilz & S. Shall), pp. 410–16. Berlin: Springer-Verlag.

Padgett, R.A., Grabowski, P.J., Konarska, M.M., Seiler, S. & Sharp, P.A. (1986). Splicing of messenger RNA precursors. *Annual Review of Biochemistry* **55**, 1119–50.

Rabinovitch, A., Quigley, C. & Rechler, M.M. (1983). Growth hormone stimulates islet B-cell replication in neonatal rat pancreatic monolayer cultures. *Diabetes* **32**, 307–12.

Rabinovitch, A., Quigley, C., Russell, T., Patel, Y. & Mintz, D.H. (1982). Insulin and multiplication stimulating activity (an insulin-like growth factor) stimulate islet β-cell replication in neonatal rat pancreatic monolayer cultures. *Diabetes* **31**, 160–4.

Swenne, I., Bone, A.J., Howell, S.L. & Hellerström, C. (1980). Effects of glucose and amino acids on the biosynthesis of DNA and insulin in fetal rat islets maintained in tissue culture. *Diabetes* **29**, 686–92.

Terazono, K., Yamamoto, H., Takasawa, S., Shiga, K., Yonemura, Y., Tochino, Y. & Okamoto, H. (1988). A novel gene activated in regenerating islets. *Journal of Biological Chemistry* **263**, 2111–14.

Tochino, Y., Kanaya, T. & Makino, S. (1983). Microangiopathy in the spontaneously diabetic nonobese mouse (NOD mouse) with insulitis. In *Diabetic microangiopathy* (ed. H. Abe & M. Hoshi), pp. 423–32. Tokyo: University of Tokyo Press.

Watanabe, T., Yonekura, H., Terazono, K., Yamamoto, H. & Okamoto, H. (1989). Complete nucleotide sequence of human *reg* gene and its expression in normal and tumoral tissues. *Journal of Biological Chemistry* (submitted for publication).

Wilbur, W.J. & Lipman, D.J. (1983). Rapid similarity searches of nucleic acid and protein data banks. *Proceedings of the National Academy of Sciences of the U.S.A.* **80**, 726–30.

Yamagami, T., Miwa, A., Takasawa, S., Yamamoto, H. & Okamoto, H. (1985). Induction of rat pancreatic B-cell tumors by the combined administration of streptozotocin or alloxan and poly(adenosine diphosphate ribose) synthetase inhibitors. *Cancer Research* **45**, 1845–9.

Yamamoto, T., Nakamura, Y., Nishide, T., Emi, M., Ogawa, M., Mori, T. & Matsubara, K. (1985). Molecular cloning and nucleotide sequence of human pancreatic secretory trypsin inhibitor (PSTI) cDNA. *Biochemical and Biophysical Research Communications* **132**, 605–12.

Yonemura, Y., Takashima, T., Miwa, K., Miyazaki, I., Yamamoto, H. & Okamoto, H. (1984). Amelioration of diabetes mellitus in partially depancreatized rats by poly(ADP-ribose) synthetase inhibitors: evidence of islet B-cell regeneration. *Diabetes* **33**, 401–4.

16

Defects of signal transduction in a tumoral islet cell line

WILLY J. MALAISSE

16.1. Introduction

Most genetic defects so far identified in pancreatic islet cells concern the pathway of proinsulin biosynthesis and conversion (Chan *et al.*, 1986; Gabbay *et al.*, 1979). The present review refers to a line of tumoral islet cells, namely RINm5F cells (Gazdar *et al.*, 1980), in which several site-specific biochemical and functional anomalies have been identified over recent years.

16.2. Anomaly in hexose transport

D-glucose represents, under physiological conditions, the major, albeit not the sole, regulator of insulin release. It is currently believed that the identification of D-glucose as an insulin secretagogue is tightly and causally linked to the capacity of the hexose to be metabolized and to augment the rate of ATP generation in the pancreatic B-cell (Malaisse *et al.*, 1979). Several rather specific features of D-glucose metabolism in normal islet cells are well suited to this glucose-sensing role. The first of these features consists of the rapid equilibration of D-glucose concentration across the B-cell plasma membrane (Hellman *et al.*, 1971). The following findings indicate that hexose transport is perturbed in tumoral islet cells.

The first indication for a deficiency of hexose transport in the RINm5F cells was obtained in a study of 3-*O*-methyl-D-[U-^{14}C]glucose uptake (Malaisse *et al.*, 1986). Results obtained at different temperatures, at various concentrations of the hexose and over different times of incubation indicated that the uptake of 3-*O*-methyl-D-[U-^{14}C]glucose represents a temperature-sensitive and saturable process, so that no rapid equilibration of hexose concentrations across the plasma membrane was reached, especially at low temperature and/or high concentrations of 3-O-methyl-D-glucose.

If the situation characterized with 3-O-methyl-D-glucose extends to D-glucose, the transport of the hexose could represent a rate-determining step in the utilization of D-glucose. The uptake of D-[U-^{14}C]glucose and the utilization of D-[5-^{3}H]glucose were indeed inhibited by 3-O-methyl-D-glucose, which failed, however, to affect D-[U-^{14}C]glucose oxidation (Malaisse et al., 1986). At variance with the situation found in normal insulin-producing cells, the transport of D-glucose into the tumoral cells may thus play a regulatory role in its metabolism in the RINm5F cells.

The latter view was further investigated by measuring the uptake of D-[U-^{14}C]glucose by the RINm5F cells (Giroix et al., 1986). Data collected over 5 min incubation at either 7 or 37° C in the presence of increasing concentrations of D-glucose (0.14–16.7 mM) again suggested that the uptake of the hexose by RINm5F cells represents a saturable phenomenon which does not result in the rapid equilibration of the extracellular and intracellular concentrations of D-glucose.

The latter radioisotopic data could not be used, however, to assess the steady-state intracellular concentration of D-glucose, since both D-glucose and its intracellular labeled metabolites are simultaneously measured. In order to overcome such a limitation, the true D-glucose content was measured by an enzymatic procedure in RINm5F cells separated from the incubation medium by centrifugation through a layer of oil into a solution of HCl (50 mM) and CsCl (0.64 M). Even after 30 min incubation at 37° C, the apparent distribution space of D-glucose (2.8–50.0 mM), although higher than that of L-[1-^{14}C]glucose, remained much lower than that of ^{3}H$_2$O (Sener et al., 1986a). As a matter of fact, the intracellular D-glucose space represented no more than 38% of the intracellular ^{3}H$_2$O volume. This implies that the intracellular concentration of D-glucose may never be sufficiently high to allow for an optimal participation of glucokinase in the phosphorylation of this hexose.

The findings so far outlined led us to investigate whether the transport of molecules structurally related to D-glucose may also be altered in the RINm5F cells. For this purpose, rat pancreatic islets and RINm5F cells were incubated for 5 min at 7 or 23° C in media containing ^{3}H$_2$O and either L-[1-^{14}C]glucose or [2-^{14}C]alloxan. In the islets, the intracellular distribution space of [2-^{14}C]alloxan represented, at 7 and 23° C respectively, 11.4 ± 1.0 and 25.5 ± 2.3% of the intracellular ^{3}H$_2$O volume. In the RINm5F cells, the distribution of [2-^{14}C]alloxan was not affected significantly by the ambient temperature and represented, after correction for extracellular contamination, no more than $5.2 \pm 0.5\%$ of the intracellular ^3H_2O volume. Preincubation for 30 min at $7°$ C in the presence of alloxan (10 mM) failed to affect subsequent D-[U-^{14}C]glucose oxidation in the tumoral cells, while causing a 70% inhibition of glucose oxidation in the islets. These results were interpreted to indicate that RINm5F cells are resistant to the cytotoxic action of alloxan, this being attributable, in part at least, to a poor uptake of the diabetogenic agent (Sener & Malaisse, 1985).

Comparable findings were collected in RINm5F cells exposed to streptozotocin (Sener et al., 1986c). When pancreatic islets were exposed for 60 or 120 min to streptozotocin (3.8 mM), both the utilization of D-[5-^{3}H]glucose and oxidation of D-[U-^{14}C]glucose were impaired (see Fig. 16.3). However, under identical conditions, the RINm5F cells appeared resistant to the cytotoxic action of streptozotocin. At this point, it should be emphasized that the RINm5F cells are not resistant to all B-cytotoxic agents. For instance, after 20 hours of culture in the presence of pentamidine (5–500 μM), a dose-related decrease in both insulin output and D-[5-^{3}H]glucose utilization was observed in pancreatic islets (Sener et al., 1986c). Under identical conditions, pentamidine also caused a dose-related destruction of RINm5F cells and, in the remaining cells, impaired D-[5-^{3}H]glucose utilization. As judged by the latter metabolic criterion, RINm5F cells appeared to be somewhat more sensitive to pentamidine than normal pancreatic islet cells.

In the last set of experiments in this series, we have investigated whether a known inhibitor of hexose transport, cytochalasin B, would affect D-glucose transport to a different extent in normal rat islets and RINm5F cells, respectively (Malaisse et al., 1987b). Cytochalasin B (17 μM) virtually abolished 3-O-methyl-D-[U-^{14}C]glucose uptake and D-[5-^{3}H]glucose utilization in the RINm5F cells. This coincided with a marked decrease in D-[U-^{14}C]glucose oxidation and suppression of the stimulant action of D-glucose upon insulin release. Cytochalasin B, however, augmented basal insulin release by the tumoral cells. The RINm5F cells appeared much more sensitive than normal islet cells to cytochalasin B, as judged from the relative magnitude of inhibition in either hexose uptake or utilization. In both cell types, the inhibitory action of cytochalasin B upon glucose metabolism seemed to be competitive, being more marked at low than high glucose concentration. Incidentally, cytochalasin B also inhibits glucose-stimulated protein biosynthesis in RINm5F cells (Valverde et al., 1988a).

These converging observations indicate that the carrier system for the transport of hexoses and structurally related molecules is indeed poorly efficient in RINm5F cells.

16.3. Anomalies in the generation and channeling of hexose phosphates

The phosphorylation of D-glucose in normal islet cells is catalyzed by both a low-K_m hexokinase and a high-K_m glucokinase (Malaisse et al., 1976b). In intact cells, the low-K_m enzyme is largely inhibited by endogenous D glucose 6-phosphate and to a lesser extent, by D-glucose 1,6-bisphosphate (Giroux et al., 1984b). Therefore, glucokinase may play an essential role in regulating the rate of hexose phosphorylation. As a matter of fact, the presence of glucokinase allows for acceleration of glycolysis in response to a rise in extracellular and, hence, intracellular hexose concentration in its physiological range. In islet cells, the phosphorylation of D-glucose could be the object of a sequential synarchistic regulation (Carpinelli et al., 1987). Indeed, the increase in cytosolic ATP concentration, which occurs in response to a rise in extracellular glucose concentration (Malaisse & Sener, 1987), may exert a positive feedback effect on glucose phosphorylation. The recent finding that a large fraction of both hexokinase and glucokinase is bound to mitochondria in normal islet cells (Sener et al., 1986b) suggests, however, that mitochondrial ATP could be used preferentially to cytosolic ATP as a substrate for hexose phosphorylation. The binding of glucose-phosphorylating enzymes to mitochondria may thus play a propitious role in stimulus–secretion coupling by sparing cytosolic ATP, which is currently viewed as the major coupling factor between metabolic and more distal ionic events in the process of nutrient-stimulated insulin release. The ambiquity of hexokinases in islet cells could itself be regulated by factors such as a change in intracellular pH (Malaisse-Lagae & Malaisse, 1988).

In several respects, a comparable situation could prevail in tumoral islet cells, which also display both low-K_m and high-K_m glucose-phosphorylating enzymic activities in both cytosolic and mitochondrial subcellular fractions (Giroix et al., 1985; Malaisse-Lagae & Malaisse, 1988). Nevertheless, the following anomalies should not be overlooked. First, at low concentrations of D-glucose, the phosphorylation of the hexose in RINm5F cell homogenates is catalyzed mainly by a type 2 hexokinase rather than the type 1 isoenzyme, which prevails in normal islet cells (Giroix et al., 1985; Vischer et al., 1987). Second, relative to the activity of these low-K_m enzymes, that of glucokinase may be somewhat lower in tumoral than in normal islet cells (Malaisse-Lagae & Malaisse, 1988).

Last, the lack of equilibration between the extracellular and intracellular concentrations of D-glucose probably prevents the full expression of glucokinase activity in RINm5F cells (Sener *et al.*, 1986*a*).

In addition to these anomalies in hexose phosphorylation, the further channeling of hexose phosphate into distinct metabolic pathways may also be perturbed, on occasion, in the tumoral cells. For instance, while the preferential orientation of α-D-glucose 6-phosphate into the glycolytic pathway and that of β-D-glucose 6-phosphate into the pentose phosphate pathway seem to be normally operative in RINm5F cells (Malaisse *et al.*, 1985*c*), an obvious difference between normal and tumoral islet cells was recently documented in the case of D-fructose metabolism. In both cell types, the phosphorylation of D-fructose is catalyzed by hexokinase, no fructokinase activity being apparently present in islet cells (Sener *et al.*, 1984*b*). Unexpectedly, in RINm5F cells exposed to D-fructose in the absence of D-glucose, virtually all the fructose 6-phosphate generated from the exogenous ketohexose is channeled into the pentose phosphate pathway (Sener *et al.*, 1987*c*). Such a preferential orientation is also observed in normal rat islets, but to a lesser extent than in RINm5F cells (Sener & Malaisse, 1988). Incidentally, in both cell types, the fate of D-fructose is dramatically altered when D-glucose is present together with D-fructose in the incubation medium. The factors responsible for the preferential orientation of D-fructose 6-phosphate into the pentose phosphate pathway in glucose-deprived cells and for the remodeling of D-fructose metabolism in the presence of D-glucose remain to be identified.

16.4. Anomaly in phosphofructokinase activation

Over recent years, we have defended the view that activation of phosphofructokinase in islet cells may be required for the rate of fructose 6-phosphate conversion to fructose 1,6-bisphosphate to keep pace with the rate of fructose 6-phosphate generation from glucose 6-phosphate (Malaisse *et al.*, 1982). Thus, the apparent role of hexose bisphosphates, especially glucose 1,6-bisphosphate and fructose 2,6-bisphosphate, as activators of phosphofructokinase was duly emphasized (Malaisse-Lagae *et al.*, 1982; Sener *et al.*, 1984*b*). For instance, it was proposed that the anomeric specificity of both D-glucose and D-mannose metabolism was attributable, in part, to the anomeric difference in the islet content in aldohexose bisphosphates (Malaisse *et al.*, 1983). Likewise, data were collected to suggest that, in islets removed from fasted, as distinct from fed, rats, phosphofructokinase was apparently not sufficiently activated to keep pace with the rate of glucose 6-phosphate formation, and this despite the well-known decrease in islet glucokinase activity during fasting (Giroix *et al.*, 1984*a*; Malaisse *et al.*, 1976*b*).

The concept that exposure of the islets to a high concentration of D-glucose results in the activation of phosphofructokinase gained further support from a recent study on the Pasteur effect in rat pancreatic islets (Malaisse *et al.*, 1988). In a previous report, it has been proposed that the Pasteur effect is not operative in pancreatic islets. However, we have now observed that hypoxia invariably increases the production of [^{14}C]lactate from D-[U-^{14}C]glucose by rat islets, whether at low or high concentrations of D-glucose. However, in the hypoxic islets, the rate of D-[5-^{3}H]glucose detritiation was increased solely at 2.8 mM D-glucose, being unaffected at 8.3 mM D-glucose and inhibited at 16.7 mM D-glucose. The latter inhibition strongly suggests that the increase in glycolysis which would be expected to occur in hypoxic islets due to the fall in ATP content and, hence, lesser inhibition of phosphofructokinase was masked and even overcome as a result of a concomitant decrease in fructose 2,6-bisphosphate and glucose 1,6-bisphosphate concentration.

When the same experiments were performed in RINm5F cells, also exposed to 16.7 mM D-glucose, the generation of ^{3}H$_2$O from D-[5-^{3}H]glucose was not decreased but, on the contrary, markedly increased by hypoxia. The contrasting response to hypoxia of normal and tumoral islet cells suggests, therefore, that phosphofructokinase fails to be fully activated in the latter cells even when they are exposed to a high concentration of D-glucose.

16.5. Anomalies in mitochondrial oxidative events

We have recently observed that, in normal pancreatic islets, the transfer of reducing equivalents into the mitochondria (as coupled with aerobic glycolysis, taken as the difference between the rate of D-[5-^{3}H]glucose utilization and the production of [^{14}C]lactate from D-[3,4-^{14}C]glucose), the rate of [1-^{14}C]pyruvate decarboxylation (as judged by the production of ^{14}CO$_2$ from D-[3,4-^{14}C]glucose) and the mitochondrial oxidation of [^{14}C]acetyl-CoA residues (as judged by the production of ^{14}CO$_2$ from D-[6-^{14}C]glucose) are all increased in response to a rise in D-glucose concentration, even when the results are expressed relative to the corresponding total rate of glucose utilization (Sener & Malaisse, 1987). It appears, therefore, that normal islet cells are organized to favour those mitochondrial oxidative reactions yielding the major fraction of the energy generated by the catabolism of exogenous D-glucose.

Further experiments (Malaisse & Sener, 1988*a*, *b*) revealed the following features. First, the preferential stimulation of glucose oxidation relative to glucose utilization was borne out by comparing the fate of D-[5-^{3}H]glucose and D-[U-^{14}C]glucose in oxygenated and hypoxic islets respectively, these experiments indicating, at variance with a prior view, the operation of a Pasteur effect in normal islet cells (Malaisse et al., 1988). Second, as judged from the ratio of D-[6-^{14}C]glucose oxidation to D-[5-^{3}H]glucose utilization, the preferential stimulation of mitochondrial oxidative events displays a sigmoidal dependency on hexose concentration and an exponential time course during prolonged exposure to a high concentration of D-glucose. Third, such a preferential stimulation can be simulated by the administration of D-fructose to islets exposed to a low concentration of D-glucose, in which case stimulation of insulin release is also observed. Fourth, the increased transfer of reducing equivalents into the mitochondria is apparently mediated, in part at least, by the glycerol phosphate shuttle. Indeed, a rise in glucose concentration in the 2.8 to 16.7 mM range causes a progressive increase in the production of ^{3}H$_2$O from [2-^{3}H]glycerol, but not from [1(3)-^{3}H]glycerol (Sener et al., 1988b). Fifth, the preferential stimulation of mitochondrial oxidative events in normal islet cells represents an unusual situation, an opposite metabolic response occurring in glucose-insensitive secretory cells, e.g. in the parotid gland. Last, the metabolic response in normal islets probably allows for the stimulation of ATP-requiring functional events by the hexose. The latter view is supported by the finding that, in islets exposed to a high concentration of D-glucose (16.7 mM), the ratio of D-[6-^{14}C]glucose oxidation to D-[5-^{3}H]glucose utilization is severely decreased when the islets are incubated in the presence of cycloheximide and absence of Ca^{2+}. The inhibitory effects of cycloheximide and Ca^{2+} deprivation are additive, suggesting that functional events such as proinsulin biosynthesis and the Ca^{2+} movements associated with stimulation of insulin release exert a feedback control on mitochondrial oxidations. This is further supported by the fact that metabolic variables are not significantly affected by cycloheximide and Ca^{2+} deprivation in islets exposed to a low concentration of D-glucose.

A vastly different situation prevails in RINm5F cells (Sener & Malaisse, 1987). Indeed, relative to total glucose utilization, the flux through the glycerol phosphate shuttle, the oxidative decarboxylation of pyruvate and the oxidation of acetyl-CoA residues in the Krebs cycle are all much lower in tumoral than normal islet cells. Moreover, in the RINm5F cells, a rise in D-glucose concentration from 2.8 to 16.7 mM fails to increase the rate of mitochondrial oxidative processes relative to the overall rate of D-glucose utilization. These results indicate that RINm5F cells have lost an essential attribute of the normal metabolic response to D-glucose, namely the capacity to adapt the rate of mitochondrial ATP generation to a rise in hexose concentration in excess of 2.8 mM.

16.6. Crabtree effect in RINm5F cells

The experimental results reviewed in the preceding section of this report led us to scrutinize the respiratory response of RINm5F cells to an increase in glucose concentration (Sener *et al.*, 1988*a*).

A rise in D-glucose concentration from 2.8 to 16.7 mM significantly augmented the production of ^3H_2O from D-[5-^{3}H]glucose and the conversion of D-[U-^{14}C]glucose to either [^{14}C]lactate or ^{14}C-labeled pyruvate and amino acids (Fig. 16.1). The rise in hexose concentration, however, decreased the rate of oxidation of both D-[U-^{14}C]glucose and D-[6-^{14}C]glucose. Whereas 2.8 mM D-glucose augmented oxygen uptake above basal value, a further rise in D-glucose concentration to 16.7 mM decreased respiration, which remained higher, however, than the basal value. The absolute value for aerobic glycolysis, taken as the difference between D-[5-^{3}H]glucose utilization and [^{14}C]lactate production, was not affected significantly by the rise in D-glucose concentration suggesting an unaltered transfer rate of reducing equivalents into the mitochondria. The oxygen uptake attributable to the catabolism of D-glucose, as judged from the rate of both aerobic glycolysis and D-[U-^{14}C]glucose oxidation, was higher at 2.8 than 16.7 mM D-glucose. However, by subtracting from such an uptake the experimental value for the hexose-induced increase in respiration, the sparing action of D-glucose upon the oxidation of endogenous nutrients appeared virtually identical in the presence of 2.8 and 16.7 mM. This was confirmed by the finding that, at the low and high concentrations of the hexose, D-glucose indeed exerted an equal sparing action upon the production of $^{14}CO_2$ from cells prelabeled with either L-[U-^{14}C]glutamine or [U-^{14}C]palmitate. Likewise, the rate of ATP generation attributable to the catabolism of D-glucose was virtually identical in the presence of 2.8 and 16.7 mM D-glucose. The rate of lipogenesis, the ATP/ADP content, the adenylate charge and the cytosolic NADH/NAD$^+$ and NADPH/NADP$^+$ ratios were all increased above basal value to the same extent at the two concentrations of D-glucose.

These results indicate that, in sharp contrast with the situation found in normal islet cells, a rise in D-glucose concentration, instead of stimulating mitochondrial oxidative events, causes, through a Crabtree effect, inhibition of hexose oxidation and oxygen consumption in tumoral RINm5F cells.

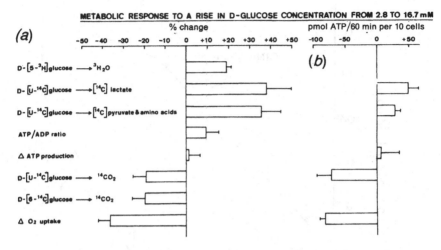

Fig. 16.1. Metabolic response of RINm5F cells to a rise in extracellular D-glucose concentration from 2.8 to 16.7 mM. (*a*) The percentage change in metabolic variables evoked by the rise in D-glucose concentration, relative to the reference value found at the low hexose level. Metabolic variables include the generation of ^3H_2O from D-[5-^{3}H]glucose, the generation of ^{14}C-labeled lactate, pyruvate and amino acids, and CO_2 from D-[U-^{14}C]glucose (or D-[6-^{14}C]glucose), the ATP : ADP ratio, the increase (Δ) in ATP production attributable to the catabolism of D-glucose, and the increase (Δ) in O_2 uptake caused by the hexose. (*b*) The changes in ATP generation rate evoked by the rise in D-glucose concentration, as calculated from and for relevant metabolic variables. In the case of Δ O_2 uptake, the P : O ratio was taken as 3.0.

16.7. Anomaly in the anomeric specificity of the metabolic and functional responses to D-glucose

A decisive support to the fuel hypothesis for insulin release stems from the observation that the greater insulinotropic action of the α- as distinct from β-anomer of either D-glucose or D-mannose coincides, in normal islet cells, with a higher rate of utilization of the α-anomers of these two hexoses (Malaisse *et al.*, 1976a; Sener *et al.*, 1982). We have therefore compared the anomeric specificity of the metabolic and functional responses in islets and RINm5F cells (Fig. 16.2). In both cell types, the rate of utilization of α-D-[5-^{3}H]glucose exceeded that of β-D-[5-^{3}H]glucose at low hexose concentrations, but this difference faded out at higher concentrations of D-glucose. Tumoral islet cells differed from normal islet cells in that the shift from an α-preferential to a non-preferential situation occurred over a much lower range of hexose concentrations

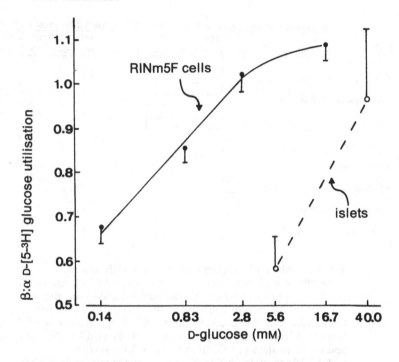

Fig. 16.2. Influence of the extracellular glucose concentration (logarithmic scale) upon the β: α ratio for ^3H_2O production by RINm5F cells and normal pancreatic islets incubated in the presence of α- or β-D-[5-^{3}H]glucose.

in RINm5F cells than in rat islets (Malaisse *et al.*, 1985*a*). It should be emphasized that such a difference could not be blamed on any obvious difference in the anomeric specificity of hexokinase in normal and tumoral islet cells, respectively (Malaisse *et al.*, 1985*b*).

In order to assess the anomeric specificity of the functional response of RINm5F cells, advantage was taken from the fact that D-glucose, although causing only a modest increase in insulin output, markedly stimulates the rate of protein biosynthesis in tumoral cells, as judged from the incorporation of L-[4-^{3}H]phenylalanine into TCA-precipitable material over 5 min incubation at 37° C (Barreto *et al.*, 1987). In the 0.15–0.4 mM range of D-glucose concentrations, the hexose-induced increment in biosynthetic activity is approximately half higher with α- than β-D-glucose (Fig. 16.3). Such an anomeric difference is no longer observed at higher D-glucose concentrations (1.4–16.7 mM). The comparison of biochemical and biosynthetic data indicates that a rise in hexose concentration exerts comparable effects upon the anomeric specificity of the

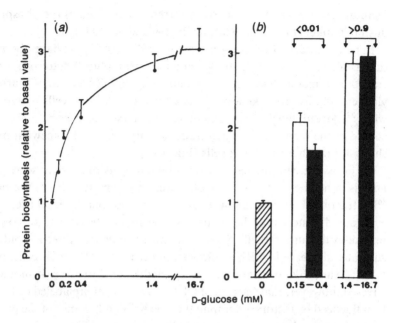

Fig. 16.3. (*a*) Effect of increasing concentrations of D-glucose, at
anomeric equilibrium, on the incorporation of L-[4-³H]phenylalanine
into TCA-precipitable material in RINm5F cells incubated for 5 min
at 37°C. (*b*) Effect of α-D-glucose (open columns) and β-D-glucose
(filled columns) on protein biosynthesis in two distinct ranges of
hexose concentrations. All results are expressed relative to the
corresponding mean basal level. Also shown is the statistical
significance of the anomeric difference.

metabolic and functional responses, respectively, of RINm5F cells to
D-glucose.

16.8. Anomalies in ionic events

In the normal process of nutrient-induced insulin release, a rise in
the cytosolic ATP concentration or ATP: ADP ratio (Malaisse & Sener,
1987) is thought to induce the closure of a class of potassium channels and,
by doing so, to result in a decrease in K^+ conductance, depolarization of
the B-cell plasma membrane and eventual gating of voltage-sensitive Ca^{2+}
channels (Cook & Hales, 1984; Rorsman & Trube, 1985; Dunne &
Petersen, 1986). In addition to these cationic changes, nutrient secretagogues also affect the movement of anions in the islet cells. In the latter
respect, the salient feature consists of a short-lived increase of inorganic
phosphate efflux occurring shortly after introduction of the nutrient

secretagogues. This phenomenon is currently described as the phosphate flush (Freinkel et al., 1974; Carpinelli & Malaisse, 1980).

In recent studies (Leclercq-Meyer et al., 1987, 1988b), we have observed that D-glucose, L-leucine, its deaminated metabolite 2-ketoisocaproate and its non-metabolized analog 2-aminobicyclo[2,2,1]heptane-2-carboxylic acid all also provoke a phosphate flush in RINm5F cells prelabeled with [^{32}P]orthophosphate and then placed in a perfusion device. The time course of the increase in effluent radioactivity appeared somewhat more sluggish than in normal islet cells (Fig. 16.4).

As far as the cationic response of RINm5F cells to nutrient secretagogues is concerned, several anomalies were encountered. First, the basal ^{86}Rb fractional outflow rate was lower in tumoral than normal islet cells. This could conceivably be related to the higher basal rate of oxygen uptake by the tumoral cells. Second, whereas D-glucose caused a rapid and sustained decrease in ^{86}Rb outflow from prelabeled RINm5F cells, both L-leucine and 2-ketoisocaproate unexpectedly failed to do so. It should be stressed, however, that the rate of ^{86}Rb outflow from prelabeled cells may be influenced by factors other than the true K^+ conductance of the plasma membrane, e.g. by an intracellular redistribution of ^{86}Rb.

In normal islet cells, nutrient secretagogues, which stimulate the net uptake of ^{45}Ca^{2+}, exert a dual effect upon ^{45}Ca^{2+} efflux from prelabeled islets (Herchuelz & Malaisse, 1980). The first of these effects consists of a decrease in effluent radioactivity which is best seen in the absence of extracellular Ca^{2+} and is thought to reflect stimulation of ^{45}Ca^{2+} uptake by intracellular organelles and/or a decreased efficiency of Ca^{2+} outward transport across the plasma membrane as possibly mediated by a process of Na$^+$–Ca^{2+} counter-transport. In the presence of extracellular Ca^{2+}, the initial decrease in effluent radioactivity is soon masked by a secondary and biphasic increase in ^{45}Ca^{2+} outflow. This secondary rise is currently ascribed to stimulation of ^{45}Ca^{2+} inflow into the islet cells, leading to a process of exchange between influent unlabeled Ca^{2+} and effluent ^{45}Ca^{2+}. In RINm5F cells stimulated by D-glucose, an increase in ^{45}Ca^{2+} net uptake, a decrease in ^{45}Ca^{2+} outflow from prelabeled cells perfused in the absence of extracellular Ca^{2+}, and an increase in effluent radioactivity from cells perfused in the presence of extracellular Ca^{2+} are all also observed. However, the relative magnitude of the changes in effluent radioactivity, especially the increase in ^{45}Ca^{2+} outflow recorded in the presence of extracellular Ca^{2+}, is much more modest than in normal islet cells. This contrasts with the dramatic stimulation of ^{45}Ca^{2+} outflow seen in RINm5F cells exposed to the non-nutrient secretagogue Ba^{2+}. We feel it

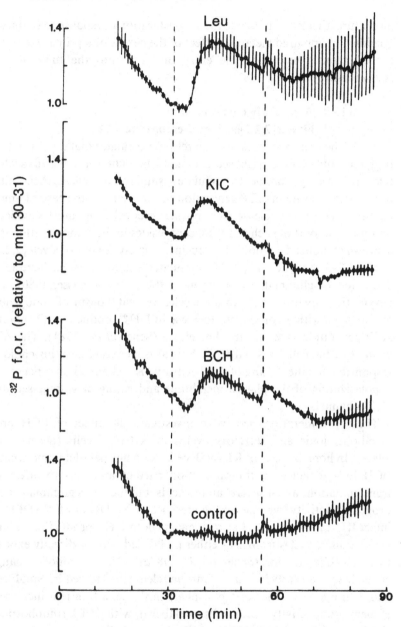

Fig. 16.4. Effects of L-leucine (Leu), 2-ketoisocaproate (KIC) and BCH, all tested at a concentration of 10.0 mM and administered during the period indicated by the vertical dotted lines, upon ^{32}P fractional outflow rate (f.o.r.) from prelabeled RINm5F cells. Control experiments, performed throughout in the absence of exogenous nutrient, are illustrated in the lower trace.

premature, therefore, to blame the sluggish cationic response to D-glucose on either an impaired metabolic effect of the hexose or a primary defect in those channels mediating the entry of Ca^{2+} into the tumoral cells (Leclercq-Meyer et al., 1987).

16.9. Anomaly in the response to bicyclo[2,2,1]heptane-2-carboxylic acid

The fuel concept for insulin release postulates that the capacity of D-glucose and other nutrient secretagogues, including amino acids such as L-leucine and L-glutamine, to stimulate insulin release reflects their ability to increase the rate of ATP generation in islet cells. In the case of amino acid-induced insulin release from normal islet cells, the most significant finding in support of such a fuel concept resides in the demonstration that a non-metabolized analog of L-leucine, namely 2-aminobicyclo[2,2,1]-heptane-2-carboxylic acid (BCH), stimulates insulin release as the result of activation of glutamate dehydrogenase (Sener & Malaisse, 1980). The enzymatic activation leads to an accelerated catabolism of endogenous amino acids with a consequent increase in NH_4^+ production, O_2 uptake, ATP generation rate and insulin release (Sener et al., 1981). The ATP content of the cells exposed to BCH is also increased and this could be responsible for the closing of ATP-sensitive K^+ channels and, hence, for depolarization of the plasma membrane and gating of voltage-sensitive Ca^{2+} channels.

These considerations led us to investigate the effect of BCH upon metabolic, ionic and secretory events in RINm5F cells (Sener et al., 1987b). In homogenates of RINm5F cells, as in normal islet homogenates, BCH fails to act as a transamination partner and fails to affect the transamination rate of several amino acids, but increases glutamate dehy-drogenase activity, whether in the presence of NADH or of NADPH. In intact RINm5F cells, BCH also augments both NH_4^+ production and the generation of $^{14}CO_2$ from cells either prelabeled with or directly exposed to L-[U-^{14}C]glutamine (Sener et al., 1987a). These metabolic changes coincide with an early increase of insulin release in perfused RINm5F cells and with the occurrence of a phosphate flush, i.e. a transient increase in effluent radioactivity from cells prelabeled with [^{32}P]orthophosphate (Leclercq-Meyer et al., 1988b). Quite unexpectedly, however, the early positive secretory response is soon followed by a severe, prolonged and irreversible inhibition of insulin release (Fig. 16.5). The latter inhibition coincides with a sustained and also irreversible increase in ^{86}Rb outflow from prelabeled cells, such an ionic effect being apparently attributable to

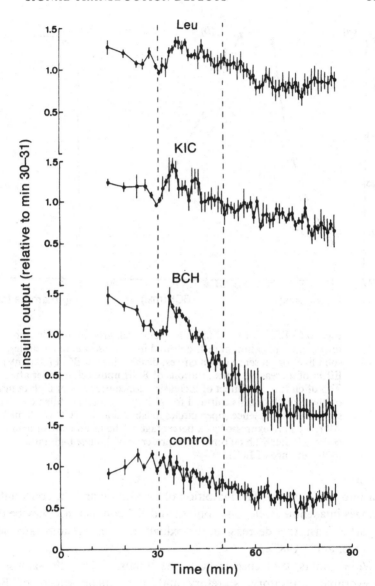

Fig. 16.5. Effects of L-leucine, 2-ketoisocaproate and BCH upon insulin output from perfused RINm5F cells. Same presentation as in Fig. 16.4.

a decrease in total adenine nucleotide content, ATP: ADP ratio and adenylate charge, as well as a decrease in oxygen uptake; these metabolic changes are all recorded during prolonged exposure of the cells to BCH. The leucine analog also impairs protein biosynthesis in RINm5F cells

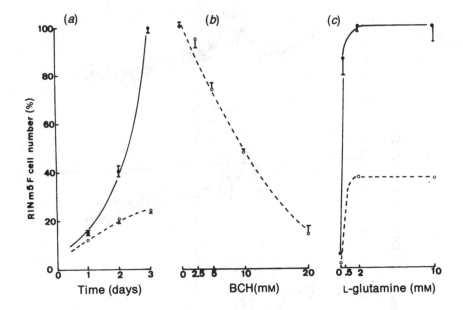

Fig. 16.6. Effect of BCH upon RINm5F cell growth. (*a*) Time course
for the multiplication of cells cultured in the absence (filled circles,
solid line) or presence (open circles, dashed line) of BCH (10.0 mM). (*b*)
Effect of increasing concentrations of BCH upon cell number after
72 h of culture. (*c*) Effect of increasing concentrations of L-glutamine
on the growth of cells cultured for 72 h in the absence (filled circles,
solid line) or presence (open circles, dashed line) of BCH (10.0 mM).
All results are expressed as a percentage of the mean control readings
collected after 72 h culture in the presence of 2.0 mM L-glutamine but
in the absence of BCH.

(Barreto *et al.*, 1989*a*). The inhibition of respiration and the concomitant
changes in adenine nucleotide content and K^+ conductance may be due,
in part at least, to a decrease in the oxidation of endogenous fatty acids
(Blachier *et al.*, 1987).

Within limits, D-glucose protects the tumoral islet cells against the
unfavorable respiratory, secretory and biosynthetic effects of BCH
(Sener *et al.*, 1987*a*).

In the light of these findings, the effect of BCH upon the growth of
RINm5F cells was eventually examined (Malaisse *et al.*, 1987*a*). As
illustrated in Fig. 16.6, BCH caused a concentration-related decrease in
the growth of RINm5F cells. L-glutamine (0.5 mM or more), the presence
of which was required to ensure optimal mitotic activity, failed to interfere
with the cytostatic action of BCH. The growth of RINm5F cells was much

Table 16.1. *Comparison of enzymatic activities, amino acid contents and metabolic flow rates in normal and tumoral islet cells*

amino acid	metabolic variable	islets	RINm5F cells
L-arginine	arginase[a]	1.48 ± 0.38	0.20 ± 0.03
L-arginine	urea production[b]	59.7 ± 4.0	5.8 ± 0.4
L-arginine	L-ornithine content[c]	15.2 ± 2.4	1.2 ± 0.2
L-ornithine	L-ornithine content[c]	14.5 ± 1.2	11.5 ± 1.4
L-ornithine	ornithine decarboxylase[a]	1.2 ± 0.2	7.4 ± 0.9
L-ornithine	di- & polyamine formation[b]	1.6 ± 0.1	11.5 ± 1.4
L-ornithine	ornithine–glutamate transaminase[a]	216 ± 31	235 ± 25
L-ornithine	L-glutamate formation[b]	8.9 ± 0.7	7.9 ± 0.8
L-arginine	amine formation: succinate oxidation[d]	0.18 ± 0.02	2.0 ± 0.7
L-ornithine	amines: glutamate formation[d]	0.18 ± 0.02	1.5 ± 0.2

[a] Enzymatic activities refer to the maximal velocity and are expressed as either nmol (arginase) or pmol (other enzymes) per μg protein per 120 min.

[b] Metabolic flow rates refer to cells exposed to 0.5 mM L-arginine or L-ornithine, as indicated, and are expressed as pmol/μg protein per 120 min.

[c] Amino acids contents refer to cell exposed to 0.5 mM L-arginine or L-ornithine, as indicated, and are expressed as pmol/μg protein under close-to-steady-state conditions.

[d] Ratios between metabolic flows refer to cells exposed to 0.5 mM L-arginine or L-ornithine, as indicated.

less severely affected by L-leucine than by BCH. The inhibitory action of BCH upon cell growth persisted for at least 72 h after removal of the L-leucine analog from the culture medium (Malaisse et al., 1987a).

Taken as a whole, these observations suggest that BCH could be used as a tool to interfere with the metabolism, secretory activity and growth of tumoral islet cells in patients with insulinomas.

16.10. Anomaly in the metabolism of L-arginine and polyamines

All the information so far reviewed in this report concerns the influence of nutrient secretagogues upon metabolic and functional events in the RINm5F cells. Non-nutrient secretagogues, such as the cationic amino acids L-arginine and L-ornithine, also stimulate insulin (and glucagon) release from these tumoral cells. In normal islet cells, the insulinotropic action of these positively charged amino acids is currently ascribed to their accumulation within the islet cells, leading to a subsequent depolarization of the plasma membrane and, hence, to the gating of voltage-sensitive channels (Charles et al., 1982). Recent work from this laboratory suggests that, in addition to such a biophysical effect, the release of insulin evoked by L-arginine or L-ornithine may also involve, to a limited extent, the de novo generation of putrescine, spermidine and spermine (Malaisse et al., 1989). We have therefore compared the metabolism of L-arginine and L-ornithine in normal rat islets and in RINm5F cells (Malaisse et al., 1989). There were obvious similarities, but also striking differences between normal and tumoral islet cells in terms of cationic amino acid metabolism. As documented in Table 16.1, four sets of comparisons deserve attention.

First, the maximal velocity of arginase in cell homogenates, as well as either the production of [^{14}C]urea or the steady-state content of ^{14}C-labeled L-ornithine in intact cells exposed to L-[U-^{14}C]arginine (0.5 mM), were all about one order of magnitude lower in tumoral than in normal islet cells, all results being expressed per microgram of tissue protein.

Second, despite a comparable cell content of ^{14}C-labeled L-ornithine in normal and tumoral cells exposed to exogenous L-ornithine (0.5 mM), the rate of di- and polyamine generation was about one order of magnitude higher in tumoral than in normal islet cells, this coinciding with a much higher activity of ornithine decarboxylase in RINm5F cells than in islet homogenates.

Third, the activity of ornithine–glutamate transaminase was roughly identical in pancreatic islet and in RINm5F cell homogenates; this coincided with a comparable rate of ^{14}C-labeled L-glutamate generation by intact cells exposed to L-[1-^{14}C]ornithine (0.5 mM).

Last, whether in the presence of L-arginine or L-ornithine, the ratio between the rate of amine generation and that of either L-glutamate formation or succinate oxidation was about ten times higher in RINm5F cells than in normal pancreatic islets.

16.11. Anomalies in insulin biosynthesis, storage and release

The metabolic and respiratory anomalies described in this chapter could well account for a defective secretory response of the RINm5F cells to D-glucose and, possibly, other nutrient secretagogues. Independently of such metabolic anomalies, the tumoral cells may also display an intrinsic defect in the biosynthesis, storage and release of hormonal products, as suggested by the following findings (Valverde et al., 1988b).

The ratio for insulin to protein content of RINm5F cells is at least two orders of magnitude lower than in normal islet cells. Moreover, since the insulin content of the tumoral cells may decrease over successive duplications to about 1% of its initial value, the ratio of insulin to protein content may, accordingly, be further decreased.

Whatever the insulin content, the basal insulin output represents, over 90 min incubation, 20–30% of such a content. The basal rate of insulin release is thus much higher, relative to cell insulin content, in tumoral than in normal islet cells. Relative to basal insulin output, the glucose-induced increment in insulin release does not exceed 20–40%, whereas, in normal islets, glucose provokes at least a 10-fold increase in hormonal output. The tumoral cells also differ from normal islet cells in their higher proinsulin: insulin ratio. The prohormone even appears to be released preferentially to its conversion product, since proinsulin represents a greater percentage of total immunoreactive material in the incubation medium than within the RINm5F cells. All of these anomalies could conceivably be attributable to a single defect, namely the inability of tumoral cells to prevent the massive release of secretory granules and, hence, to store adequate amounts of insulin.

The high fractional turnover of insulin in RINm5F cells could theoretically be considered as a favorable situation to monitor the synthesis, conversion, and release of newly synthesized proinsulin and insulin. Such is not the case, however. Indeed, the extremely low content of proinsulin and insulin in tumoral cells severely hampers the qualitative and quantitative characterization of newly synthesized proinsulin and insulin, the amount of which is virtually negligible compared with that of non-hormonal radioactive peptides in cells exposed to a suitable labeled tracer amino acid (e.g. L-[4-^{3}H]phenylalanine). Nevertheless, the incorporation of the tracer amino acid into TCA-precipitable material was found to be increased manifold in cells exposed to D-glucose. Hence, the much higher relative magnitude of the biosynthetic, as distinct from secretory, response to the hexose provides the opportunity to investigate with greater accuracy the metabolic determinants of the functional response of RINm5F cells to nutrient secretagogues. For instance, this approach was used to assess the anomeric specificity of the functional response to D-glucose (see above).

16.12. Anomalies in glucagon secretion

Tumoral cells of the RINm5F line contain not only insulin but also glucagon. The release of glucagon by the RINm5F cells displays several and severe alterations, when compared to the situation found in normal islet cells (Barreto et al., 1989b).

First, the rate of glucagon release is, relative to the cell content in the hormone, much higher in tumoral than in normal islets. For instance, over 90 min incubation in the presence of 1.4 mM D-glucose, the release of glucagon represents $50.6 \pm 5.8\%$ of the paired final cell content (i.e. 1.41 ± 0.27 ng/10^6 cells). L-arginine (10 or 20 mM) only causes a modest increase in glucagon output, the amino acid-increment in glucagon output above and relative to the control value not exceeding $37.3 \pm 5.7\%$, $26.0 \pm 5.8\%$ and $16.8 \pm 6.6\%$ respectively in the absence of D-glucose and presence of 1.4 and 2.8 mM D-glucose.

Second, the tumoral cells release a rather large fraction of their immunoreactive material in the form of peptides with a molecular mass in excess of that of true glucagon. Thus, as judged by immunoreactive criteria after chromatographic separation, the ratio of high molecular mass immunoreactive glucagon (M_r 9000) to true glucagon (M_r 3500) was close to 0.51 ± 0.01 and 0.47 ± 0.04 in RINm5F cell homogenates and incubation media, respectively.

Last, RINm5F cells release in the incubation medium a peptidase-like factor able to degrade both unlabeled porcine glucagon and [^{125}I]glucagon. Incidentally, the peptidase-like factor also degrades, but to a lesser extent, [^{125}I]insulin, but not [^{125}I]proinsulin.

16.13. What is normal in RINm5F cells?

Although we have so far emphasized, in the present review, differences in the metabolic and functional pattern of normal and tumoral islet cells, respectively, it should be realized that RINm5F cells do not always display an abnormal behavior. Several situations could be mentioned in which the response of these tumoral cells is qualitatively, if not quantitatively, close to normal. The following examples are illustrative.

First, in RINm5F cells perfused in the nominal absence of Ca^{2+} and glucose, the administration of Ba^{2+} (2 mM) and theophylline (1.4 mM) provokes a dramatic stimulation of both $^{45}Ca^{2+}$ outflow and insulin release (Fig. 16.7), suggesting that the intracellular redistribution of Ca^{2+} provoked by the Ba^{2+} cation and the subsequent activation of the Ca^{2+}-responsive effector system for insulin release is essentially preserved in tumoral islet cells (Leclercq-Meyer et al., 1987, 1988a).

Second, RINm5F cells display a positive secretory response to cholinergic agents, at least in the presence of D-glucose (Blachier & Malaisse, 1987). This coincides with activation of phospholipase C and increased production of tritiated inositol phosphates in cells preincubated with myo-[2-^{3}H]inositol. In the tumoral cells, the occupation of muscarinic receptors is apparently coupled to the activation of phospholipase C at the intervention of a novel GTP-binding regulatory protein, distinct from both Ns and Ni (Blachier & Malaisse, 1987).

Last, we recently observed that, in RINm5F cells like in normal islet cells, exogenous ATP or its stable analog stimulates the hydrolysis of inositol-containing phospholipids (Blachier & Malaisse, 1988). This observation suggests that the purinergic pathway for stimulation of islet cells is also operative in RINm5F cells. Further work is required, however, to characterize the secretory response of these tumoral cells to ATP and its stable analog.

16.14. Conclusions

The present review illustrates that tumoral islet cells differ from normal islet cells in several distinct, albeit interrelated, biochemical, biophysical and functional respects. These, possibly in part genetic, anomalies could serve as a model in the search of comparable alterations in the B-cells of diabetic subjects.

References

Barreto, M., Sener, A., Malaisse, W.J. & Valverde, I. (1989a). Inhibition by a nonmetabolized analogue of L-leucine of protein biosynthesis in tumoral pancreatic islet cells. *Hormone and Metabolic Research*, in press.

Barreto, M., Valverde, I. & Malaisse, W.J. (1987). Anomeric specificity of glucose-stimulated protein biosynthesis in tumoral pancreatic islet cells. *Medical Science Research* 15, 1103–4.

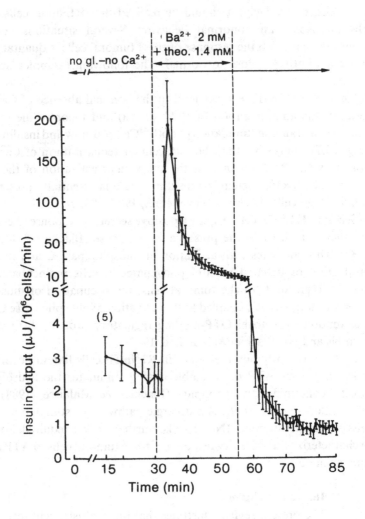

Fig. 16.7. Effect of the association of Ba^{2+} (2.0 mM) and theophylline (1.4 mM) upon insulin release from RINm5F cells perfused in the absence of Ca^{2+} and glucose.

Barreto, M., Valverde, I. & Malaisse, W.J. (1989*b*). Glucagon storage, release and degradation by tumoral islet cells (RINm5F line). *Diabetes Research*, in press.

Blachier, F. & Malaisse, W.J. (1987). Possible role of a GTP-binding protein in the activation of phospholipase C by carbamylcholine in tumoral insulin-producing cells. *Research Communications in Chemical Pathology and Pharmacology* **58**, 237–55.

Blachier, F. & Malaisse, W.J. (1988). Effect of exogenous ATP upon inositol phosphate production, cationic fluxes and insulin release in pancreatic islet cells. *Biochimica et Biophysica Acta* **970**, 222–9.

Blachier, F., Sener, A. & Malaisse, W.J. (1987). Interference of a nonmetabolized analogue of L-leucine with lipid metabolism in tumoral pancreatic islet cells. *Biochimica et Biophysica Acta* **921**, 494–501.

Carpinelli, A.R. & Malaisse, W.J. (1980). The stimulus-secretion coupling of glucose-induced insulin release. XLIV. A possible link between glucose metabolism and phosphate flush. *Diabetologia* **19**, 458–64.

Carpinelli, A.R., Sener, A. & Malaisse, W.J. (1987). Positive feedback effect of ATP on glucose phosphorylation in pancreatic islets. *Medical Science Research* **115**, 481–2.

Chan, S.J., Nanjo, K., Miyano, M., Lu, Y.J., Welsh, M., Nielsen, D., Haenda, M., Kwok, S.C.M., Tager, H.S., Rubenstein, A.H. & Steiner, D.F. (1986). Abnormalities of insulin gene structure and expression – A minireview. In *Diabetes 1985* (ed. M. Serrano-Rios & P.J. Lefèbvre), pp. 486–94. Amsterdam: Excerpta Medica.

Charles, S., Tanagawa, T. & Henquin, J.C. (1982). A single mechanism for the stimulation of insulin release and $^{86}Rb^+$ efflux from rat islets by cationic amino acids. *Biochemical Journal* **208**, 301–8.

Cook, D.L. & Hales, C.N. (1984). Intracellular ATP directly blocks K^+ channels in pancreatic B-cells. *Nature* **311**, 271–3.

Dunne, M.J. & Petersen, O.H. (1986). Intracellular ADP activates K^+ channels that are inhibited by ATP in an insulin-secreting cell line. *FEBS Letters* **208**, 59–62.

Freinkel, N., El Younsi, C., Bonnar, J. & Dawson, M.C. (1974). Rapid transient efflux of phosphate ions for pancreatic islets as an early action of insulin secretagogues. *Journal of Clinical Investigation* **54**, 1179–89.

Gabbay, K.H., Bergenstal, R.M., Wolff, J., Mako, M.E. & Rubenstein, A.H. (1979). Familial hyperproinsulinemia: partial characterization of circulating proinsulin-like material. *Proceedings of the National Academy of Sciences of the U.S.A.* **76**, 2881–5.

Gazdar, A.F., Chick, W.L., Oie, H.K., Sims, H.L., King, D.L., Weir, G.C. & Lauris, V. (1980). Continuous, clonal insulin- and somatostatin-secreting cell lines established from a transplantable rat islet cell tumor. *Proceedings of the National Academy of Sciences of the U.S.A.* **77**, 3519–23.

Giroix, M.-H., Dufrane, S.P., Malaisse-Lagae, F., Sener, A. & Malaisse, W.J. (1984a). Fasting-induced impairment of glucose 1,6-bisphosphate synthesis in pancreatic islets. *Biochemical and Biophysical Research Communications* **119**, 543–8.

Giroix, M.-H., Sener, A., Dufrane, S.P., Malaisse-Lagae, F. & Malaisse, W.J. (1985). Glucose metabolism in insulin-producing tumoral cells. *Archives of Biochemistry and Biophysics* **241**, 561–70.

Giroix, M.-H., Sener, A. & Malaisse, W.J. (1986). D-glucose transport and concentration in tumoral insulin-producing cells. *American Journal of Physiology* **251**, C847–51.

Giroix, M.-H., Sener, A., Pipeleers, D.G. & Malaisse, W.J. (1984b). Hexose metabolism in pancreatic islets. Inhibition of hexokinase. *Biochemical Journal* **223**, 447–53.

Hellman, B., Sehlin, J. & Täljedal, I.-B. (1971). Evidence for mediated transport of glucose in mammalian pancreatic β-cells. *Biochimica et Biophysica Acta* **241**, 147–54.

Herchuelz, A. & Malaisse, W.J. (1980). The dual regulation of calcium efflux from pancreatic islets. *Hormone and Metabolic Research* (suppl. 10), 116–21.

Leclercq-Meyer, V., Giroix, M.-H., Sener, A., Marchand, J. & Malaisse, W.J. (1987). ^{32}P, $^{86}Rb^+$ and $^{45}Ca^{2+}$ handling by tumoral insulin-secreting cells (RINm5F line). *Hormone and Metabolic Research* **19**, 525–31.

Leclercq-Meyer, V., Giroix, M.-H., Sener, A., Marchand, J. & Malaisse, W.J. (1988a). Dynamics of insulin release by perfused insulin-producing tumoral cells: effects of glucose, forskolin, leucine, barium and theophylline. *International Journal of Pancreatology* **3**, 17–31.

Leclercq-Meyer, V., Marchand, J., Sener, A., Blachier, F. & Malaisse, W.J. (1988b). Effects of L-leucine, its 2-keto acid metabolite and its non-metabolized analogue on rat tumoral islet function. *Journal of Molecular Endocrinology* 1, 69–76.

Malaisse, W.J., Blachier, F., Mourtada, A., Camara, J., Albor, A. & Valverde, I. (1989). Stimulus-secretion coupling of arginine-induced insulin release. Metabolism of L-arginine and L-ornithine in normal and tumoral islet cells. *Biochimica et Biophysica Acta*, submitted.

Malaisse, W.J., Giroix, M.-H., Dufrane, S.P., Malaisse-Lagae, F. & Sener, A. (1985a). Environmental modulation of the anomeric specificity of glucose metabolism in normal and tumoral cells. *Biochimica et Biophysica Acta* 847, 48–52.

Malaisse, W.J., Giroix, M.-H., Dufrane, S.P., Malaisse-Lagae, F. & Sener, A. (1985b). Anomeric specificity of hexokinase in rat, human and murine tumoral cells. *Cancer Research* 45, 6376–8.

Malaisse, W.J., Giroix, M.-H., Malaisse-Lagae, F. & Sener, A. (1986). 3-O-methyl-D-glucose transport in tumoral insulin-producing cells. *American Journal of Physiology* 251, C841–6.

Malaisse, W.J., Giroix, M-H. & Sener, A. (1985c). Anomeric specificity of glucose metabolism in the pentose cycle. *Journal of Biological Chemistry* 260, 14630–2.

Malaisse, W.J., Giroix, M.-H. & Sener, A. (1987a). Inhibition of insulinoma cell proliferation by a nonmetabolized leucine analogue. *Medical Science Research* 15, 1533–4.

Malaisse, W.J., Giroix, M.-H. & Sener, A. (1987b). Effect of cytochalasin B on glucose uptake, utilization, oxidation and insulinotropic action in tumoral insulin-producing cells. *Cell Biochemistry and Function* 5, 183–7.

Malaisse, W.J., Malaisse-Lagae, F. & Sener, A. (1982). The glycolytic cascade in pancreatic islets. *Diabetologia* 23, 1–5.

Malaisse, W.J., Malaisse-Lagae, F. & Sener, A. (1983). Anomeric specificity of hexose metabolism in pancreatic islets. *Physiological Review* 63, 321–7.

Malaisse, W.J., Rasschaert, J., Zähner, D. & Sener, A. (1988). Hexose metabolism in pancreatic islets: the Pasteur effect. *Diabetes Research* 7, 53–8.

Malaisse, W.J. & Sener, A. (1987). Glucose-induced changes in cytosolic ATP content in pancreatic islets. *Biochimica et Biophysica Acta* 927, 190–5.

Malaisse, W.J. & Sener, A. (1988a). Mitochondrial oxidative events and Pasteur or Crabtree effects in pancreatic islet cells. *Abstracts of the 7th Annual Meeting of the Endocrine Society* (New Orleans), p. 181.

Malaisse, W.J. & Sener, A. (1988b). Dissociated regulation of glycolytic and mitochondrial oxidative events in pancreatic islets. *Diabetes* 37 (suppl. 1), 192A [abstract].

Malaisse, W.J., Sener, A., Herchuelz, A. & Hutton, J.C. (1979). Insulin release: the fuel hypothesis. *Metabolism* 28, 373–86.

Malaisse, W.J., Sener, A., Koser, M. & Herchuelz, A. (1976a). The stimulus-secretion coupling of glucose-induced insulin release. XXIV. The metabolism of α- and β-D-glucose in isolated islets. *Journal of Biological Chemistry* 251, 5936–43.

Malaisse, W.J., Sener, A. & Levy, J. (1976b). The stimulus-secretion coupling of glucose-induced insulin release. XXI. Fasting-induced adaptation of key glycolytic enzymes in isolated islets. *Journal of Biological Chemistry* 251, 1731–7.

Malaisse-Lagae, F. & Malaisse, W.J. (1988). Hexose metabolism in pancreatic islets. Regulation of mitochondrial hexokinase binding. *Biochemical Medicine and Metabolic Biology* 39, 80–9.

Malaisse-Lagae, F., Sener, A. & Malaisse, W.J. (1982). Phosphoglucomutase: its role in the responses of pancreatic islets to glucoses epimers and anomers. *Biochimie* 64, 1059–63.

Rorsman, P. & Trube, G. (1985). Glucose dependent K⁺-channels in pancreatic β-cells are regulated by intracellular ATP. *Pflügers Archiv European Journal of Physiology* 405, 305–9.

Sener, A., Blachier, F. & Malaisse, W.J. (1988a). Crabtree effect in tumoral pancreatic islet cells. *Journal of Biological Chemistry* 263, 1904–9.

Sener, A., Giroix, M.-H., Hellerström, C. & Malaisse, W.J. (1987a). Influence of D-glucose upon the respiratory and secretory response of insulin-producing tumor cells to 2-aminobicyclo[2,2,1]heptane-2-carboxylic acid. *Cancer Research* **47**, 5905–7.

Sener, A., Giroix, M.-H. & Malaisse, W.J. (1984a). Hexose metabolism in pancreatic islets. The phosphorylation of fructose. *European Journal of Biochemistry* **144**, 223–6.

Sener, A., Giroix, M.-H. & Malaisse, W.J. (1986a). Impaired uptake of D-glucose by tumoral insulin-producing cells. *Biochemistry International* **12**, 913–19.

Sener, A., Leclercq-Meyer, V., Giroix, M.-H., Malaisse, W.J. & Hellerström, C. (1987b). Opposite effects of D-glucose and a nonmetabolized analogue of L-leucine on respiration and secretion in insulin-producing tumoral cells (RINm5F). *Diabetes* **36**, 187–92.

Sener, A. & Malaisse, W.J. (1980). L-leucine and a nonmetabolized analogue activate, pancreatic islet glutamate dehydrogenase. *Nature* **288**, 187–9.

Sener, A. & Malaisse, W.J. (1985). Resistance to alloxan of tumoral insulin-producing cells. *FEBS Letters* **193**, 150–3.

Sener, A. & Malaisse, W.J. (1987). Stimulation by D-glucose of mitochondrial oxidative events in islet cells. *Biochemical Journal* **246**, 89–95.

Sener, A. & Malaisse, W.J. (1988). Hexose metabolism in pancreatic islets. Metabolic and secretory responses to D-fructose. *Archives of Biochemistry and Biophysics* **261**, 16–26.

Sener, A., Malaisse-Lagae, F., Giroix, M.-H. & Malaisse, W.J. (1986b). Hexose metabolism in pancreatic islets: compartmentation of hexokinase in islet cells. *Archives of Biochemistry and Biophysics* **251**, 61–7.

Sener, A., Malaisse-Lagae, F., Giroix, M.-H. & Malaisse, W.J. (1986c). Sensitivity to pentamidine but resistance to streptozotocin of tumoral insulin-producing cells (RINm5F line). *Pancreas* **1**, 550–5.

Sener, A., Malaisse-Lagae, F., Lebrun, P., Herchuelz, A., Leclercq-Meyer, V. & Malaisse, W.J. (1982). Anomeric specificity of D-mannose metabolism in pancreatic islets. *Biochemical and Biophysical Research Communications* **108**, 1567–73.

Sener, A., Malaisse-Lagae, F. & Malaisse, W.J. (1981). Stimulation of islet metabolism and insulin release by a non-metabolizable amino acid. *Proceedings of the National Academy of Sciences of the U.S.A.* **78**, 5460–4.

Sener, A., Malaisse-Lagae, F. & Malaisse, W.J. (1987c). Fructose metabolism via the pentose cycle in tumoral islet cells. *European Journal of Biochemistry* **170**, 447–52.

Sener, A., Rasschaert, J., Zähner, D. & Malaisse, W.J. (1988b). Hexose metabolism in pancreatic islets. Stimulation by D-glucose of [2-³H]glycerol detritiation. *International Journal of Biochemistry* **20**, 595–8.

Sener, A., Van Schaftingen, E., Van de Winkel, M., Pipeleers, D.G., Malaisse-Lagae, F., Malaisse, W.J. & Hers, H.G. (1984b). Effects of glucose and glucagon on the fructose 2,6-bisphosphate content of pancreatic islets and purified pancreatic B-cells. *Biochemical Journal* **221**, 759–64.

Valverde, I., Barreto, M., Blachier, F., Sener, A. & Malaisse, W.J. (1988a). Effect of hexoses upon protein biosynthesis, respiration and lipid metabolism in tumoral islet cells (RINm5F line). *Diabetes, Nutrition and Metabolism* **1**, 193–200.

Valverde, I., Barreto, M. & Malaisse, W.J. (1988b). Stimulation by D-glucose of protein biosynthesis in tumoral insulin-producing cells (RINm5F line). *Endocrinology* **112**, 1443–8.

Vischer, U., Blondel., B., Wollheim, C.B., Höppner, W., Seitz, H.J. & Iynedjian, B.P. (1987). Hexokinase isoenzymes of RINm5F insulinoma cells. Expression of glucokinase gene in insulin-producing cells. *Biochemical Journal* **241**, 249–55.

Index

340